The Conceptual Foundations of Contemporary Relativity Theory

THE MIT PRESS, CAMBRIDGE, MASSACHUSETTS, AND LONDON, ENGLAND

FOREWORD BY JOHN ARCHIBALD WHEELER

JOHN COWPERTHWAITE GRAVES

The Conceptual Foundations of Contemporary Relativity Theory

Set in Linotype Baskerville and printed and bound in the United States of America by Halliday Lithograph Corporation.

ISBN: 0-262-57049-1 (paperback)

First MIT Press paperback edition, 1977
Second printing, 1978

Library of Congress catalog card number: 77-122257

Contents

On 21 February 1870 twenty-four-year-old William Kingdon Clifford put before the Cambridge Philosophical Society the view that a particle is nothing but a kind of hill in the geometry of space. On 4 November 1915 Albert Einstein presented before the Prussian Academy of Science at Berlin his paper on general relativity where one for the first time had a law for the dynamic development of geometry with time. This geometry was for Einstein in the beginning primarily a means to describe gravitation: matter here curves space here. Curvature here inevitably brings in its train curvature there. Curvature there affects matter there. Thus arises action of matter here on matter there! The ensuing decades and not least the advances in astrophysical measurements brought by the laser have given and continue to give impressive tests of the correctness of this geometric theory of gravitation.

If Aladdin called forth from the bottle a genie who was unwilling to return, Einstein gave to geometry a life and independent existence, which he equally was astonished to see taking shape before his eyes. Geometry, conceived first as the slave of matter, transmitting influence from mass here to mass there, turned out to possess dynamical degrees of freedom of its own. These degrees of freedom manifested themselves to Einstein in the very first months of his "geometrodynamics" as gravitational waves, not in the least disturbing to Einstein's views, and today actively under exploitation by ingenious experimental physicists seeking to open a new channel of communication with violent events taking place in supernovae and in galactic nuclei. Only after a few years had passed did Einstein discover that the genie that was geometry, in addition to such small-scale quivering and waves, manifested a large-scale dynamics. In brief, his geometrodynamic law applied to the simplest model universe that one could readily conceive (a closed spherical universe endowed with a roughly uniform density of mass and energy), predicted that such a universe should expand from a point ("big bang"), reach a maximum dimension, recontract, and go to complete gravitational collapse. No prediction seemed to him more preposterous. No one who had ever looked at the universe before had ever thought of its possessing any such large-scale dynamics and he saw no evidence that any such dynamics was present. Much against his will he felt it necessary to search for some way to change his basic geometrodynamic law. The difficulty is great because the arguments of simplicity and correspondence with Newtonian gravitation theory left no alternative. The only possibility he could see was to add an artificial so-called cosmological term to the equation, as means to get a static universe. Less than a decade later, in 1929, Edwin Hubble proved that the universe is not static; it is undergoing expansion in qualitative accordance with the original Einstein predictions. Thereafter Einstein abandoned the cosmological term, calling it "the greatest mistake of my life." The observed expansion became and remains the most dramatic single proof one has today for the correctness of the idea the "Geometry is Dynamic, and Einstein is Its Prophet."

Like a poor man suddenly put in possession of a mansion with twenty servants and feeling at a loss how to measure up to his possibilities and responsibilities, the investigator also feels at a loss who unexpectedly finds this powerful genie standing at his right arm waiting to serve. What questions shall he ask? What tasks shall he undertake? What obligations to be of service has he acquired? How does one talk sensibly about initial conditions for this geometry? How does one come to terms with the inevitability of gravitational collapse, collapse not only at the scale of the universe but even collapse of a star to a so-called black hole? What does the quantum principle have to contribute toward an understanding of this "collapse"? This collapse is a crisis in the theoretical physics of our times. Out of the resolution of this crisis surely decisive new insights must emerge. Out of those insights, yet to be won, new understanding has to be anticipated about the origin and fate of the universe and even, one can believe, about man's position in the universe. A truer vision about the constitution of particles surely will emerge from the clearing of this crisis, far though it is from the simple picture of Clifford's hills in the geometry of space.

"No progress without a paradox" was the guiding principle in the life work of Niels Bohr and at his Copenhagen School of Physics. The paradox stands out clear today: the universe exists but it is postulated to have been born out of a condition of impossibility—infinite compaction. Surely this paradox will resolve itself as has every paradox in the past, not by revolution in the laws of physics but by evolution. A larger unity must exist that includes both the quantum principle and general relativity. Surely the Lord did not on Day One create geometry and on Day Two proceed to "quantize" it. The quantum principle, rather, came on Day One and out of it something was built on Day Two which on first inspection looks like geometry but which on closer examination is at the same time simpler and more sophisticated. One looks toward the future as deeply challenged as one has ever been in the whole history of physics. One is as hopeful of new insights as one has ever been. At the same time one knows that any hope of progress depends most of all on a solid understanding of what the situation is today and how it came into being.

Until today it was not easy to know where to look for a penetrating overview of general relativity. The subject has experienced important developments in recent decades. Time has come to be understood more clearly than ever as a many-fingered entity, not least through the achievements of

Tomonaga and Schwinger. The initial value problem—what must be given to predict the future—has been the subject of active investigation in recent years. Mach's principle, according to which inertia here originates from mass-energy there, is seen in a new perspective. New insights have been won on the many ways in which geometry serves as a building material to construct geometrodynamic objects of the most diverse kinds. Nowhere has there been a place where one could turn for an account of the philosophical foundations of general relativity with special emphasis on modern developments, until this book by John Graves. Nowhere better than here can one clearly see the historical continuity of thought along these lines, from Plato's famous "All is geometry" through Descartes' vortices up through Einstein's standard 1915 geometrodynamics in all its richness.

If by reading this book one gains a new feeling of gratitude to the workers of the past for the wonderful patrimony they have won for us, one also acquires a new appreciation of how a philosopher deals in detail with a physical theory. May this book serve as a happy reminder that no one can break the ties that unite philosophers and scientists and all who have a love of learning in one of the greatest of enterprises: to understand this beautiful and mysterious universe.

John Archibald Wheeler
January 1971

I am indebted to many individuals for their help at various stages of the long period of preparation of this manuscript. Most especially, I wish to thank Professors Carl G. Hempel and John A. Wheeler of Princeton University. They have gone to a great deal of time and trouble in reading an earlier version of this work and providing valuable detailed comments and criticisms. I am also very grateful to Professors Thomas S. Kuhn, Richard Rorty, and Charles C. Gillispie, also of Princeton, for reading parts of the manuscript and providing further evaluations and conferences.

Many other people have given me ideas or suggestions which are reflected in one way or another in this treatise, or have encouraged me to proceed by their interest in the project. Among them I would like to mention and thank Professors Gregory Vlastos, Hilary Putnam, Abner Shimony, James E. Roper, Lawrence Sklar, Richard Grandy, Dieter Brill, John Stachel, Adolf Grünbaum, Peter Achinstein, Philip Morrison, and J. J. C. Smart. In some cases their help took the form of an answer to a direct question; in other cases it simply reflects my experience of hearing them put forward their ideas. I wish to give especial thanks to my colleague Sylvain Bromberger, for his sustained interest and gentle prodding during the latter stages of the work.

Students in two of my courses, "Philosophy of Science" and "Tradition and Innovation in Twentieth Century Physics," have had a chance to hear much of the material of this book, since it served as an important basis of my lecture notes. I thank them for their patience, enthusiasm, and occasional helpful criticisms during those dry runs. I am also grateful to my former graduate student, Caroline Whitbeck, who forced me to clarify the ideas of geometrodynamics in my own mind so that I could clarify them in hers.

Finally, I am deeply indebted to Mrs. Sheila Toomey, who spent the hot summer months diligently typing the final draft of this very lengthy manuscript.

Acknowledgments

The Conceptual Foundations of Contemporary Relativity Theory

PHILOSOPHY 1

Introduction 1

This book is intended to be a comprehensive account of the philosophically significant aspects of a particular physical theory, general relativity or "geometrodynamics," as its contemporary version is now called. I firmly believe that the theory should be taken seriously, even if it has not yet been perfectly confirmed, and that it has important and sometimes surprising implications for both physics and philosophy. At least some theories, and this one in particular, include among their explanatory features the task of trying to construct a coherent and intellectually satisfying picture of the world, and philosophy should be concerned to explicate and evaluate this picture. In any case, general relativity is recognized to be one of the most impressive intellectual accomplishments of the twentieth century: massive, complex, and in its own way beautiful, but with a range and power still largely unknown. Such a theory must touch on many things outside itself, not only on the state of science at the time of its creation and development, but also on the rest of our intellectual history. My desire for comprehensiveness stems from my conviction that no theory of this sort can be fully understood until all these external factors are taken into account.

In order to delineate the world-picture and to evaluate the philosophical significance of a theory, a philosopher, or physicist qua amateur philosopher, must have a definite philosophical framework of his own. Often this is only implicit, or assumed to be the same as that of the reader, but it is desirable to spell it out explicitly in advance, particularly if one uses a special set of technical terms. Without such a framework and overall position one can hardly know what to look for, what aspects are important, and what criteria of evaluation should be used. Philosophers who disagree on these issues are likely to interpret the theory in very different ways. But this overall position need not be a set of specific dogmas; indeed, the philosopher can hardly be said to take the theory seriously if it is. He then can only distort the theory by singling out certain features which he interprets to serve as grist for his mill, and rejecting the rest. His position should be a flexible outline which supplies general principles of interpretation, but allows the theory to supply the specific details. To be sure, these principles may influence his development of the theory and reveal new and unexpected aspects, and on the other hand the exigencies of the theory may cause him to modify these principles; but the point is that both are needed and function together. Furthermore, one should recognize that the ideas in a theory are not created ex nihilo, as the study of any discipline will reveal. Rather, they evolve by a gradual process; their full implications are not always fully understood at the time when they first appear, and their development reflects the overall increase in human knowledge and awareness of new possibilities. Therefore it is essential to look for historical antecedents for the ideas of the theory, to see how the theory modifies or makes them different from those antecedents, and whether the new versions hold out greater prospects for successful explanatory power. Finally, it is essential to understand the physics of the theory, the mathematical symbolism used, the techniques of application, and the emphases of the physicist. This may require delving into the theory in considerable detail.

With these ideas in mind, I envisage this book as the confluence of three main streams. In the first major section I describe my own philosophical position, which I call "scientific realism." This includes the following main points: (1) scientific theories do have ontological import; if they are considered adequately confirmed, we should assume that the entities and processes to which their 'theoretical terms' purport to refer really do exist; (2) this conclusion does not rule out the existence of anything else, since the world as a whole consists of levels which cannot be completely reduced to one another, and individual theories can cover only one level; and (3) models are essential to scientific theories, and can function as 'ontological hypotheses', enabling us to determine just what is included at the level of a given theory. To bring out the character of my scientific realism, I criticize two basic assumptions which I call the "empiricist error" and the "logicist error," which have been accepted too readily by many philosophers of science. Because of the difficulty of arguing for or against one's most general philosophical assumptions, and the fact that such debates would take me too far away from geometrodynamics, I am less concerned to argue the merits of my own position above all others than to present it as a plausible approach, and see how well it is able to interpret geometrodynamics. If a person wishes to try this same task from a different starting point, he is welcome to do so, though I believe that I present some arguments for why he would be unlikely to succeed. In this section of the work I emphasize very abstract and general issues, and introduce a special terminology, which I can later apply to the specific case of geometrodynamics. Examples taken from other theories could not be developed at sufficient length within the confines of this work, and in abbreviated form might only confuse the issue.

In the second, or historical, section I introduce a notion of 'conceptual geography' which I use to compare the concepts of thinkers discussing the same general issues, but in different "intellectual milieux." Since I believe that, in their different ways, Plato and Descartes both took the same point of view on the relation of space to matter as Einstein, Wheeler, and contemporary geometrodynamicists, I discuss their positions here in considerable detail. This detail is important in evaluating the degree to which they may really be considered predecessors of geometrodynamics, or whether there is only a superficial resemblance. This comparison is important for another reason: since the theories of Plato and Descartes were subject to criticism which made their influence on physics short-lived, it is important to see whether general relativity can avoid these same criticisms. I also mention the work of Newton and others in order to show the alternative theories available then, so that I can consider whether they are still available now.

Finally I come to general relativity proper. I try to present the main physical ideas without using detailed mathematics except where it is necessary to present the result of a calculation. I also envisage the physics as formed from three substreams of empirical data, mathematical techniques, and awareness of conceptual possibilities, which must come together at the right time to produce a new and revolutionary theory. I present first the 'classical' version of the theory, with its postulates, triumphs, weaknesses, and already revolutionary features. Then I turn to its recent reinterpretation, Wheeler's vision of physics as geometry and of a space-time that encompasses everything material and needs no other 'foreign entities', analyzing how this can resolve some old problems though at the cost of raising new ones. I have chosen the order of philosophy, then history, then physics so that in each section I can refer back to the ideas and results of previous sections without presupposing subsequent ones.

Finally, the three streams join together to form the mighty river of a completely monistic and self-contained level of the universe. Here finally I can tie the earlier material together, locating geometrodynamics as a *philosophical* theory and giving its picture of the world. I consider what monism is in general, how it can be derived from physics, how it can account for the observed diversity and plurality in the universe, and what its overall ontological significance is. Once this has been done, the river can at last flow into the sea of human knowledge.

A physicist who is hoping to find in this work a textbook for learning how to work within the theory will be disappointed, for my purpose is primarily philosophical. My goal will always be to shed light on the more general philosophical issues concerning the construction and use of theories in physics and the status of the theoretical entities which they postulate, *through showing how these problems arise concretely in the context of a particular physical theory.* For it is my fervent conviction that unless philosophers of science constantly compare their generalizations with the particular theories actually developed and used by scientists, they run the grave danger of making a priori pronouncements which hardly mirror actual scientific practice. The danger then is that philosophy of science may become a coterie discipline like Medieval Scholasticism, which produced many clever analyses and subtle distinctions, but gradually lost contact with the reality with which it purported to deal, by remaining too closely within a set framework of questions, general assumptions, and methods of analysis.

Philosophers should always remain consciously aware of the essentially 'metadisciplinary' character of their subject. As each of the various special sciences broke away from philosophy by choosing a particular level of abstraction and finding paradigmatic concepts, methods, and principles appropriate to it, philosophers have sometimes been tempted to isolate some subject matter as uniquely appropriate to philosophy, and try to achieve 'progress' in this area in a manner paralleling that of the sciences. I contend, on the contrary, that philosophy as a subject for serious study has a fully adequate rationale and justification in its analytic function of investigating the logical structure and presuppositions of some other discipline, and in its synthetic function of giving that particular discipline a perspective with respect to the others and seeing whether it can produce a coherent picture of some aspect of the world. But these tasks require an extensive knowledge of what actually goes on within a field.

Most philosophers of science have considered themselves empiricists, often rather self-consciously. My recommendation, then, is that they practice in philosophy what they preach to science. As I see it, the task of the philosopher of science should be this: to study actual scientific theories to see both how and why they have evolved and how they are presently understood by their users. It is natural and reasonable to hope that true statements about theories in general will result from such a study. But I would insist that any such statements have the character of inductive generalizations, advanced tentatively. One must always allow the possibility that the things which the scientific community calls theories, even the more restricted class called 'good theories', are related only by 'family resemblances', in Wittgenstein's sense. Even if, as appears more likely, there are indeed statements which seem to be universally true of all theories, there is no guarantee that such statements will shed much light on the nature of any *particular* theory. For the features which a theory has in common with all others may only be a small and insignificant part of that theory, certainly not enough to characterize it. The truly interesting features, those which make it most useful and raise the most problems, may be the ones which differentiate it from other theories. Philosophers who take a more formalistic approach to theories in general run the risk of overestimating the importance of the common features. The only corrective is to see just what role these aspects play in a particular case, and to discover which ones the scientist finds most relevant, and why he does so. Geometrodynamics is as yet only a small part of physics, just as physics is only one of a multiplicity of special sciences, but it has important lessons to teach anyone who proposes to philosophize about science as a whole.

2 Scientific Realism versus the Empiricist Error

In this treatise I intend to approach contemporary relativity theory from the standpoint of what Smart has called "scientific realism."[1] This position I take to include the following claims: (1) There is an external world, independent of anyone's sensory perception of it. (2) This world may contain entities and processes radically different from those which seem to be directly disclosed in sense perception. (3) We can, however, attain some degree of knowledge of this external world, even though we may never be entitled to claim perfect knowledge. It does not consist of noumena, or things-in-themselves that by their very nature must remain unknowable. (4) Science is the attempt, and indeed the most legitimate attempt, by human beings to understand the structure of this external world, both in its general outline and in all the complexities and details of its operation. It thus seeks to transcend the contents of immediate experience, while at the same time explaining them. (5) Scientists do this by developing theories, which postulate "theoretical entities" named by "theoretical terms." In doing so, they hope to achieve as close a correspondence as possible between their postulated entities and the actual elements of the real (or external, or objective) world.

Thesis (1) is common to all realisms. To deny it, and thus to make reality dependent upon some perceiving subject, is to advance some form of idealism. To accept (1) but deny (2) is probably the typical attitude of the "man on the street," but it is philosophically gratuitous, for it lacks the explanatory power of other realisms, and requires that the perceptions of any two individuals must be similar, if they are simply mirrorings of such a readily accessible reality. The denial of (3) is characteristic of Kantianism, though Kant himself tempered it by allowing sufficient scope to the activity of the mind, so that the world which is the object of science is far more than the passive accumulation of sensations. In this respect he clearly went beyond Hume, who could not accept such notions as 'substance' and 'causality' insofar as they could not be derived from immediate perceptual data. Kant might be regarded as accepting (4) while rejecting (3). Thesis (4) gives the job of investigating the external world specifically to science, rather than to some special discipline such as pure metaphysics or phenomenology. I am here taking "science" as a collective name for the historically evolved disciplines of physics, chemistry, biology, etc. This position may be controversial. It might be argued that only a "rationally reconstructed" unified science is capable of such an investigation, but I will argue against such a notion below. Thesis (4) would be rejected by, for example, scholastic metaphysicians, who believe that they have their own key to reality. Thesis (5) is primarily a corollary to (4). However, it goes somewhat further in emphasizing that the intended correspondence between science and reality is one of detail, and includes the material as well as the formal aspects of the postulated entities. Eddington might be taken as denying (5) while accepting (4).

We may make an additional distinction between what may be called 'aggressive or dogmatic realism' and 'hypothetical or programmatic realism'. The former confidently asserts the truth of each of the five theses above, and claims that we can be sure that science is in fact revealing more and more of the actual structure of reality. The latter is more tentative. It recognizes that the sciences, especially physics, have postulated radically different things as ultimate elements at different stages in their histories, and that we have no assurance that the basic entities currently fashionable will continue to enjoy such a preferred status in the future. However, this consideration need not drive us to complete relativism. In the first place, by any criteria of coherence, adequacy in representing a wide variety of empirical facts, and explanatory power science does show some sort of progress, even though such progress cannot be regarded as a linear accumulation of new facts, as Kuhn has brilliantly demonstrated.[2] Secondly, anyone who accepts theses (1)–(3) must give some indication of how we can learn anything about the world, and what the purpose and intellectual value (aside from its successful technological applications) of science is. Accepting (4) enables him to do both things at once. Finally, the belief that he is actually discovering the way the world really is can have significant heuristic value for the practicing scientist, in giving him a sense of what he is doing and guiding his future research. Realism can thus function as a regulative idea or ideal, in Kant's sense.[3] Körner has rightly emphasized the need for such regulative ideas in organizing and directing research.[4] Regulative principles may come from a variety of sources. They may reflect cosmological beliefs, independent of any scientific theories, about the structure of the universe as a whole. They may reflect the technological standards or limitations of their era, which in turn are crucial in providing empirical tests. They may reflect esthetic or moral values, and may be governed by a variety of religious and political influences.[5] Nevertheless, some such regulative principles will always be present to provide constraints on scientific theorizing. Once accepted, they then determine what Toulmin calls the "ideals of natural order" for the scientific community, which in turn give criteria for deciding what sorts of phenomena need to be explained, and what counts as an adequate explanation of them.[6]

Some may appear rather specific and arbitrary, and thus will readily be rejected by a scientist who does not accept the underlying assumptions; others, like the principle that the entities with which physical science deals are independent of any particular observer, or the principle that all actual events are related to antecedent events by principles of causality or sufficient reason, are harder to give up while still trying to analyze phenomena rationally.* It may indeed be the case that some can never be given up, because they reflect the fixed and unchanging structure of our cognitive apparatus. If so, these principles will then be constitutive. This of course was Kant's position, and one might also try to justify it by appeal to universal features common to any language. However, we need not *assume* that any particular regulative principles are also constitutive. It is the mistaken interpretation that such principles must be constitutive of science (which dogmatic realism might accept) that seems to underlie Pap's denial that science has undemonstrable metaphysical presuppositions.[7]

My approach in this treatise will therefore be to take an attitude of hypothetical scientific realism toward geometrodynamics as a specific physical theory. I will assume that it purports to say something about a real external world, and that the referents of its basic concepts should be the elements of such a world. This is especially important in that geometrodynamics is one of the most abstract and comprehensive—indeed, cosmological—theories ever developed seriously in science. But as hypothetical rather than dogmatic realists, we should be prepared to admit the possibility that the theory may not contain such a coherent picture of the world, or else one that cannot be connected with any of those features which not only everyday experience but also other accepted physical theories have led us to take as a proper ontology.

In this case the philosopher of science is confronted with a trilemma: either (1) he must give up the regulative principle of realism altogether, in which case he must find some other principle that can replace it with at least equal heuristic value; or (2) he can hold on to the principle in general, but insist that it does not apply to this particular theory, in which case he must give a more than ad hoc account of what distinguishes this theory from others to which the principle does apply; or (3) he can hold on to the principle in general and use it as a ground for criticizing the theory, either by rejecting it outright or recommending that further refinements be made. I maintain, though a detailed marshaling of supporting evidence would be beyond the scope of this work, that the history of science would reveal many examples of all three approaches, not only for realism but for most of the other regulative ideas which have been em-

*It should be noted that both of these are challenged in the "Copenhagen interpretation" of quantum mechanics. See Niels Bohr, *Atomic Physics and Human Knowledge,* John Wiley & Sons, New York, 1958; N. R. Hanson, "The Copenhagen Interpretation of Quantum Theory," in Arthur Danto and Sidney Morgenbesser, eds., *Philosophy of Science,* pp. 450–471, Meridian Books, New York, 1960.

ployed in science. Indeed, on particular questions the scientific community has often shifted from one attitude to another. In any case, I will try to see just how far one can go with a realistic interpretation of geometrodynamics, and what sort of picture of the physical world can be elicited from the theory.

I believe that most scientists—and in particular Wheeler, the leading developer of contemporary geometrodynamics—have adopted realism in the conduct of their research (which after all is the area to which regulative principles apply). They have presumed that their theoretical terms referred to real (though sometimes very odd!) things, rather than being mere calculating devices. If this is so, it should be surprising that so many philosophers, notably the "empiricists" and their more extreme cousins the "positivists," have never accepted scientific realism at all. Indeed, they have offered what they consider cogent reasons for rejecting even its more temperate claims. I wish to argue that much of contemporary philosophy of science, and the criticism of realism that it offers, is based on two related assumptions or biases, which I claim are not well founded. I will refer to them as the "empiricist error" and the "logicist error." Taken together, they constitute the core of logical empiricism, though in dubbing them errors I do not mean to suggest that logical empiricism has not contributed many significant insights to the philosophy of science.

In criticizing what I am calling the "empiricist error," I am identifying empiricism with the set of theses given below. I do not pretend that every philosopher who has called himself an empiricist would subscribe to all these statements, but when taken together they constitute at least a representative example of contemporary empiricism. (1) There is a certain kind or class of statements, which will henceforth be called observation statements, whose truth or falsity can be determined by direct, unaided observation. (2) On the strongest version of the principle, a version certainly not accepted by all members of this school, the truth or falsity of the observation statements, once determined in this way, is forever fixed and not open to further challenge. They thus provide an absolutely certain class of empirical truths. (3) These statements can be formulated in a language which is neutral, in the sense of not containing any terms which refer to unobservables nor expressing any commitments to a particular theory, even one of a very low order. It may be called the (or an) observation language. (4) General statements can be formulated in the observation language. (5) But more complex generalizations can be expressed in a more compact and convenient way if one introduces terms which do not refer to observables, thus taking us outside the observation language. These may be called theoretical terms. (6) Scientific theories are sets of statements which relate these 'theoretical terms' to each other as well as to observation statements by means of what are called 'theoretical laws' and 'correspondence rules'. (7) Let us now assume that we have established pretheoretically what is to count as an observation statement in a given area, and furthermore that we have found some subclass of them to be empirically true (and presumably others false). The basic goal of scientific theories is then to reproduce the true observation statements as accurately as possible, rule out the false ones, and then make specific predictions as to which of the remaining ones will be true, all by deduction from the principles of the theory. These predictions can then be tested by new observations and experiments. The observation statements thus have ontological as well as epistemological priority, in that theories have no empirical content above and beyond them. The situations that they, rather than the theoretical statements, describe are taken to be the basic realities of the world.

It should be noted that I am emphasizing the linguistic aspects of empiricism, in particular the notion that it is possible and desirable to construct a pure observation language as a basis for physics. This of course differs from traditional (Humean) empiricism, which emphasizes the concept rather than the statement and looks at concept-formation either psychologically, in terms of the association of atomistic sense impressions in a sort of "mental chemistry," or in terms of logical constructions from such data.[8] Logical constructs can of course always be analyzed back into the original data, just as in my version of empiricism the content of any theoretical statement can always be analyzed back into a set of observation statements. The linguistic version is somewhat broader in scope since it is not restricted to phenomenalism, which is hard pressed to avoid subjective idealism and give an account of the intersubjective testability of scientific statements. Carnap has emphasized the possibility of using either psychological or physical predicates and phenomenalistic or physicalistic languages as a basis.[9] The important assumptions for our purposes, and the ones that I shall criticize, are that a *single* language, whose basic terms all belong to some definable class, is adequate to express the content not only of physics, but of the whole of science; and that such a language involves no commitment to any particular theory.

Pure phenomenalism must be able to give a complete and adequate characterization of exactly what are the basic ingredients given to us in experience. Furthermore, such a description must be immediate, in the sense of not involving any background knowledge or expectations about the sources, objects, or referents of such data. But there is certainly no agreement on just what these elements are. Most commonly they have been interpreted atomistically, as something like a small red patch in a portion of one's visual field. But such notions have been challenged by Gestalt psychologists, who claim that we directly perceive wholes in a way at least partially determined by psychological expectation or 'set', and from them we later abstract out the 'atomic' parts in interpretative analysis. Even Carnap, in the *Aufbau* system, chose as his element the total instantaneous visual field.[10] In any case, any claim to offer a pure description of 'the given in experience' should be part of a sophisticated theory of sense perception, and this is a task for empirical psychology, not introspective armchair philosophy.

Sellars has argued persistently and effectively against 'the myth of the given'.[11] He rightly stresses the role of language, as a learned intersubjective medium of communication, in determining the very character of our perceptions. We cannot think at all (and here I take 'describing' as a kind of 'thinking') without some sort of language with more or less definite concepts and rules for combining them into judgments. These provide classifications and distinctions for the ordering of experience. Thus an infant's senses may be stimulated, but he cannot be said to have any experience until he has learned at least the rudiments of a language. Kant was quite correct in this criticism of Hume, that the mere passive reception of sense data without any concepts to categorize them cannot be called experience. If this requirement restricts the possible range of the term 'experience', at least it includes everything that is relevant to empirical science, when the latter claims to be based on 'experience'. Kant's mistake was to assume that these categories could be derived from the unchanging structure of the human cognitive process, rather than being in large part a product of the social and historical experience of other people which has become embodied in an actual language. The actual categories used, in particular the choice of ordinary physical objects as the basic substances, have demonstrated a certain survival value, though this does not imply that they must always be the most convenient for all purposes, including those of a particular science—a mistake which some ordinary-language philosophers tend to make.

Sellars has emphasized that, unless one brings in God as a permanent perceiver (Berkeley's option), one must allow possible as well as actual sense contents if one is to be able to make generalizations that can serve as ground-level scientific statements, or even reproduce the content of physical-object statements.[12] But in this case severe problems arise. It is extremely difficult to formulate generalizations which do not include terms referring to features of the spatio-temporal framework of physical objects; and even if such generalizations could be made (as assumption (4) of my version of empiricism requires) they would have to be only 'accidentally autobiographical', to use Sellars's phrase. But Sellars shows that one can formulate only essentially autobiographical generalizations about one's private sense contents, in that one can separate out the peculiarities of one's environmental history only by using the language of 'public' physical objects. He continues,[13]

And a generalization which is an expression of the contingencies of my existence can scarcely be one of the generalizations which, in the intersubjective conceptual heritage of the race, support the phenomenally conditional sense contents postulated by the phenomenalist.

These arguments all tend to the conclusion that any sense-datum language must be parasitic on physical-object language and cannot function as the observation language which empiricism requires, although sense data might be introduced as theoretical entities in a proposed *explanation* of the world of physical objects.[14] Sellars,[15] Smart,[16] and others have given additional cogent arguments against various forms of phenomenalism, but it would lie outside my purposes to discuss them here. It thus appears that the language of ordinary physical objects is the leading candidate for our basic observation language.

Let us consider the language of ordinary, macroscopic physical objects as a possible observation language. This now appears as the last bastion of empiricism, if its assumption of a universal observation language as a basis for all of science is not to be considered a mistake. Here Sellars offers the striking argument that the very reasons which led one to reject phenomenalism are also reasons for rejecting the physical-object language. The crucial assumption for empiricism is here the fourth, that there are lawlike generalizations, whether universal or statistical, which adequately describe the behavior of these macro-objects; and that such generalizations can be formulated without introducing any theoretical terms. But Sellars asks, "Is the behavior of macro-objects even statistically lawful in a way which leaves to theories only the job of deriving these laws from its postulates and correspondence rules?" and answers in the negative.[17]

In order to evaluate the force of Sellars's argument, we must reconsider the nature and function of the scientific enterprise. I will make the following claims: (1) science is a *public,* interpersonal activity, rather than the construction of a personal *weltanschauung;* (2) its goal is to *explain* observed phenomena; and (3) explanation consists in making these phenomena *intelligible* to the scientific community, in terms of the standards and ideals which it holds at a given time. Thesis (1) should not be controversial. Indeed, it was largely to preserve (1) that the logical positivists were led from phenomenalism to physicalism. To be sure, it has been denied by Bridgman in his later writings.[18] But Bridgman is led to his 'methodological solipsism'—with his insistence that there is only 'my science', 'your science', etc., rather than something which may be called 'science' underlying them all—not by any previous metaphysical commitments but by attempting to apply the principles of his operationism consistently. Whereas in his earlier books[19] Bridgman had stressed the physical, manipulative operations of the experimentalist, he later emphasized the fact that the logic of the term 'operation' requires that it be performed by only one person, so that any interpersonal comparison of operations and their results depends on more or less arbitrary conventions. Thus unless one follows Bridgman in identifying concepts, scientific or otherwise, with sets of operations, one need not accept his subjectivist conclusions.

Thesis (2) would also be accepted by almost all philosophers of science, but the problem here is that 'explanation' can mean many different things, as Toulmin has pointed out.[20] Thesis (3) is therefore offered as a means of unpacking the content of (2), and it is certainly a controversial claim. What I wish to stress is that explanation, *as understood by the scientific community* (whose ideals philosophers of science ignore at their peril), is primarily a *pragmatic,* rather than semantic or syntactic notion. To make a phenomenon intelligible is to make it accord, or fit in, with the standards of human rationality which prevail at a given time. It is easy to misunderstand this position. In the concrete situation of explaining something to some particular person, the pragmatic aspect is that reasons are offered until that person is satisfied. What satisfies an individual may depend strongly on subjective, emotional, and irrational grounds. If he wishes strongly to believe, he may take the merest shred of supporting evidence as conclusive; if, on the other hand, he is dead set against the conclusion, he may shut his mind to the most cogent arguments. Clearly a scientific explanation must not be one that depends on these individual peculiarities. But the important fact is that a scientific community does exist, and does set up interpersonal standards which are taken to be objective and rational. Among these standards are the regulative principles mentioned earlier. I do not mean to suggest that philosophers (as being outside the scientific community) are not entitled to criticize these standards. But when the standards do change, the changes are usually initiated by the scientific community itself. Such changes may introduce new restrictions, such as the requirement that laws be given a rigorous mathematical form; or subtract them, as when physicists stopped requiring an account of the 'real nature' of such forces as gravity.

Among the most important of such standards for science is that all natural phenomena are governed by laws, and can be subsumed under them as special cases of certain *kinds* of phenomenon. This implies that there are no miracles or 'loose ends' in the universe, whose existence can be explained only by supposing unique acts of divine intervention. It is this particular principle which the Hempel-Oppenheim theory of explanation serves to explicate.[21] The explanandum must be deducible or inductively inferrable from universal or statistical lawlike statements, which classify the phenomenon as of a certain kind in terms of essential properties or causal antecedents, and initial conditions which particularize the application of the law. As long as the scientific community retains this standard (which, after all, is of fairly recent vintage in Western intellectual development) the Hempelian analysis may be considered a *necessary* condition for scientific explanation. But it is not *sufficient;* it would be sufficient only if universal lawfulness were the only regulative ideal held by the scientific community, and in fact this is not the case. Mere conformity to the Hempelian model would not guarantee the adequacy of a proposed explanation for the scientific community. Material conditions relating to the sorts of entities represented by terms in the laws and initial conditions are just as important as the formal condition of inferrability, and these must be determined by other regulative ideals, conditions of intelligibility, ideals of natural order, or however else they may be described. There is great diversity in the kinds of laws that are actually used in scientific explanations. To call them all 'laws' may be useful in some contexts, but it may distract attention from more important differences.

We are now ready at last to complete the argument that empiricism, as represented by the conjunction of the seven theses above, is really based on an error. For the question now becomes, Can we formulate laws in an observation language (for which the physical-object language is the only remaining candidate) which will satisfy the scientific ideals of natural order? We might try to interpret the question as follows: Granted that there are such things as physical objects, does the world which they constitute have sufficient internal coherence to satisfy the demands of rationality of the scientist? For it is just this hypothesis which Sellars is denying. Science demands a certain perfection and exactitude even in statistical laws, and this is never achieved if one stays at the level of inductive generalizations in the observation language, or at least not to a degree which could serve as an adequate basis for sophisticated science. And it was this very inadequacy that was our reason for rejecting phenomenalism.

A traditional empiricist view of the nature of scientific explanation proceeds as follows:[22] We begin with pure descriptions of observable phenomena, and then give a first low-level explanation of them by showing that they can be inferred (deductively or statistically) from inductive generalizations formulated in the observation language. Theoretical terms, and laws involving them, may be introduced at this point to explain the inductive generalizations. But the relation between the two kinds of statement is the same as at the first level of explanation: the theoretical laws are simply premises from which the inductive generalizations (which together express their content) can be deduced. They do not refer directly to the particular phenomena, which must therefore be presumed to be covered adequately by the generalizations. The process can be repeated, but Feigl claims that the various levels of explanation differ from each other only in terms of degree of generality, like a series of Chinese boxes, each of which encompasses all its predecessors.

The basis of Sellars's criticism, and mine, is that the low-level inductive generalizations do *not* explain particular phenomena, insofar as the latter are presumed to be consequences of them. At best there is only approximate agreement, which may be adequate for some purposes but not others. In a somewhat different context, Scriven has pointed out three major kinds of inaccuracies which may appear: (1) observable data do not satisfy the degree of precision which they are taken to have in the generalizations; (2) the data are not related to each other in the relatively simple way expressed by the generalization; and (3) no absolute claim to completeness in including all possible factors which might be relevant can safely be made.[23] It is a fact to be explained that there are reasonable approximations but so few exact laws at this level. Sellars therefore argues that "theories about observable things do not explain empirical laws (since these are not even true!); they explain why observable things obey, to the extent that they do, these empirical laws."[24] Thus it is unreasonable to expect theories to reproduce the putative empirical laws exactly; if they did, not only would they fail to give us any new information, but they would even be false, insofar as these 'laws' are!

I consider the foregoing arguments to have established that the empiricist belief in a universal observation language and its adequacy for all of science is indeed fallacious. The Chinese boxes account of theoretical explanation is unable to get started, and in any case cannot account for the "to the extent that they do" clause quoted above. Having cleared the way for allowing to theoretical statements some additional content and function above 'explaining' empirical laws, let us see how in fact they might account for observable phenomena. Sellars says that if we realize that a gas *is* a cloud of molecules behaving in theoretically defined ways, we can understand why the Boyle-Charles law (a typical example of an inductive generalization) does describe the gas's behavior adequately for some purposes, *and* why under certain conditions the law breaks down completely. Furthermore, it suggests sources and interpretations of a second approximation, the Van der Waals law. Schematized, Sellars's argument seems to run as follows: a supposedly observable thing *O* (the gas) is not really as it appears but is rather some theoretical entity *T*; and because *T* is postulated by some definite theory, we know laws governing its behavior in sufficient detail and to a sufficient degree of precision. We also know that an entity of kind *T* will manifest itself to our senses in the form *O*, in terms of which we have tried to make inductive generalizations. From the known properties and relations of *T* we can predict the degree to which *O* should be expected to obey regularities couched in the terms appropriate to it. But this regularity is rarely, if ever, exact, simply because these are not the terms appropriate to *T*, the genuine reality of which *O* is only an appearance. Sellars suggests that his position is Kantian, but only to the extent that the 'noumenon' *T* is knowable, indeed even more knowable than the 'phenomenon' *O*.[25]

Sellars is thus taking a quite radical step. Not only is he claiming that theoretical entities are real, or have an independent existence, but also that ordinary physical objects do not thus exist, that in fact they are mere appearances. He states,[26] "Correspondence rules appear in the material mode as statements to the effect that the objects of the observational framework *do not really exist—there really are no such things.* They envisage the *abandonment* of a sense and its denotation." At this point I must part company with Sellars's version of realism. For he shares with the empiricists (and for that matter, the Platonists) the assumption that only one level of being can be truly real, and that all others must therefore be mere appearances or manifestations of this underlying level. I will hold that (1) this assumption is not required, nor is it methodologically desirable; and (2) if accepted, it weakens the case for scientific realism.

As an alternative to Sellars, I propose what I will call 'the levels hypothesis'. By this I mean the hypothesis that reality should be presumed to consist of a multiplicity of levels, each characterized by a distinctive class of entities, their properties, and the relations which these entities bear to each other. It is important to note that each level has both material and formal aspects, and either or both of these may differ radically in kind from one level to another. Ideally, each level should be characterized by a language that is uniquely appropriate to it. Its terms should correspond to the actual things, events, and processes which are the material constituents of the level, and its syntax to the specific kinds of relations or laws which connect them with each other. A level may very well exist, and its existence may provide a ground for explaining a certain kind of phenomenon without anyone's knowing it; or its existence may be postulated, as in Bohm's 'hidden variable' approach to quantum mechanics, without any of the details of its structure or mode of function being known.[27] But insofar as a level can be appealed to for genuine explanations, it must be presumed to include both material and formal aspects, and these must be adequately articulated. In some cases the kinds of things which serve as elements for the level may be postulated before an elaborate theory of their interrelationships is developed, as was the case in classical (Democritean) atomism; in others a sophisticated mathematical theory may precede a generally acceptable ontological interpretation of its theoretical terms, as is the case in quantum mechanics with its continuing debates on the meaning of the probability amplitude. Formalists and positivists who believe that the formal role alone is a necessary and sufficient condition for scientific explanation claim support from the situation in quantum mechanics, but such a move seems merely a counsel of despair in the face of an already unsatisfactory situation.

The levels hypothesis provides an analysis of two distinct kinds of explanation. Insofar as one insists on remaining within a particular level, explanation can proceed only in terms of statements of greater generality or scope, as determined by the amount of diversity possible within that level; and as long as such generalizations are well confirmed, this is a perfectly legitimate move. However, there is another kind of explanation, which consists in carrying the problem to a different level, and coordinating a part of that level to the explanandum. (The nature of the coordinating relation will be discussed below.) Since, at least as a start, there may be an indefinite number of such levels, there are indefinitely many kinds of such theoretical explanation, depending on the particular levels involved. Sellars's approach must ultimately allow only one such kind of explanation, since the explanans must come from the one and only level which is allowed genuine reality, i.e., that of theoretical entities. This might not be an objection if *all* the entities postulated by physical theories could be taken to constitute a single level. But I will argue that this cannot be done; nor can it be done in the case of 'observables'.

My last objection to Sellars could be established completely only in terms of a more precise conception of what constitutes a separate level. Furthermore, if the coordination relation in theoretical explanation is to be asymmetrical, we must have some criterion for determining which level is greater in explanatory power, so that, given an explanandum from a certain level, we can at least limit the sources of a possible explanans. Most philosophers who have adopted the levels hypothesis, like Bunge,[28] have assumed that the levels of reality constitute a single linear hierarchy. Sometimes this is done in terms of size and resulting complexity, as in the series: elementary particle, atom, molecule, crystal, tangible physical object, etc. The rationale for this approach is that each range seems to have a set of properties and laws peculiar to it. As one goes down the list, new laws and properties come to the fore at each stage, as manifestations of the greater complexity of structure, and others, while not disappearing altogether, at least lose their importance or centrality. This approach seems most useful for distinguishing between levels within physics. A second order for this hierarchy emphasizes the evolution in time of one level out of another. This is common in biology and the social sciences. Yet a third distinguishes the levels in terms of degree of abstraction, ranging from concrete particulars, with their indefinitely many 'accidents,' in Aristotle's sense, to the most abstract and universal features common to everything, such as (say) extension. This often helps to distinguish between the various sciences, e.g., physics, astronomy, chemistry, biology, etc. My contention is that no single definition of a level can be given, though each of these three criteria proposed for establishing a hierarchy of levels calls attention to important distinctions, especially since they seem to draw the lines in the same places. In practice, scientists have little difficulty in recognizing when a levels transfer, to new kinds of entities and laws, is being made. It is also possible that a single theory, together with its extension or range of application and interpretation, may be capable of introducing a new level by itself, and in effect creating a subject matter. I will argue that this is what has taken place in the case of geometrodynamics.

It may be thought that I have been guilty of a confusion between the ontological and epistemological meanings of 'level.' For a realist, the explanatory power gained by a levels transfer derives from the fact that the new level really exists, and in addition explains why the old one exists, or seems to exist. However, I have been characterizing levels in epistemological terms, in particular suggesting that a theory, which is certainly an epistemological device, can define a level in terms of its range of validity. Following Bunge, let us distinguish between cognitive and ontic levels. The various criteria mentioned above may distinguish among cognitive levels, but what guarantee have we that such distinctions correspond in some way to ontic levels, levels in the actual structure of the real world? In particular, how do we know that there is not just one ontic level (in which case Sellars would be correct), on which the various cognitive levels are merely perspectives? In keeping with his assumption (2), a realist must certainly allow the possibility that the levels structure assumed by his science may not agree with the actual levels structure of reality. Furthermore, cognitive levels are fluid, with new ones being discovered or developed and old ones being assimilated, in a way that could not be postulated of ontic levels except by an idealist, who collapses the distinction.

In order to answer this question we must turn to the problem of theoretical reduction. I have so far avoided this question because I do not intend the levels hypothesis as a denial of the claim that (at least sometimes) one level can be reduced to another, at least as long as reductionism is taken as a hypothesis, rather than a pontification about the way the world really is. I will assume, along with Carnap, Nagel, and most other philosophers of science, that a theoretical reduction of one level *A* to another level *B* requires the following two things: (1) Each term that appears in *A* should be definable in terms of the language of *B*. At best such definitions should be explicit, but if this fails one can use something like a system of 'reduction sentences.' Putting the same point in more ontological terms, each entity in *A* should be fully characterized in terms of the properties and relations of the entities in *B*. This would constitute a reduction of the *material* aspects of *A* to those of *B*, and provide a translation of the language of *A* into the language of *B*. (2) Once this translation has been carried out, we can express the laws of *A* in terms of the language of *B*. But at this stage these *A*-laws will appear only as unjustified assertions within *B*. To complete the reduction, we must also show that the *A*-laws, as expressed in the language of *B*, can indeed be derived from the basic laws of *B* by some deductive procedure. Success here would constitute a reduction of the *formal* aspects of *A* to those of *B*:[29] In principle, it is certainly possible that such a complete reduction could indeed be carried out for a given *A* and *B*. However, five cautions should be borne in mind by anyone claiming the success of such a reduction.

1. The reduction of one level to another must always be carried out in full detail before it may be assumed to have occurred. Every term of *A* must be defined, and every law must be derived. Too often such reductions are claimed on the basis of a partially completed program. A few terms and laws are accounted for, and then, by what can only be described as a spectacular inductive leap, it is assumed that the rest will fall out in a similar pattern. Quite the contrary: it is likely that initial successes reflect examples well chosen because they do work, while the more difficult cases are never accounted for at all. The program is certainly worth trying, but there is no guarantee of ultimate success.

2. Each level to be reduced must be articulated in the fullest detail possible before we concentrate on carrying out the reduction. Certainly it is important to know just how much we have to reduce, or we can never know whether the reduction is complete. Each new discovery in *A* poses a problem for the reductionist. To avoid biasing the case, *A* must be studied in its own right, as a genuine level, rather than merely as a candidate for reduction. If we take the latter attitude, we may very well fail to appreciate the full complexity or significance of *A*. In the attempt to reduce psychology to neurophysiology, such philosophic disciplines as ordinary language analy-

sis and phenomenology can play a very important role in revealing this complexity of human phenomena, even though they need not be considered the last word on the subject. Sellars has made a similar point in a somewhat different context. He points out that the 'manifest image' of man, as derived from the study of human language, modes of perception, and general cultural heritage, differs radically from the image which emerges from contemporary scientific theory, and suggests that it may be methodologically advisable to enunciate all the details of each image separately before attempting to tie them together in a way which might prove premature.[30] Of course Sellars is here recognizing only two levels, the 'manifest image' and the 'scientific image.'

3. We may, however, speak of partial reductions in which a part (which is recognized not to be the whole) of one level can be reduced to a part (or the whole) of another. Likewise we may speak of the *degree* to which one level appears reducible to another, on the basis of the amount of success already achieved along those lines. Such partial reductions may be extremely significant. However, each must be evaluated in terms of its particular accomplishments, which may differ greatly from case to case, rather than as mere parts of a more general enterprise which ought to be carried out. Their importance may be more scientific than philosophical.

4. A level is characterized by means of categories and concepts appropriate to it. Any reduction of such a level should reproduce these classificatory distinctions, or at least explain why it is natural or appropriate to make them at the places where they are made. (This does not preclude the possibility of recommending some changes, but if changes were made on a wholesale basis it would no longer be clear just what was being reduced.) In the case of Eddington's two tables,[31] it may be correct to say both that "tables are really just swarms of atoms" and (separately) "chairs are really just swarms of atoms," but this does not solve the problem of explaining, at the level of atoms, how tables differ from chairs.

5. The relation of reduction has been assumed to be transitive by philosophers like Carnap, who claim that if biology can be reduced to chemistry, and chemistry to physics, then biology has effectively been reduced to physics. If the reduction is complete at each stage, then the relation is indeed transitive. However, if the reductions are only partial, which at the present stage of science still seems to be the case, we cannot assume transitivity. The reduction of biology may require specifically chemical terms, laws, and distinctions. Unless all of these can be formulated in the language of physics proper, in such a way that both the referent of the chemical term can be isolated and its syntax reproduced by the corresponding expression from the language of physics, then the biological has *not*

been reduced to the physical. Furthermore, even a single biological concept, like that of an 'egg,' may require a tremendous amount of chemistry for a complete explanation. The more chemistry we need, the less likely it is that all of this chemistry will have already been included in a partial reduction of chemistry to physics.

Unless one accepts completely the Kantian viewpoint that the noumenal world is unknowable, these difficulties for the reduction of cognitive levels must also hold for ontic levels. If the world *is* really simple in containing only relatively few kinds of things and processes, all of which can be connected into a unified level under a single set of laws, it is difficult to explain why it should appear to have the many-dimensioned diversity and complexity reflected by the existence of so many different cognitive levels. (I take it to be an undeniable empirical fact that we now have and use many cognitive levels, whose mutual compatibility is at least unproven.) Each such cognitive level makes its categoreal distinctions in radically different places (and for good reasons in each case), and this fact appears inexplicable if there is only one ontic level whose distinctions, wherever they may lie, are thus the only really significant ones.

These remarks must not be taken to imply that there is a neat one-to-one correspondence between cognitive and ontic levels, with the former simply reflecting the latter. Bunge has argued decisively against such an interpretation.[32] He claims that the lack of perfect correspondence results from the fact that even a single concrete particular (considered ontologically) may include indefinitely many aspects of which we can never know, because of the limitations of our faculties of perception; and that the very nature of our cognitive process, as well as its objects, determines the results of our cognitions. This essentially Kantian attitude includes the fact that human cognition is mediated by certain abstract concepts and idealizations which are never presumed to have any ontological referents, though they may be part of a theory which does when taken as a whole. In addition, Smart has emphasized the importance for a realist (though more especially for a materialist in his sense) of thinking of man, with his perceptions and cognitions, as a part of nature, so that cognition may be understood as a particular kind of natural process. But there is no reason why this, or any other such process, should mirror the totality of nature better than some other.[33]

Bunge now makes the very interesting suggestion that the multiplicity of ontic levels does not simply limit human knowledge, but is the fact which makes partial knowledge and explanation possible. If we explain a phenomenon by relating it to a variety of different cognitive levels, each such explanation throws light on certain aspects and thereby makes us understand the whole phenomenon better. It allows us to neglect, in a particular explanation, all factors except those from a particular level of interest. If there were only one level, we could not isolate these factors and would therefore find ourselves in the dilemma of Bradleyan idealism: we would have to know everything (about this one and only level) before we could know or explain any particular thing, since otherwise we would have no assurance that we had not neglected any essential factors.[34]

I have mentioned above that most philosophers, especially those with reductionist leanings, have assumed that the levels form a single, one-dimensional hierarchy, in which any pair can be ordered unequivocally by a relation of 'higher' or 'deeper'. The actual criterion proposed for this job may be the degree of generality or comprehensiveness, certainty, simplicity, abstraction, or anything else; the point is that these philosophers believe a single criterion can do the job. But this assumption is not really warranted. Feyerabend has argued that the laws and concepts which appear at various levels are strictly incompatible with each other, rather than somehow including each other. The facts which they are designed to explain are radically different as well, even if they are understood as observables. Since, as was argued above, there is no such thing as a pure observation language perfectly adequate for all descriptions, our descriptions will use terms which help assure their relatedness to the terms of the level in which they are being explained. While some of the 'facts' expressed in one language can be directly correlated with those expressed in some other language, other 'facts' cannot be—and in any case there will be striking differences of emphasis.[35]

It is generally recognized that not every aspect of a given level may be reproduced at all, or at any particular one of the other levels. Insofar as this irreducible residue refers to certain (usually qualitative) features of the level being explained or reduced, it has been the backbone of the case for epiphenomenalism. On the other hand, some aspects of the explaining level may not appear in the level being explained, although this is usually considered to be a matter of legitimate approximation, as when the relativistic effects on high-speed elementary particles are neglected in the observation of slow-moving macroscopic objects, even though the latter are assumed to be composed of the former.

Since each cognitive level is assumed to have a uniquely appropriate language (at least ideally), we may pose the problem of coordination in terms of the degree to which such languages are mutually intertranslatable. The language of physics and the language of sense data seem to be at opposite ends of any hierarchy of cognitive levels, but the claims of physicalism and phenomenalism, respectively, were that the content of every legitimate statement could be completely translated into the one language or the other. But in fact this can be done only in a way which biases the case, by defining 'content' in the terms appropriate to whatever is taken as the ground level. On the contrary, it seems more reasonable that the content of a statement at any given level should be expressed by reference to the terms of its own level; namely, what it says about the things and processes which comprise that level. Once this is done the translatability thesis becomes problematic. Feyerabend, and to a lesser extent Kuhn,[36] have gone so far as to suggest that there is no real translatability at all. Like most relativisms, however, this seems a mere counsel of despair, and it loses the advantages of the levels hypothesis. Its effect is to suggest that there is only one acceptable explanatory level at any given time, since other levels cannot be compared with it. During scientific revolutions this level is simply jettisoned and replaced by a new one.

If the ontic levels are not completely independent of each other, the cognitive levels should not be independent either. There must be some connections between them, which would result in at least a degree of translatability that preserves content. I have previously used the term 'coordination' to describe the relation holding between different levels. Insofar as level B is used to explain phenomena originally described in the language of level A, I will now say that level A is *grounded* in level B. It might be asked how my notion of 'grounding' differs from Sellars's notion of 'identifying with', which is essentially the relation of an appearance to its underlying reality. But the point which I wish to stress in the strongest possible way is this: the coordination is not carried out on a piecemeal basis, with parts of A identified with the 'corresponding' *parts* of B in succession. For an adequate explanation of even a small part of A may in fact require anything up to the *whole* of B; and conversely, two phenomena which are quite distinct in A may appeal to the same parts or aspects of B. The reason for this is the fourth objection to reductionism mentioned above: that B cannot (in general) reproduce all the distinctions important to A or explain why they should be made where they are. If they did fall in the same places, we would have a true reduction and Sellarsian identification. But while it may be part of a perfectly good explanation of tables to appeal to collections of molecules and point out that the being of tables, as well as that of chairs and other macroscopic physical objects, is grounded in that of molecules (and so on down to elementary particles and below), tables are *not just* collections of molecules, as the identification metaphor would lead one to say. The organization, the wholeness, the specific identity, or what Aristotle would call the 'form' of the table are what do not appear at the level of molecules.

A degree of translatability is therefore achieved in this way: a statement is made in the language of a level *A*, and has a content expressible in *A*; to the extent that *A* is grounded in *B* or, more generally, that *A* and *B* are coordinated, the statement also has a content in *B*, expressible in the language of *B*. Now this notion of coordination does not require that for any two levels, selected at random, every aspect of each must have a coordinate in the other. For example, one aspect of *A* may have a coordinate in *B* but not in *C*, while another has a coordinate in *C* but not in *B*. Assuming that *B* and *C* likewise have coordinate aspects, I would conclude that *A, B,* and *C* are mutually (though not perfectly) coordinated. It is very rare (i.e., it happens only in the case of a genuine reduction) that the full content of a particular level, or the sum of the contents of the totality of statements in the language of that level, can be expressed in the language of another level without residue. A more serious difficulty arises if a particular statement appears to have no translation into *any* other level. The ontological significance of such a statement is questionable, especially since, as mentioned above, any language makes use of abstractions and idealizations which arise primarily from the human cognitive process rather than from the external world. Such a statement may thus do nothing more than reflect the peculiarities of our chosen symbolism. If a statement seems to have an invariant significance between at least two levels, it is then more plausible to assume that it says something about the real world.

In fact, the various levels have many common aspects, and this is reflected in language. I have insisted that each level may be characterized by a particularly appropriate language, with distinctive terms and relations (laws). However, this by no means implies that these languages will have no terms and relations in common. Indeed, I contend that the great majority of terms applicable at a given level are cross-level terms, applicable to a variety of levels, and in some cases, to all. This point was recognized by Aristotle, who insisted that along with the concepts and principles peculiar to any given science, there were some which cut across all the sciences. These Aristotle considered to be the proper object of first philosophy, or metaphysics. Since Aristotle is normally considered a paradigm of an anti-reductionist, it is noteworthy that he sees in the individual term a source of unification. It may be suggested that even though the same word may be used at two levels, it means different things in each case. However, in an important sense the meaning of a term and its functional role within a level is the same, and a study of the reasons for sameness and difference of meaning or role between levels can shed light on many interesting philosophical questions.

The notion of cross-level terms provides a useful basis for criticizing the observational-theoretical dichotomy. Putnam has pointed out that this distinction was first intended to apply to terms in a language, but later extended to statements, where an observation statement is taken to be one that contains only observation terms and logical vocabulary, while a theoretical statement is one that contains at least one nonobservational term.[37] Now 'terms', as used in the distinction, may refer to things and processes (entities) on the one hand and properties and relations on the other. The following questions then arise:[38] Can observable things and processes have unobservable properties? And can unobservable things and processes have observable properties? In my terminology, one might rather ask, Can entities in level *A* have properties appropriate to level *B*, where *A* and *B* are any two distinct levels? If the languages for each level were completely isolated, the answer to these questions would have to be no. But in fact they are by no means isolated. And if I am correct here, this is a further argument in favor of the levels hypothesis. It might be argued that if one allows all the distinct cognitive levels suggested by scientific theories, then of course there are cross-level terms connecting them. But if 'observational' and 'theoretical' terms each define a tightly closed and self-sufficient realm, with the connection between them mediated only by 'correspondence rules' which are largely arbitrary and conventional, then one could argue that the observational-theoretical distinction has a far greater ontological significance than that between any two levels. Consequently, the observational and the theoretical are the only two real (ontic?) levels.

Putnam offers four good objections to the observational-theoretical dichotomy:[39] "(1) If an 'observation term' is one that cannot apply to an unobservable, then there are no observation terms. (2) Many terms that refer primarily to what Carnap would class as 'unobservables' are not theoretical terms; and at least some theoretical terms refer primarily to observables. (3) Observational reports can and frequently do contain theoretical terms. (4) A scientific theory, properly so-called, may refer only to observables (Cf. Darwin's theory)." For our purposes (1) and (3) are the most interesting. With regard to (1), we may say that the properties and relations which a theoretical entity is taken to have are determined by the theory in which it appears. This point should be conceded, especially by anyone who insists that the axioms of a theory are 'implicit definitions' of the nonobservable terms which they contain. There is no reason why a theory should not state explicitly that color predicates, say, are applicable to the entities postulated by the theory. A theory always represents a degree of abstraction from the totality of properties of any concrete phenomenon. A molecule, inasmuch as it is extended in space, may reasonably be assumed to have some of the other properties of macroscopic physical objects. But some of these properties may not be needed for the explanatory purposes of the theory; in this case the entities are not assumed to have them, in order to avoid unnecessary complications in the theory. Furthermore, the assumption of some of these properties may actually contradict the principles of the theory, a situation that has arisen most strikingly in quantum mechanics. Through the theories in which they appear, we find that molecules need not—and electrons cannot—be considered colored. But other kinds of unobservables might be, as Newton supposed when he postulated that red light consists of red corpuscles.[40]

Before proceeding further, we must analyze the very concept 'observable'. If it is to be contrasted with 'theoretical', it must be understood as meaning 'directly observable', without the mediation of any theories. If this interpretation is taken in a phenomenalistic or even physiological sense, it should be justified by a very highly confirmed psychological theory. In any case, there are grave difficulties. Maxwell has argued that if we refuse to allow looking through a microscope as a case of direct observation, we may ask why this restriction should not also rule out the use of eyeglasses, window panes, and even the thermal currents in the ambient air.[41] At least there appears to be a continuum from immediate to mediated observation. If all observation is mediated to a greater or less degree through a physical medium, we must have some sort of theory about the influence of that medium. (I include as a limiting case the theory that certain kinds of medium have no influence, though this seems to be largely a matter of convention.) But then every observation report is mediated by concepts as well as by media and instruments, the concepts of the theory of that medium. Maxwell and more especially Hanson[42] and Kuhn[43] have pointed out that we never 'just see' any phenomenon, in the sense of 'seeing' as a human *activity* involving awareness rather than mere passive stimulation of the visual organs. What we sense, the features in the continuous sensory flux to which our attention is called, is determined by our expectations, and these depend on the state of our knowledge, as expressed in a language which we consider capable of permitting an adequate description of the phenomenon. Popper has pointed out that even infants have certain expectations (which may be called 'low-level theories' for them), which are constantly being revised to the extent that they fail in helping adjust to present and future events.[44] The Copernican and the Ptolemaist simply see different things in watching the same sunrise.

One great value of science is that it gives us new ways of seeing things that we could not see before. Molecules were certainly unobservables when they were first introduced into scientific theories, but it is because they were believed to be real and distinct physical entities that the principles of molecular physics and chemistry were conjoined to those of optics and other theories in developing the principles for the construction of electron microscopes with sufficient magnifying power to make them visible. It is difficult to understand the rationale of such activity if molecules were assumed to be mere calculating devices, forever unobservable. Maxwell has emphasized that it is unreasonable to assume that any theoretical entity introduced into science must, as a matter of logical necessity, remain unobservable. On the contrary, scientists are likely to start trying to find a way in which it might be observed. The means finally suggested may require extending the *criteria* for what is to count as 'seeing', but if this is done for good reasons it need not change the *meaning* of 'seeing' thereby, a point which Putnam has emphasized.[45]

Thus Putnam's contention (3) is established. Since there is no such thing as a pure observation language, any attempt to describe what we observe in a particular situation must make use of a language that involves some theoretical commitments and thus may be expressed by using 'theoretical terms'. But in the light of (2) and (4), which state in effect that the classes of theoretical and nonobservational terms do not coincide, nor is one included within the other, this is no longer startling. But it indicates that the observational-theoretical distinction has no profound ontological significance.

Putnam has also criticized Carnap's notion of a partially interpreted calculus, where theories are given a partial interpretation in an observation language, and the latter is the only one assumed to have genuine ontological significance. The main basis for his criticisms is the great asymmetry which Carnap believes to hold between the theoretical and observational levels. Without this bias, the notion of partial interpretation becomes quite close to my notion of 'imperfect coordination' between levels, which *can* be a symmetric relation. In accordance with this viewpoint, Putnam has insisted that[46] "Justification in science proceeds in any direction that may be handy—more observational assertions sometimes being justified with the aid of more theoretical ones, and vice versa." If a (more) theoretical language is partially interpreted in terms of a (more) observational one, then the observational also has a partial interpretation in terms of the theoretical. Sellars seems to be moving closer to my viewpoint in his recent writings. For he insists that there is no fundamental difference between explanation of observables by means of unobservables, and explanation of one theoretical framework by means of another. The 'correspondence rules' between the levels involved have the same logical and functional status in both cases, although Sellars still interprets them as reductionistic identifications, rather than my coordinations or groundings.[47]

3 The Logicist Error

We are now ready to consider what I will call the logicist error. I mentioned above that the two errors were not really separate and distinct. Most twentieth-century logical empiricists have held them together, and the requirements of the one have influenced the development of the other. I wish to make it quite clear at the outset that I am not claiming that the use of formal logic in a physical theory is *per se* a mistake, or the even more absurd idea that theories are (or should be) alogical or even illogical. What I am opposing is the notion that a theory must have the same structure as a highly developed formal system, or "uninterpreted calculus."

In order to see the problem, let us consider the possible relations between logic and ontology. Here I will take Aristotle and contemporary neo-Aristotelian logicians like Veatch as one extreme and the logical atomism of Wittgenstein's *Tractatus Logico-Philosophicus* as the other. Aristotle built up his logic from an analysis of the term to the statement, as a grammatical subject-predicate combination of terms, to the argument, as a logical sequence of statements. This final step appears in the formal theory of the syllogism in the *Prior Analytics*. But Aristotle could never consider this a stopping point. His goal was always a scientific explanation of the real world, in the sense which he laid down in the *Posterior Analytics*. For this purpose logic can be no more than an 'organon', a tool or instrument. As long as logical relations mirror relations in the natural world, thereby enabling scientists to move correctly from one true statement about the world to another, well and good; if they fail in this task, they must be modified or replaced. In Veatch's terminology, the whole nature of logical terms and relations is to be intentional, to be *about* something other than themselves.[1] They are nothing in themselves, mere conceptual shadows cast by the real world upon our language. It follows that the terms 'correct' and 'incorrect' are properly and unambiguously applicable to any proposed logical system; and since there is only one real world, there can be only one correct logic. The crucial assumption in this whole argument is that we must have nondiscursive (intuitive) knowledge of the structure of reality prior to the construction of any logical system. This knowledge is what gives us a basis for comparison and criticism of logical systems. Aristotle is well aware of this need. In order to satisfy it he postulates the faculty of the active intellect, which is so constituted as to give us this 'insight' into the nature of reality and leave to logic only the task of unpacking the details of its content. But this assumption has an ad hoc character which makes it the weak link in an otherwise plausible account.[2]

To this approach we may contrast that of the *Tractatus* and its successors. Whereas Aristotle had claimed that we have an adequate knowledge of the general features of the world prior to any language or logic, Wittgenstein insisted in the *Tractatus* that we can have no knowledge of the world at all without a language, where by 'language' he means a formal system with an explicitly articulated logical structure. It follows that we cannot transcend language and appeal directly to 'the world' for advice in constructing and correcting our logical system. But if this is so, it is quite reasonable to construct our logic or ideal language in a purely a priori fashion, guided only by such 'internal' features as consistency, ease of manipulation, economy in use of axioms and rules of inference, and fruitfulness in proving interesting theorems. Since any axioms have a certain arbitrary character, rather than being taken as ultimate truths about the world, it is desirable to have as few independent axioms as possible. (An Aristotelian who holds that we can know a great many independent truths with equal certainty would not be bothered by this matter.) Now the young Wittgenstein was obviously impressed by the fact that an extremely powerful formal system had just been created, the *Principia Mathematica* (*PM*) of Russell and Whitehead. From relatively few assumptions a system had been constructed with far greater power than any previously proposed logics. Its culminating achievement was the reduction of all of classical mathematics to this logic. Today, mathematics; tomorrow, the whole world. The use of mathematics is certainly an important part of scientific investigation, and its applicability to problems of physics could be guaranteed if the world was taken to have the logical structure of *PM*. What grounds could we then have to consider this assumption false?

Thus I take Wittgenstein to be asking an essentially Kantian, presuppositional question in the *Tractatus*. Kant had asked, "What must the (phenomenal) world be like if Euclidean geometry and classical mechanics are to be true of it?" If his analysis is correct, and geometry and mechanics are accepted without question, we must accept Kant's conclusion. Likewise Wittgenstein asks, "What must be the logical structure of the world if the system of *PM* is to be true of it?" where the validity of the latter is taken for granted. If we have no other legitimate method of metaphysical investigation, and no other system comparable to *PM*, I see no way of avoiding some form of logical atomism as the answer. Aristotle's conclusion is turned on its head; since there is only one true logic, there can be only one kind of world, the world whose structure is given by that logic. I will call any claim that the world, understood as the object of scientific investigation, must have a structure conforming to a previously developed logic, and cannot be known except through this logic, "philosophical logicism." It is a stronger claim than mathematical logicism, which contends only that mathematics is structurally reducible to logic.

Logical atomism as a systematic metaphysics began to die out after 1925.[3] Among the reasons for its decline may be cited criticism of the putative reduction of mathematics to *PM* logic and developments within formal logic which indicated that *PM* was not the last word, both of which undermined the basic premise of logical atomism; as well as the extreme reaction against all metaphysics taken up by Wittgenstein's logical positivist heirs. However, most of the latter continued to accept and use the presuppositions of logical atomism, although refusing to carry them to their logical conclusions.

One might ask why so many philosophers of science during the 1920s and 1930s were attracted by the legacy of logical atomism, with its emphasis on logical syntax, focussed on the structure rather than the particular content of statements, and the use of highly elaborate formal systems. The answer seems to be twofold. On the one hand, they believed that the main philosophical problems within the sciences stemmed from a lack of clarity and precision in the language of science, and in the relation of scientific terms to their observation language. If the language actually used were replaced with an ideal formalized calculus, they hoped that complete clarification of the foundations of science could be achieved. On the other hand, having rejected the notion that philosophy could achieve some special ontological insights independent of science, they had to find an intellectually respectable task for the discipline of philosophy. By concentrating on the analysis of syntactical relations, they hoped to make philosophy a science in its own right, which would deal with material at a single level in the same way as do the special sciences, and ultimately achieve comparable progress.

But philosophy has no single level of its own. Its synthetic function is rather to stand outside all the levels and see them in their relations to each other. As Aristotle emphasized, its special province is the cross-level terms, and the modifications such terms can undergo while retaining a basic identity of meaning from one level to another. Even if the sciences are taken together, they do not exhaust all the legitimate questions that can be asked. There is no assurance that at least one science will allow any given question in the belief that some of its chosen methods will be adequate to solve it. Thus philosophy may be called 'the garbage can of the sciences'. It is left to handle the fractious residue that the sciences all leave behind. If none of their methods seem applicable, the presumption arises that the problem is not susceptible of a definitive solution (which may be considered a step of progress), but this does not make the problem less legitimate or important.

It was precisely this residue of problems for which there were no criteria for a definitive solution that the logical positivists tried to reject with their notion of meaninglessness. If, given the knowledge available at a given time, one could not think of a standard way of answering a particular question (at least in outline), the question was to be ruled out as meaningless. Their confusion seems to be this: it is perfectly legitimate and even essential for any particular science to rule out certain questions as meaningless *for that science,* but the positivists jump from this to a notion of 'absolute meaninglessness'. This can be done only by changing the meaning of 'science' from 'the totality of the various special sciences' to 'knowledge in general', or 'Science with a capital S'. Unlike physics, biology, etc. this 'Science' cannot be characterized in terms of its subject matter or degree of abstraction. Therefore it would have to be defined in terms of method or logical structure, independent of content. What I am claiming, on the other hand, is that there is no such thing as 'Science', except as a collective name for 'the sciences'. Furthermore, since each science is confined to a particular level instead of trying to include them all at once, the totality of the sciences cannot exhaust the totality of legitimate and important questions.

Methods of investigation and systems of logic are simply tools developed by human beings in their attempt to understand the world. As a realist, I must insist that natural processes go on whether or not anyone knows about them. But if the external world is independent of our cognitive processes, only divine revelation could guarantee that a single method is always the correct one, or that a single kind of system embodies The Form of the world. If my levels hypothesis is correct, the very fact that the levels are not perfectly coordinated is a good reason for seeking diversity of methods and systems.

Thus I conclude that philosophical logicism is a form of idealism and is subject to the various objections thereto. It can be defended only by denying that there is an external world with an independent structure of its own which must exert some influence on any human attempts to grasp its nature. But this need not drive us to the Aristotelian position. In fact, I have been insisting throughout (contra Aristotle) that all knowledge is mediated through a particular language and is never just a matter of nonlinguistic intuition. However, I here take 'language' in a much broader sense than do the logicists; in particular, I wish to emphasize that it must have intensional content as well as logical form. The actual situation seems rather to be this: without having any nondiscursive overall grasp of reality, we can acquire some direct knowledge of its particular features. We develop methods and formalisms which both express and help to extend this knowledge, but this knowledge still functions as a continuous corrective to any proposed system. We can in fact become aware of the limitations and inadequacies of any of our tools through using them. Thus I must disagree with Kant that our conceptual tools can never change. We must start with something; but our tools can be modified, refined, and sometimes even transcended in the course of bringing them into a closer correspondence with reality. Since they are always so intimately bound together, any question about the relative priorities of direct knowledge and the development of methods and formalisms must have an unanswerable chicken-and-egg character. I believe that a careful study of the history of science would substantiate my claim.

If I am correct in maintaining that logicism is a form of idealism, it is understandable that the logicists should emphasize the logical form above the content of particular theories. For idealism has always asserted that the content of a term, in the sense of the nature of its referent, is largely 'mental' or at least dependent on a knowing and perceiving mind. It thus has no extramental ontological significance. Likewise, in unpacking the belief that nature is rational, the idealist must give an a priori criterion of rationality which is somehow determined by the structure of our cognitive apparatus. The realist considers it a problem for scientific investigation to find out just what kind of rationality does apply to nature.

There are two immediate dangers resulting from this emphasis on form above content: (1) Within a theory, philosophers may put too much stress on its syntactical features. Positivists did distinguish among the pragmatics, semantics, and syntax of any language, but they tended to concentrate on the syntax, since this was the part most amenable to treatment by formal logic. This led to the suggestion that logical syntax was self-sufficient in that it could be studied without reference to the other two aspects. On the contrary, I would hold that all three are so intimately conjoined that they cannot be separated out in the case of any living and growing theory. (2) Formalists have attempted to give definitions of such central notions as 'law' and 'theory' in formal syntactical and semantical terms. Once this is done, they can then be treated as free variables in a calculus, and one can say things like "Let T be a theory and L be a law within that theory . . . ," or "Pick a theory, any theory; pick a law, any law. . . ." But this requires that all theories from relativity to evolution to psychoanalysis, insofar as they may be considered scientific, must be alike in all philosophically important respects, and must be composed of laws in the same way. The fact that even a superficial study of known theories reveals such striking differences is not considered important; the aspects in which they differ are taken to be philosophically irrelevant. This assumption is utterly unjustified. Any criterion of relevance, in whatever context, is essentially evaluative and should make reference to some goal or ideal. Now one would expect the goal here to be an explication of what aspects practicing scientists have found most important in increasing our knowledge of nature. But in fact the actual goal of such philosophers seems to be simply ease and convenience in the manipulation of symbols. The net result has been a proliferation of clever displays of technical virtuosity in the use of mathematical logic which a scientist finds quite useless. I would suggest rather that theories and laws, when all their features are taken into account, bear what Wittgenstein calls 'family resemblances' to each other, rather than the strict identities required by formal logic. The logicist error has also led its proponents to underestimate the importance of analogy in science, as will be seen in the next chapter. The problem is that analogy involves a notion of 'likeness without identity', and thus seems to depend on an unformalizable intuition.

Equally misleading is the assumption that the internal structure of theories and the relations among their concepts must be like those which hold in a formal calculus. This assumption is implicit in the claim that any legitimate theory contains such a calculus as a component, with statements in this calculus related to observation statements by 'correspondence rules'. Now a formal system is supposed to have primitive (uninterpreted) terms, rules for producing well-formed formulas, axioms within the system, unchallengeable rules of inference, etc. These can be discussed without regard to any possible interpretations, which are to be made only later. It is possible to find formally different axiomatizations that will generate equivalent classes of theorems. If the system is recursive, there is an effective procedure for determining whether any well-formed formula is a theorem, and consequently a sharp dichotomy can be made between theorems (including axioms) and all other well-formed formulas. Terms which are not primitive must be defined explicitly by means of the primitive terms.

Do actual scientific theories have such a structure? In the first place, a calculus is never developed as such without any interpretation. Instead it is developed along with a 'model' which suggests the lines of its further elaboration and the addition of new concepts and principles when necessary. I will discuss this in more detail below. Braithwaite concedes this point as a matter of historical fact, but suggests that once this job has been done the calculus should be "disinterpreted" into a mere formal system and later "reinterpreted" in terms of an observation language.[4] However, Braithwaite's method operates only after the real work of the theory has been done. Not only does it add nothing to our knowledge, but it prevents us from developing the theory further in useful directions. Within an uninterpreted calculus, all theorems are equivalent as equally analytic; there is no way of telling in advance which may prove to be physically significant, as the use of the model enabled us to do. Thus it freezes the theory at an instant and petrifies it forever. It is "dropping the pilot," with disastrous consequences for the scientific ship of state. In addition, we do not have an indefinite number of fully developed calculi just lying around and waiting to be given an interpretation by some otherwise arbitrary choice of correspondence rules. Calculi in physics are developed when and as they are needed, with a specific interpretation in mind.

I maintain that the truly interesting scientific theories are not those whose principles are fixed, static, and codified, but those which are constantly being modified and added to in an effort to achieve a more perfect correspondence with reality. The best example of a static system seems to be classical mechanics. It *can* be put on a completely axiomatic basis, and indeed, a variety of equivalent axiomatizations have been put forth by Newton, d'Alembert, Lagrange, Hamilton, Hertz, Mach, etc.[5] Classical mechanics is the first full-scale physical theory with which a student comes into contact, and he *ought* to be impressed with the simplicity, elegance, and power of its logical structure. It is naturally tempting to seize on it as an ideal to which all scientific theories should try to conform, and I think this is precisely what many formalists have done. But to claim that as a matter of fact all theories do contain such an isolable calculus would certainly be an unwarranted induction if based on classical mechanics alone, and a careful study of other theories would show it to be false. And to claim that theories should have this form, and therefore try to fit them into it, can only have the stultifying effect described above. For classical mechanics is a dead theory. By this I do not mean that it has been discredited or discarded. Within its proper sphere of application (weak fields and small velocities) it is as useful as ever. But it has been dead for some time in that no serious attempts have been made to introduce even slight modifications in its principles. The latter were simply taken for granted and applied as far as they could go. When they finally broke down in electromagnetism, astronomy, and atomic theory they were simply replaced wholesale by relativity and quantum mechanics, which are not yet perfectly settled.

Likewise a rigorous analytic-synthetic distinction does not hold in contemporary scientific theories, although it is required by the logicists' notion of a calculus. I do not wish to go into the details of this controversy, except to indicate that I am in agreement with Quine[6] and Putnam[7] that the distinction has no profound significance for science. Partial axiomatizations may be useful at a given time in demonstrating the coherence of a theory. They also provide useful fixed points, and suggest a reasonable order for trying out revisions in response to anomalous empirical data. But the most basic axioms may themselves be changed (and in fact often have) when this appears necessary to effect a closer correspondence with reality. There need not be any statements which are absolutely secure from possible revision in the light of future experience by being analytic, nor are the other (synthetic) statements simply given up at the first sign of anomaly. Maxwell, although supporting scientific realism, has nevertheless argued that within any theory one should be able to distinguish a class of 'meaning postulates' for the terms of the theory. These need not be any sort of translations into an observation language; but they differ from the other intratheoretic statements in which the term occurs by serving to isolate and fix the meaning of the term. These meaning postulates thus provide a basis for a rigorous analytic-synthetic distinction.[8] Now, Maxwell admits that we cannot unambiguously locate such a class of statements within actually given scientific theories; but he takes this as a reason for recommending a 'rational reconstruction' of science which will provide this supposedly needed exactness of meaning.

I will argue against 'rational reconstructionisms' later. But in any case I think this view of meaning is unreasonable for a realist. I would argue instead that theoretical terms normally refer to entities and their properties within the level being described by the theory in question. I construe 'the meaning' of a term as the totality of the relationships of its referent to the other entities which are taken to comprise that level, although these relationships may have different degrees of importance. None of these is necessarily prior to any other. It may be claimed that this is true only of ontic levels, and I have certainly phrased it in ontological terms. Cognitively, some features of a level will be better known and considered more important in characterizing that level than others. It may therefore be suggested that priorities should be established within cognitive levels. But any such evaluations must be made at a given time, and since as our knowledge increases we may come to regard other features as more significant, a proper theory of meaning should allow us to account for these changes.

I have mentioned before that not every term in a theory is taken to refer to an existing entity. There are always some abstractions and idealizations, and it may seem that names for these should be introduced by explicit stipulative definitions. But as Quine has pointed out, "Stipulation is not a lingering trait" but rather a datable action.[9] Putnam has shown brilliantly in the case of 'kinetic energy' how a term first introduced by a precise definition comes to acquire multiple connections in the theory through a variety of laws, which give the term a larger but more precise extension.

In order to preserve these other laws and to avoid changing the extension of the term, it may ultimately be necessary to reject the original definition as universally valid, thus converting a supposedly analytic statement into a synthetic one. I agree that most theoretical terms, understood cognitively, are or at least *can become* law-cluster and hence multiple-criterion terms.[10] Shapere has likewise argued for the intricate interdependence of concepts within a theory. He contends that[11]

> The geometrical concepts within a theory function partly to determine what can count, within that theory, as a fact, what can count as an entity, what can count as the behavior of an entity, and what can count as an explanation of the behavior of an entity; and shifts in the spatial concepts of the theory entail corresponding shifts in the concepts of "fact," "entity," "behavior of entities," and "causal explanation" within that theory. If this is so, then reducing one such concept to another will lose most of the richness of the connections that exist between them within the theory; it will even falsify the way those concepts work together to form what we call a "theory."

This account seems to me absolutely correct and essential for a philosopher of science to recognize. Any attempt to reduce the relations obtaining in actual scientific theories to the definitional relations of formal systems can only result in a distortion of the structure and function of a theory at a given time, and its potential for future growth.

I have shown above that a major part of the empiricist error was the insistence on having a ground floor of absolutely certain statements expressed in an observation language, whose truth-value could be determined by direct observation. But this requirement is also part of the logicist error. Wittgenstein had taken as the basis of the Tractarian system a set of 'atomic sentences' which had no meaningful sentential parts. All other scientifically legitimate sentences were 'molecular' in the sense of being truth-functional combinations, no matter how elaborate, of the original atomic sentences. Two other features were essential: (1) Each atomic fact was absolutely self-contained and independent of all the others. It could be true or false whatever the truth value of any others, and its truth-value could be determined independently, by looking to see whether the 'atomic state of affairs' which it described did in fact obtain. (2) The actual truth-values of all the atomic sentences together, and consequently the set of existing atomic states of affairs, was an adequate characterization of 'the state of the world'. Every complete assignment of truth-values to atomic sentences represented a possible state of the world, and logically all of these were equally probable; conversely, any real description about the state of the world must be analyzable into the set of atomic facts which express its content. This desire for an exhaustive enumeration of possible states of the world reappears in attempts to develop an 'inductive logic' which, like Carnap's, introduce a measure-function as a numerical determination of what proportion of such states are included or ruled out by any given hypothesis. It is this need to have something corresponding to atomic sentences that led the positivists to introduce their "Protokollsätze." Phenomenalism and physicalism were simply attempts to find the most appropriate correspondant for these Protokolls. Ayer has shown how the search for suitable Protokolls reflects a supposed need for an absolutely certain fixed basis.[12]

All this seems to me completely unjustified. In the first place, there is no reason to suspect that any proposed candidate for an atomic fact is really thus independent of all other similar facts. Certainly we expect it to be *causally* related to others. If one insists that although physically dependent they are still *logically* independent, it is reasonable to question whether this sort of logical independence has any significance for physics, or even whether it simply reflects a particular and inadequate conception of logic. Secondly, the observation languages for different theories may not be fully comparable, as I have suggested above. Feyerabend has pointed out that logicism is highly undesirable methodologically and can lead to dogmatism rather than genuine empiricism.[13]

I will finally criticize that cherished dogma of most contemporary philosophy of science, the sharp distinction between the contexts of discovery and justification, as being based in part on the logicist error. A standard example of the supposed irrelevancy of the mode of discovery of a hypothesis to its justification is Kekulé's dream of a snake biting its tail, which led him to think of the ring structure of the benzene molecule (cf. the apocryphal story of Newton's conceiving the law of universal gravitation after being hit on the head by a falling apple). The conclusion is that scientific hypotheses can be and are arrived at in all sorts of nonrational ways, and the interesting philosophical problems do not arise until one considers the degree to which scientists are justified in accepting such hypotheses. The dismissal or referral of the study of how hypotheses are actually conceived to empirical psychology (which seems surprisingly loath to take it up)[14] probably appears most strongly in Popper,[15] but it is also common to logical empiricism. Popper uses it to reject induction (at least in the Baconian sense) as a problem for philosophy.

Philosophers have tended to dismiss any device that seems useful for thinking of fruitful hypotheses but does not confirm the hypotheses as having "mere heuristic value." Such a phrase sounds very strange to practicing scientists. Popper's account of scientific method as an attempt to falsify a succession of hypotheses, one after the other, suggests that there are an indefinite number of such hypotheses simply lying around and awaiting their respective turns. But this is clearly not the case. Hypotheses that are worthy of extensive (and often expensive) tests are hard to come by, and to say that any device has genuine heuristic value should be to give it very high praise. Scientists can hardly sit around waiting for inspiration; and if they are rational, they will want a procedure that not only produces hypotheses but orders them in terms of some notion of initial plausibility, coherence with the rest of their scientific tradition, promise of novel and interesting (deductive) conclusions, and various other criteria. These considerations reflect the fact that scientists have only limited time and money, and thus want some principle that will enable them to test hypotheses in an *efficient* way. I believe that the possibility of finding such principles, which are understood to operate *before* any actual tests of the hypotheses by comparing them with observable phenomena are made, should be an important problem for philosophers of science.

The problem is often phrased as the question, "Is there a logic of scientific discovery?" I think that this formulation is somewhat misleading. For logic, as understood by virtually all contemporary philosophers, deals only with *timeless* relations. A logical truth cannot be true now but possibly false or indeterminate at some other time in the past or future; it must be equally true at all times. In evaluating the confirmation of a hypothesis by means of inductive logic, one deals with a timeless relation between the hypothesis and the data that is being cited as evidence. It is irrelevant that *this particular set* of data happened to be the one available at a particular time. From the standpoint of inductive logic, it is just one of a (perhaps infinite) number of logically possible combinations of evidential data, each of which would confer a similar degree of confirmation on the hypothesis. (I have suggested above that this position requires the unwarranted assumption of a list of all logically possible states of affairs.) But discoveries are datable events or processes, although Kuhn has pointed out that there are difficulties in dating them precisely.[16] They do take place in time, and thus in a total context of assumptions, mathematical techniques, paradigms of good theorizing, available experimental devices, background knowledge, individual ideals, and even such sociological facts as the prestige of the discoverer in the scientific community. All of these are bound up together and evolving in time, and they are contributing factors in an account of why a certain kind of hypothesis should be thought up and tested. It is certainly reasonable to evaluate, criticize, and attempt to justify the achievements in putting forth any particular hypothesis, quite independently of whether subsequent research proves that hypothesis to be right. Kuhn has raised the very important point that in scientific revolutions, the evidential basis for the revolutionary new hypothesis is likely to be far smaller than the previously well-confirmed old one which it seeks to replace, if only because the possible elaborations of the new hypothesis have not as yet been worked out. Acceptance of the new hypothesis represents a commitment on the part of a scientist which cannot be justified by the principles of inductive logic.[17] Let us assume that philosophers of science, like ordinary mortals, are 'glad' that such revolutions did in fact occur. Then they should be confronted with a dilemma: either they should say that the 'conservatives', who had more evidence in their favor at the time, were really the more justified; or they must give up the notion that timeless inductive logic is the sole source of justification.

The most obvious historical example of a 'logic' whose relations were essentially temporal is of course Hegel's dialectic, which was introduced for the very purpose of overcoming the supposed inadequacies of Kant's static logic. In its general outline, dialectic may be able to shed some light on the way that theories have actually succeeded each other, but this may stem from the vagueness of the operation of dialectic, which can be twisted to cover a variety of developments after the fact. In any case it is hardly adequate as a logic of scientific discovery. Hegel, however, was at least correct in insisting that scientific thought (or any other kind) never operates in a vacuum, but that new ideas are engendered by the existence of anomalies alongside of previously adequate points of view, and that such new theories must somehow combine the insights of the new and the old, though perhaps in a radically different way which could not be predicted in advance and which requires the action of real scientific genius.

The great danger in Hegelianism is the genetic fallacy, that the actual mode of discovery and development to the point of acceptance by the scientific community can itself constitute a justification for a hypothesis. This fallacy, which may be the Scylla to the Charybdis of the logicist error, was what led Hegel to his identification of the rational with the real, and the later contention of the evolutionists that mere victory, by whatever means, in the 'struggle for survival' among rival hypotheses was adequate evidence for the correctness of any one. However, I would insist that more independent reasons must be given for the opposite requirement, that justification must be a timeless logical relation, reasons which do not founder on Kuhn's objections. In any case I will argue that an understanding of the way in which a theory was actually conceived and developed is essential for an understanding of its content, rationale, function, and modus operandi, as well as the hopes which its users have of possible future accomplishments. New theories have been developed to replace old (and often crude) ones which have proved inadequate for a variety of reasons, not all of them assimilable to inductive logic. Scientific advance must always be seen in this total human context.

Certain positivistically oriented philosophers have contended that although the foregoing considerations may be correct as matters of historical fact, they nevertheless have no place in a 'rational reconstruction' of science. But by now it should be clear just what is wrong with the notion of a rational reconstruction. Any advocacy of a reconstruction presupposes a belief that the original construction was somehow defective or insecure; in this case, that it was irrational or at least less rational than it might have been. But this is an extremely paradoxical position to take. As I have suggested before, it is science as it has actually developed that is generally taken to be paradigmatic of a rational approach to nature. And this rationality certainly does not mean simply that the great scientific ideas which have later proved so fruitful and which have revolutionized the history of science have been more highly confirmed than the hypotheses with which they were competing, according to some system of inductive logic. In fact, if their development is seen in an historical setting, this criterion is often not even satisfied. It is reasonable to ask any reconstructionist whether his method, if followed strictly, could have led to all the scientific accomplishments which we now value. I believe that it could not.

In the first place, once a theory has been codified into a canonical form, one can simply take the results (usually understood in terms of sets of observation statements) and find logically equivalent ways of deducing them from different sets of 'axioms'. Depending on the criteria used, any one of these might be considered the rational reconstruction. But this is really more like a reformulation than any sort of construction, and depends on the original construction having been effected by other means. Like Braithwaite's notion of 'disinterpretation', it is a classic example of the logicist error. In the second place, the reconstruction usually involves relating all theories to the one universal observation language. But we have criticized this in connection with the empiricist error. To the degree that the languages in which observations germane to different theories are expressed are not perfectly coordinated, it may not be possible to express the total content of one theory in terms of the observation language for the other. In the third place, the primary motivation for rational reconstructions seems to be the quite unfounded belief that science should be strictly cumulative, so that there can be no future scientific revolutions which will reject ideas that (for good reasons) had previously been accepted. This belief appears most strongly in Bridgman's almost neurotic reflections on the need to make future Einsteins unnecessary.[18] But if Kuhn and Feyerabend are correct, revolutions are needed to open up new and unsuspected cognitive levels, which in turn may be keys to new and unsuspected ontic levels. Only a dogmatist can reject these in advance. In conclusion, I submit that a reconstructionist may be able to replace an existing scientific edifice with another; and to his esthetic sensibility it may be more elegant and have cleaner lines. But it will not be functional; indeed, it can only be a monument to two great errors.

Having laid these two bugbears to rest at last, we may now turn to a more positive account of the methodology of scientific realism. In particular, I will discuss the nature and function of models in science. Nagel has emphasized that a complete account of the operation of theories cannot refer only to a formal system and a set of correspondence rules; there must be some third factor to mediate between them, and this he calls the model for the theory.[1] As already mentioned, we cannot remain entirely within the formalism if we hope to get significant physical results from it; we need some guide to give direction to our deductions, to suggest that if we combine some set of the axioms in a certain way, we will be more likely to prove theorems which have interesting physical interpretations. More importantly, we must have some means of selecting correspondence rules. These are clearly different from statements within the formalism, for they relate terms of the formalism to observable entities. But presumably this correspondence could be effected in a number of ways, even within the requirement that the observational predictions resulting from a particular choice of correspondence rules be mutually consistent. This is unsatisfactory. It leaves the correspondence a completely external and arbitrary relation, while we require that it should be in some sense natural. I contend that this can be achieved through the use of a model.

I have previously argued that formalisms do not appear without an interpretation. Now I will claim that there is a double interpretation involved: first, the formalism is interpreted in terms of a model; and second, the model is interpreted in terms of observable phenomena. But the relation called 'interpretation' differs considerably in the two cases. The first kind is not to be understood as a separate act, performed after the development of the formalism has been completed; it should rather be understood as the material aspect of a single cognitive process, of which the succession of deductions is the formal. Both support each other: by regarding the terms of the calculus as referring to conceivable entities, one can give a content to the relations among them which appear as theorems in the calculus, and find promising lines for further development; on the other hand, the actual deductions place limitations on the kinds of entities which may be considered, and in doing so refine our knowledge of what sorts of things they might be. The net result is to locate the theory as belonging to a particular cognitive level, or even to require that the theory define a new level. If it is legitimate to regard the material aspect of a pure (i.e., not yet empirical) theory as an 'interpretation' of the formal aspect, it should be equally legitimate to consider the formal as an interpretation of the material. The former considers what sorts of things can behave in the way described by the formalism; the latter describes how a postulated

The Role of Models 4

set of things does in fact behave. For realism requires that a theory is an articulation of part of a cognitive level, and every level has both formal and material aspects. The emphasis on the former at the expense of the latter has usually been considered a form of Platonism; but since the only properties and relations susceptible of complete treatment by formal logic are logical and quantitative (mathematical), this move is really a retreat to Pythagoreanism, and is motivated by the same desire for tidy rational perfection as was Pythagoras's own position.

On the other hand, the 'second interpretation' is more like a coordination relation between the level of the pure theory (formalism *cum* model) and a level which is taken to be observable. As I have argued above, the notion of 'observable' does not define a single universal level, and that actually chosen is likely to depend on the nature of the pure theory. Degrees are possible here; depending on the criteria used for deciding what is to count as an observation of a kind of entity, one level may be more or less observable than another. It is also possible that 'second interpretation' may occur in steps, so that A, the level of the pure theory, is 'interpreted' in terms of B, and B in terms of the relatively more observable C, rather than going from A to C directly. It may be thought that I am suggesting that theories are properly applicable only to models, whereas in fact they should be about observable phenomena. This is not intended; but in view of the objections of Sellars and Scriven, we must recognize that their application to observables may be only imperfect or inaccurate. The model then becomes "that to which they apply perfectly." And this perfection results from the fact that the model can be constantly refined in response to the demands of the formalism, rather than simply reflecting a fixed pattern of observations. The model can contain ideal elements, and while these may be extrapolated from experience, they need have no actual correlates in experience. If a theory is to be empirical (in the good sense) a second interpretation must indeed take place; but to refer to either the first or the second as "The Interpretation" would be to make a gross oversimplification. It is possible that the distinction between the two 'interpretations' may collapse, and the two thus become identical, as in the case of 'phenomenological' theories like classical thermodynamics, but such a circumstance can equally well be considered as merely a limiting case, a 'self-coordination' of the level of the pure theory with itself, and this must of course be perfect coordination.

Sellars has argued that 'correspondence rules' (second interpretations) should be such as to provide an ontological explanation of observed phenomena, e.g., that gases behave the way they do, and the classical 'gas laws' are valid to the extent that they are, because gases are *really* collections of particles whose behavior is governed by the kinetic theory and statistical mechanics.[2] Having rejected Sellars's notion of identity here, I would hold that second interpretations should show the observable level to be *grounded* in that of the pure theory. Sellars then goes on to argue that the theoretical level must provide an explanation of why phenomena do not obey perfectly the inductive generalizations formulated in the observation language being used. This contention is in keeping with Sellars's belief that there can be only one truly real level, and it is certainly a worthwhile ideal for the theoretical explanation. But if taken strictly, it must rule out a great many systems which have generally been called theories, including phenomenological ones which may be the best presently available for a certain class of phenomena. Returning to Bunge's point, I would argue instead that since levels may be only imperfectly coordinated, we may (as yet) know of no single level which can provide a complete grounding for another, and therefore we must draw on several at once for a complete explanation. Those observable features which cannot be perfectly coordinated with the level of the pure theory may be explicable in terms of some other; and this deficiency need not detract from the usefulness of the theory. Nevertheless, it is always reasonable to hope (and therefore this can be a regulative ideal) to find some one level that can do the job by itself, as long as this is not taken to preclude the reality of other levels.

It would be helpful if the foregoing description of its functions could serve as an 'implicit definition' of the term 'model'. But this would be dangerous, for I am using the term in a rather special way, and I wish to prevent unintended connotations from other uses of the term from prejudicing the argument either way. In the first place, models are certainly devices employed to make observable phenomena intelligible. But this statement is purely formal until the notion of 'intelligibility' is analyzed. Following Toulmin, I have suggested above that intelligibility in science reflects 'the ideals of natural order' that are held by the scientific community at a given time. These can change in time, and the great mistake of most advocates of models has been to seize on a contemporary ideal and offer it as forever fixed and constitutive, in an attempt to give an explicit criterion of what makes a good model. Thus Kelvin insisted that only explanations in terms of a mechanical model could be truly intelligible.[3] Such a standard was ridiculed by Duhem, when he contrasted the French mind, which he claimed could think in abstract logical terms, with the English mind, which could understand the logical structure of scientific theories only in terms of mechanical gadgets made of "springs, pulleys, and vulcanized rubber."[4] Nagel[5] and Bunge[6] have discussed the nature of mechanical explanations, and the fact that the notion can be interpreted in many different ways. I certainly do not want to require that models should be mechanical, either in terms of the formal structure of the relations obtaining within them or of the sorts of entities whose behavior they describe. I am not even requiring that they be visualizable. The latter is a fairly extreme claim. In practice, almost every proposed model will indeed be visualizable, presenting itself to our imagination, if not to our perception. Furthermore, the model is almost sure to have features analogous to perceptible qualities, especially if it is set in a spatio-temporal framework. But the important thing is that the entities in the model need not behave in the same way (obey the same laws) as perceptible physical objects. The choice of a model for a scientific theory is thus limited only by the scientist's powers of conception, including nonvisual conception, on the one hand, and his ideals of natural order on the other. Models must have both material and formal aspects, though the latter are expressed by the 'already interpreted' formal calculus of the theory. The material aspects are the entities to which it purports to refer, like the 'billiard-ball' particles of classical mechanics. Their correspondence with the formalism must be perfect. With observed phenomena the correspondence is imperfect, but sufficiently close to determine a natural set of correspondence rules with the observations that give the theory its 'empirical content'. The problem is somewhat similar to Kant's, who inquired how the pure a priori categories of the understanding could become applicable to the contents of our sensory experience. To unite them he introduced the schematized categories, which share with perceptions the property of being manifested in time. Our model-interpreted formalisms share material and qualitative features with actual perceptions, and through these the connection is effected.

One mistake made by adherents of the logicist error was to identify theories with sets of statements (in the limiting case single statements) characterizable in such a way that they could be called 'laws'. The 'concepts' of the theory were given implicit definitions by such laws, but their real definitions came from the correspondence rules, whether these were formulated in terms of operations, reduction sentences, or anything else. I hold on the contrary that there are no such things as isolated concepts, defined adequately without reference to others on the same level, or isolated laws, of which a theory is simply a collection. For what is really desired in a theory is not a compilation, but a systematic unification of the laws and concepts which provides an 'organic' coherence for the theory. Coherence requires logical consistency as a minimal necessary condition, but this is not enough. What a model can do is to represent these concepts and laws as aspects or features of a single underlying reality, a kind of process characteristic of a particular level. Such a process must be intelligible in terms of the ideals of natural order of the scientific community, with a 'surplus content' of interpretations, distinctions, and classifications beyond those phenomena for whose explanation the model was originally proposed. The totality of functional linkages and roles of a concept in the model of a theory is the theoretical meaning of that concept, and it is only through these that we can determine the correspondence rules that complete the meaning by providing an empirical content.

Operationism, at least in the extreme form advocated in Bridgman's early writings,[7] is now generally recognized to be an inadequate philosophy of science. However, it provides a nice contrast to the view propounded above. In order to avoid complete chaos from which no science could ever get started, Bridgman was forced to have some (also operational) criterion for identifying the concepts defined by distinguishable operations. For this he used mere numerical equality (to a reasonable approximation) of measures—a contingent, empirical fact which does not justify extrapolation. But such a criterion can be neither necessary nor sufficient. Our experience is always limited to a relatively small class of phenomena. Within that class, it may well happen that two properties which on theoretical grounds we would want to keep distinct have proportional measures; then on Bridgman's grounds we should (tentatively) identify them. Likewise if two devices presently considered different measures of the same property cannot both be applied to the same phenomena in this limited experience, we would have no reason to identify them. In fact, our distinctions could only be made in terms of the characters of the operations themselves, and this could lead to a radical reclassification of concepts. For example, the use of rulers or diffraction gratings involve quite different pieces of apparatus and rules for using them. If one did not already have some notion that they could both be used to measure lengths, he would probably assume that they referred to categorically distinct concepts, and certainly not think of trying to apply them to the same things, even if this were possible. It might in fact be possible to develop some sort of science on the basis of these reclassifications, but it would be very different from what we have today and, I suspect, far less well developed or confirmed. Bridgman's mistake, then, is to seek the source of unity and coherence in operations as the manipulatory devices of a *single* individual, rather than in the independent natural processes which theories seek to describe with the help of models.

We have said that models reflect the actual world, though with the greater degree of precision, perfection, or rationality required by the formalism of the theory. But insofar as they are cognitive instruments, they need not be static, and can be constantly refined to achieve a closer correspondence with our observations, thus further rationalizing or 'saving' the phenomena. But such a refinement will presumably require an increase in the complexity of the model, and in doing so decrease the coherence which the model was created to give, by introducing new, apparently unrelated or ad hoc elements. Thus in proposing a model the scientist must try to slip between two dangers; he must not lose the systematic unifying power of his theory by including within it a variety of relatively unrelated, ad hoc assumptions; but he must not desert the phenomena for an ideal world of his own. The latter is essentially Platonism, which restricts possible knowledge to the pure forms, on the grounds that observable phenomena can never exemplify them adequately. It is for this reason that it is often desirable to hold on to a variety of different theories, each of which explains only certain aspects of what can actually be observed, rather than trying to reduce them all to one.

I have argued that a model gives a theory coherence by representing its concepts and laws as features of a single underlying process. To this it might be objected that if we already understood this process well enough so that merely introducing it would count as an explanation, we would not need a theory. This is quite true; but what the model does is to point to some process which is better known or understood than the relatively unknown process being theorized about. It points to analogies or isomorphisms between the known and the unknown (or imperfectly known), in such things as the mathematical structure of the equations in their respective formalisms, or the fact that certain terms in the formalisms must stand for the same sorts of entities. From this more or less limited domain of resemblances an analogy is drawn and then extended tentatively. This extension suggests both new developments within the formalism (new theorems worth trying to prove) and new interpretations of specific statements to provide more empirical tests. If these lead to false predictions, the formalism must be modified accordingly, and this in turn requires modification of the model, if the two are to remain in perfect correspondence. Thus the model and the formalism influence each other.

Hesse has made a useful distinction between two senses of 'model'.[8] Models_1 are the actual representational devices of a theory, in perfect correspondence with the formalism. Models_2 are the *other* natural processes from which the analogy is first drawn. They must be known to exist and to have certain very definite characteristics. But aside from the positive analogies with the theory being interpreted, models_2 will in general include a negative analogy, a set of features which can be known to have no counterpart in the process which the new theory is seeking to describe.

A $model_1$ cannot have any negative analogy. In any case, until the theory is completed, codified, and until its $model_1$ is interpreted so that it can in turn provide a $model_2$ for some other theory, there will be a neutral analogy, a set of features in the $model_2$ which may or may not have counterparts. These then provide the extensions of the theory and its $model_1$. For the $model_1$ initially includes the neutral, as well as the positive analogy taken from the $model_2$ and then by empirical and mathematical tests divides this up into additional positive and negative analogies, keeping only the former. The point is again that science never works in a vacuum. We always have some knowledge, imperfect though it be, of 'how things work'. (Of course, our standards of what constitutes knowledge here may change along with our ideals of natural order.) We focus on certain of these processes (motions, forces, propagations, transitions, or whatever else) and describe them by means of a system of concepts that has proved itself useful. When confronted with a new class of phenomena about which we are trying to theorize, it is natural and reasonable to draw our first approximation from these tried-and-true concepts, which therefore give us a $model_2$. But we can transcend this starting point by what may be described as a method of successive approximations, in which features may be added, subtracted, or reinterpreted. It is possible that so many features may have to be subtracted in order to eliminate the negative analogy from the original $model_1$ that the latter may be considered incomplete in the sense that it is assumed to have one of a class of features, but every choice within this class leads to an incompatibility with the principles of the theory. As a result, the $model_1$ may come to look like a pure mental construction or mere collocation of whatever properties the scientist chooses to put in, rather than something which may be conceived to exist in its own right in that having certain properties requires that it have certain others. I have stated above that a $model_2$ is taken to have such an independent existence, and that what was originally a $model_1$ for one theory may later be used as a $model_2$ for another. But our notions about which properties must go with which (in all contexts) can and do change in time. It was undoubtedly a radical conceptual innovation to suggest that atoms, in the sense of small but still extended particles, did not have any color, although all macroscopic physical objects did. The crucial notion again is that of level. Any $model_2$ comes from a particular level, and its use as a first approximation represents an attempt to locate the new theory at that same level. But required combinations of properties are relative to particular levels. If in fact we cannot satisfy these requirements without introducing negative analogies, we should consider this a strong argument for believing that the theory really belongs on another level. A different initial $model_2$, taken from some other level, may work here; or we may be forced to conclude that the theory belongs to an entirely new level.

Critics of the importance of models in scientific investigation have tended to concentrate on $model_2$ as the only legitimate sense of the term. They have emphasized the fact of negative analogies, which both spoil the logical structure of the theory and limit its interpretation by means of the $model_2$. I have been concentrating on $model_1$, and mean 'model' in the sense of $model_1$ whenever I use the term without a subscript, especially for a reason which I will present below. But I do not wish to imply that $models_2$ are not tremendously important. In the first place, one cannot begin the construction of a $model_1$ without them. In the second place, they do reveal extensive positive analogies, and this is a considerable accomplishment in its own right. The ability to link together previously unrelated observations by seeing them as analogous to kinds of behavior of a single $model_2$ is a mark of scientific genius. Black has argued that there is an important continuity between the notions of 'model' and 'metaphor'. He asserts,[9]

> A memorable metaphor has the power to bring two separate domains into cognitive and emotional relation by using language directly appropriate to the one as a lens for seeing the other; the implications, suggestions, and supporting values entwined with the literal use of the metaphorical expression enable us to see a new subject matter in a new way. The extended meanings that result, the relations between initially disparate realms created, can neither be antecedently predicted nor subsequently paraphrased in prose. . . . Much the same can be said about the role of models in scientific research.

Of course models must be far more precise, complex, and organized than poetic metaphors, which capitalize on the vagueness and ambiguities of ordinary usage. "Systematic complexity of the source of the model and capacity for analogical development are of the essence."[10] Certain of the concepts used in science have been found to have such power that they have acquired the status of "archetypes," which function both as elements within a vast range of models and as regulative ideals for the construction of new models. It is not necessary that a $model_1$ must be constructed from a single $model_2$. For in fact the $model_1$ for a theory may incorporate two or more of these (provided that they are on the same level), possibly modifying them in the process. Harré calls this "development" as opposed to "deployment," which is simply an elaboration of the possibilities implicit in a particular $model_2$, or "paramorph."[11] However, I believe that Harré is confused, especially in his claim that description of a model can by itself serve as a theoretical explanation.[12] For Harré's paramorphs are only what Black calls "analogue models." They have only formal or structural isomorphism to the phenomena that the theory is trying to explain, without any suggestion that the *material* elements of the two might also be similar. But even if this condition were satisfied, the theoretical explanation must be based on the $model_1$, if it is to use the full power of the formalism, whereas in fact Harré's paramorphs can only be $models_2$, with all the negative analogies which vitiate their use in explanations. They suggest the outline of an explanation without providing it themselves.

It may still be thought that all the above establishes is that models have heuristic value. I have suggested previously that this alone would be enough to make them an important subject for philosophers of science. Campbell has argued that this is misleading. He distinguishes between theories and laws (or collections of laws) by the criterion that scientifically valuable theories must "display an analogy."[13] There are indefinitely many sets of laws (understood as isolated statements) which can predict a class of phenomena by logical deduction. The problem is rather to find that set of laws whose deductive steps can be interpreted as formal aspects of a process introduced by the analogy. We may use various heuristic devices in trying to find such a theory, or selecting it from a group of logically equivalent formalisms which may be available, but the analogy itself is the criterion for 'theoryhood', and not something to be discarded as soon as a single correct formalism has been found. I think that Campbell is absolutely correct in his contention that formalisms without models are somehow incomplete, and that we have good grounds for discriminating among formalisms, each of which equally well allows the deduction of a class of particular facts or inductive generalizations. However, he goes too far in claiming that models do *not* have heuristic value. For in general it is not so easy to find *consistent* formalisms which allow us to deduce the explananda, except in the trivial sense of simply taking the statements and replacing their words by logical symbols (constants or variables). The use of a model, even a $model_2$, can be the very thing that gets us started in the construction of a formalism.

Nagel has pointed out that there can be danger, or negative heuristic value, in the unrestricted use of models.[14] It is essential that a model have surplus content, going beyond the phenomena which suggested it. But it can go too far. Nagel suggests that a $model_2$ can waste valuable time leading a theory into blind alleys which delay the arrival at a correct interpretation. But this need not be harmful: I would claim that such false starts are an essential part of the conversion of a $model_2$ into a $model_1$. In exposing the limitations of the proposed $model_2$ these investigations can perform a valuable function in determining its actual range of validity; and furthermore, since interpretation cannot occur in the absence of knowledge, it is unlikely that the correct interpretations could have been pulled out of the blue until the way had been cleared by the elimination of the others. More serious is the fact that there may (at least for a long time) be no empirical correlate for a concept in the model, or even a coordinate at any other level. Then it is not clear whether statements using that concept, which may become a major part of the development of the formalism, have any empirical significance or not. I would say that they may in fact refer to real things in nature, which however appear only at one level; but we have no guarantee that this is so, and that we are not simply being misled by the nontestable features of the model. This will be an important point in my discussion of the wormhole model of geometrodynamics.

While on the subject of heuristics, I should mention two further notions: 'paradigm' and 'plan'. 'Paradigm' was first introduced by Kuhn. His use of the term is somewhat vague, but it refers originally to the concrete scientific accomplishment which can be held up as an example of the way research in a field should be carried out.[15] The point which Kuhn wants to emphasize is that paradigms are logically prior to formalisms, models, correspondence rules, and any other aspects which can be abstracted out of actual scientific theories. In vivo they are all essentially conjoined, and this is a fact which any later analysis must take into account. A good paradigm suggests a way of looking at phenomena (and a mode of theoretical explanation appropriate to them) that holds out a good prospect of success, since it has worked well in other situations. I have suggested that each particular science chooses its own level of abstraction and discusses only such problems as it considers solvable through its chosen methodology. I agree that it is the construction of a successful paradigm that delineates such a field of investigation, by the fact that it includes formulations of both the questions and the techniques allowed for answering them.[16] Kuhn has pointed out that in fact no field of investigation can get started as a science without a clear paradigm which can serve as an example to be followed. The paradigm then governs the later experimental work, giving a new evaluation of what sorts of features in our experience are important or relevant for science, and thus enabling us to see the world differently. It does have the effect of limiting empirical work to experiments that the paradigm suggests are especially important (though another paradigm might select a different class), those which allow very accurate tests of its predictions, and those which further articulate the paradigm and fill in its gaps. I have mentioned that Feyerabend has shown that such limitations, though useful for making progress within the 'normal science' developed out of the paradigm, can be stifling in the long run. I will consider some of these points in more detail in the next chapter. Wheeler gives

J. L. Synge credit for introducing the notion of 'plan'.[17] The term is never defined, though it seems closely related to 'model', in that both may be considered to be interpretations of a formalism. But whereas a model is concerned with providing a world-picture implicit in the formalism by representing its relations as aspects of some natural process, the plan is more pragmatic. It is concerned with how the formalism may be used or applied in such a way as to exhibit its full power. Given the formal relations, it proposes a logical sequence for using them, suggesting what sorts of things can be specified as initial data and how the theory can work from them to predict new phenomena. It therefore indicates an appropriate formulation for specific problems before they can be subsumed under the theory, and a language for expressing its potential content.

I hope to have established by now that even if models did have only heuristic value, they would not thereby lose their importance for a philosopher of science; and that the interpretation of their role as primarily heuristic is at least misleading. I now wish to suggest that for a scientific realist they have another and possibly more important role: they are at the same time ontological hypotheses. Sellars has advanced the very striking suggestion that if we consider a theory to be a *good* theory, we should be prepared to accept its ontology, and to say that the things and processes to which it purports to refer really do exist.[18] I contend that this notion is essentially correct, and indeed that a scientific realist cannot reject it without giving up premises (4) and (5) mentioned on p. 7. However, when stated in this bald form the principle raises grave difficulties, and sorely needs clarification before it can be effectively applied.

In the first place, it is not clear what 'accepting an ontology' involves. I have argued in Chapter 2 contra Sellars that it does not include rejecting other ontologies with which it may be coordinated whether drawn from common sense (as expressed in 'ordinary language') or from other accepted theories. It consists rather in locating the ontology of the theory at a particular level, thereby further elaborating the nature of that level, or in extreme cases outlining a new level.

More importantly, it is not always clear just how, for any given theory, one can determine just what its ontology is, by somehow digging it out of the formalism. Existing formalisms can be and have been successfully applied to classes of phenomena without any real 'first interpretations', and this fact has undoubtedly comforted philosophers who do not believe that models are essential intermediaries between formalisms and phenomena. In conceding this point I may appear to be denying my earlier claim that formalisms are never uninterpreted. But this applies only to the first appearance of a formalism, the time when it is first developed, and there the 'first interpretation' may be in terms of quite abstract mathematical entities. But *once developed* the formalism can be reapplied to new domains that are sufficiently similar in the appropriate structural features to the original. This similarity need be only partial, if the theory treats only certain limited features of a domain, and it certainly does not imply that the entities obeying these formally similar relations must also be materially similar. Nagel[19] and Bunge[20] have emphasized this point in discussing the difficulties in characterizing a mechanical explanation. It is certainly not enough to show that a theory can be given a Lagrangian formulation, even though this was first developed as a reformulation of classical Newtonian mechanics. What a Lagrangian representation of electrodynamics shows is not that Maxwell's theories can be adequately interpreted in terms of a swarm of mass particles moving around together, but rather that it is possible to look at the content of the theory in a new way which involves giving greater emphasis to certain concepts and laws at the expense of others, and this reformulation may in turn suggest unexpected and interesting lines for further development and application of the theory.[21] It is true that the tensor calculus was fully developed before Einstein gave it its first extensive application in general relativity; but this was so because it had been interpreted in terms of the concepts of differential geometry.

I have suggested that models can play an important role as ontological hypotheses. It is of course undeniable that hypotheses are essential to any empirical science. But it is not obvious that a method which involves advancing various hypotheses tentatively as possible candidates for scientific laws, and then testing them and their implications against 'the facts' that can be observed, is reasonable in ontology. There seems to be no way of verifying or falsifying an ontological hypothesis by seeing whether or not it corresponds with observable facts, since its very nature is to transcend our experiences by exhibiting them as merely the sensible manifestations of an underlying reality. If, like Popper and the verificationists, one holds this requirement of testability as a demarcation criterion between science and metaphysics, ontological hypotheses must certainly be considered metaphysical, with whatever pejorative connotations that much-maligned term may still arouse. The scientific realist cannot be satisfied with this sharp dichotomy between physics and metaphysics, such that they bear no relation to each other in their methods or conclusions. For him the creation of a coherent world-picture is an essential (perhaps even the most essential) part of science. Metaphysical questions are legitimate and important; and if science is to be our primary source of knowledge, it would be unreasonable if it could not shed light on such questions.

It should be remembered that what I am advocating is hypothetical realism, not dogmatic realism. This position recognizes that the formal structure of a given theory may have no obvious model, or there may be more than one, all of which seem equally (if at all) appropriate. Thus no final claim can be made that a particular ontology must be the correct one even within a given theory, and when this is compared with theories at the same or other levels, the claim becomes even less probable. Yet it must be considered a start in the right direction. Models may be distinguished from ontologies in that an ontology refers to an entire level, the parts of which may be described by many different theories, while a model is essentially associated with a single one of these theories. The construction of an ontology for a level is therefore likely to require the convergence of several different models, in the course of which these models may need modification and thereby suggest modifications in their corresponding formalisms. Thus the use of a model does not involve any 'ontic commitment' in Quine's sense, but must always include the possibility and perhaps the hope that it might turn out to be a vital part of that world-picture that, as far as we can tell, gives the most adequate reflection of the actual structure of the world.

Let us now return to Sellars's argument. Sellars's criterion for accepting the model of a theory as an ontology, i.e., as a (not necessarily proper) part of the ontology of the level with which the theory deals, is simply whether or not we consider the theory to be a 'good' theory. Unfortunately, Sellars does not tell us what he means by 'good' in this context. The positivists have a workable criterion for the goodness of a theory; they simply consider the totality of its predictions of particular phenomena, and see how many agree, and how many disagree, with our actual experience. However, the positivists' approach is based on an explicit denial of the relevance of ontological considerations in determining the goodness of theories. It would hardly be reasonable (or fair!) to use the results of the positivist criterion, then suddenly take an interest in the previously ignored ontology to the extent of accepting whatever it turns out to be. It is certainly possible to construct theories of very high predictive power whose implicit ontologies may not only be unusual or hard to express with our present stock of concepts, but may even be quite incoherent. Indeed, it may turn out that the 'best' theory available at a given time to cover a certain class of phenomena may turn out to have this defect. It has in fact been argued that this is exactly the situation which prevails in quantum mechanics, and this in turn has been used as an argument for the positivistic over the realistic interpretation of theories. Can Sellars really be asking us to accept an incoherent ontology, one which does not hang together in a unified way, without hesitation?

We may well ask what is the alternative for a scientific realist. One approach might be to accept a certain kind of ontology in advance, and then try to find a model for the theory that will be at least consistent with this ontology. This line is typified by the attempts in the late nineteenth century to find a mechanical interpretation of the electromagnetic ether, which were motivated by the successes of mechanical models in representing and extending other physical theories. This is at least an improvement over the Aristotelian requirement that our ontology must be drawn from general metaphysics, rather than specific physical theories. However, in the history of science such preassigned ontologies have ultimately proved to be strait-jackets to later theorizing, even though they have given some stability to the scientific world-picture and have inspired worthwhile research in their own right.

On the other hand, is it ever legitimate to rule out an ontology which may be quite unlike what seems to be found in our ordinary experience, but which is internally coherent and seems to be the only reasonable model for a powerful and apparently accurate theory? If one claims that we cannot simply accept whatever ontology seems to emerge, then we must have grounds for criticism independent of the theory in question. But what could such grounds be, and how might they be justified? The answer, I believe, must lie in the regulative ideals accepted by the scientific community. These are not fixed and immutable, and indeed such ontological questions are a frequent reason for modifying them. The ontologies proposed with new theories, and the regulative ideals governing their construction, are intimately bound together and are harmonized by what is essentially a process of successive approximations. In this connection one of the most important ideals is that of the increasing unification of science by means of common principles of explanation of the greatest possible power and generality. If the model proposed for a particular theory seems so completely different from that of the other theories at that level that there appears to be no prospect of unification, it is reasonable to look for another model, at least as a first move. Insofar as some regulative ideals are needed at any stage in the development of a science, the criticism which they provide is justified. Nevertheless, one must always remain aware of the dangers. Too close attention to ontological considerations on the part of a practicing scientist might stifle his research by leading him to give up theories which might later prove to be the best, and thereby lead to a modification of those ideals. This point has been repeatedly stressed by Feyerabend,[22] as well as by Kuhn in his criticism of inductive logic. If regulative ideals are to be regulating ideals, they must at least be somewhat lax and tolerant in their administration.

A further problem is to determine when models or ontologies are inconsistent with each other in a way that rules out the possibility of their simultaneous validity. This is of course a crucial problem for quantum mechanics, some aspects of which have been interpreted by means of a particle model, others by a wave model. Both waves and particles are classical entities. Used in different theories, both are highly successful, and as I have suggested above, it was reasonable to make them the first approximation in interpreting the new theory. But when taken together, as being apparently simultaneous aspects of the same sorts of entities, they appeared quite inconsistent. Now if the phenomena with a wave model and those with a particle model were taken to be on different levels, or if wave and particle explanations could each cover overlapping ranges of phenomena so that they could be considered coordinations of this class of phenomena with these two different levels, there would be no problem about compatibility. The whole point of the hypothesis of imperfectly coordinated levels is that no one may be assumed capable by itself of accounting for all the explainable features of the world. But the problem in quantum mechanics is that both the wave and particle models seem to be on the same level, i.e., they seem to be accounting for phenomena on the same level. The regulative ideal that a theory should have a single coherent model to cover its entire range, or at least a set of mutually consistent ones, thus militates against the Copenhagen complementarity interpretation of quantum theory. In taking the classical notions of wave and particle as final, despite the fact that they were developed only through another physical theory, the Copenhagen school seems to be taking the attitude criticized above, that ontologies must be determined in advance and cannot be changed. They prefer to reject the ideal of a single model for a theory rather than the ideal that any model at that level must be based on waves and/or particles. I would suggest rather that the situation requires a new model which reflects the formalism of quantum mechanics more accurately, rather than a 'complementary' amalgam of foreign models. These considerations on the need for new kinds of models will come to the fore more strongly when I discuss the model of geometrodynamics.

Sellars has suggested that although contemporary physics uses an atomistic ontology, there are some features of experience, for which he offers such as the 'ultimate homogeneity' of an ice cube that is pink all through its three-dimensional extent, that can be explained only by a theory that postulates an underlying continuum, and even goes on to suggest that the present notion of space-time may provide just such a model.[23] Nevertheless, his suggestion is quite out of place in this context. In the first place, his assumption that some continuous (nonatomistic) model will eventually be needed in explanations is quite unfounded. For it assumes that if there is one level with continuous features, it must be grounded in another similar one, whereas in fact it may be adequately grounded in an atomistic one—though of course this must be demonstrated in the particular case. More importantly, the argument for ultimate homogeneity is based on an extrapolation from sensory experience which involves envisaging smaller and smaller regions of a given thing, the pink ice cube, with the assumption that this can go on without limit. But this is precisely how Kant argued in the antithesis of the second antinomy.[24] Kant concluded that if this proof was valid, so was the proof of the thesis that matter is composed of atoms. Therefore neither argument was valid. The great value of Kant's achievement here, which Sellars seems to overlook, is that he showed that the question of atomism versus continuity could not be resolved by armchair extrapolations, and thus must be answered by the customary methods of scientific investigation, *if at all.*

Even if it is in fact the case that a level with a continuum ontology is needed for some explanations, there is no guarantee that we can find one adequate for this purpose, and in particular we cannot assume that the first one to appear (i.e., geometrodynamics) can do this job. We cannot simply create a complementary model (presumably at a different level) by listing all the desired features which are not in the original model, and somehow putting them together in an ontological tossed salad. Any model must arise from a particular theory, not from the miscellaneous residue of a set of theories. Geometrodynamics was developed for purposes quite other than explaining pink ice cubes. It may in fact be able to give a better account of them than atomistic theories can; but we certainly have no a priori guarantee of this. Since a model must be developed in conjunction with the formalism that allows for testable inferences, we can not know a priori what its final content will be.

HISTORY 2

I believe that the central *conceptual* problems with which geometrodynamics, as developed primarily by Einstein and Wheeler, is trying to deal concern the relationship of space to matter. Somewhat more specifically, one may ask: can space and matter be identified, or must they remain forever ontologically distinct entities? If they are thus radically different, by what criterion can they be consistently distinguished? If they can be, but need not be identified, are there any grounds other than conceptual economy for doing so? If they are identified, how rich must the single concept covering them be in order to account for all the features that have traditionally been assigned either to one or to the other? These are hardly novel questions. On the contrary, I maintain that every philosopher seriously concerned with cosmological problems has been forced to deal with them in one form or another. But I will make the further and certainly more controversial claim that the answer proposed by Einstein and developed in more detail by contemporary geometrodynamics is similar in important respects to those given by Plato and Descartes, and that in any case, a great deal can be learned about the theory by considering exactly where and why they agree and disagree. I also claim that the ontology common to Plato, Descartes, and Einstein is sharply opposed to that which was developed out of Newtonian mechanics; but on the other hand, that the ontology of geometrodynamics depends on a basic methodological insight which Newton introduced in sharp opposition to Descartes.

Now it might be argued that it is illegitimate to compare the answers of such men as Plato, Descartes, and Einstein, Wheeler, and other contemporary geometrodynamicists on the grounds that they are not and cannot be addressing themselves to the same question. The reasoning here would probably proceed as follows: In the centuries that have elapsed between Plato's time and Wheeler's, the semantics and syntax of such concepts as 'space' and 'matter' have changed so radically that they have nothing in common but the names of the concepts (or rather their supposed translational equivalents). Therefore, the bald statement that Plato and Descartes are alike in that they identified space and matter is misleading, if not completely false. They certainly differ in how and why the identification was made and what its results and implications were taken to be. There is a great deal to be said for this argument, especially as a weapon against those who like to organize the history of philosophy by lining up philosophers according to which of a limited number of simple answers they are taken to have given to some question. At least in certain contexts, it might be quite useful to use subscripts, such as '$\text{space}_{\text{Plato}}$', to make sure that actually different concepts are not confused. But if carried too far, this view is fatal to an understanding of the history or evolution of any sort of discipline. For concepts are not, as it seems to assume, cut from whole cloth, with precise meanings and syntactical rules for use fixed by a set of unchanging analytic statements introduced at the same time as the concept itself. Concepts are essentially tools, instruments developed by rational human beings in an attempt to organize their experience in order to provide a complete and coherent picture of the world. As such they have the dual function of unifying otherwise distinct particular experiences and allowing for new distinctions. These depend on what classifications are considered relevant or important to make at a given time, and this can change in response to new knowledge.

I suggest that concept formulation in science may profitably be likened to cartography. Toulmin has stressed the analogies between theories and maps,[1] but his conclusions are misleading because they relate primarily to road maps. These are essentially devices showing various possible routes from one particular place to another, and accuracy in reproduction of the terrain may be sacrificed to convenience in presenting this 'travel information'. My concern is rather with those maps which strive only for accuracy, rather than any possible practical uses. Reality then corresponds to the territory to be mapped, and concepts to the regions of the map, whose boundaries may be more or less sharp. The different cognitive levels correspond to the different kinds of maps that can be made, maps of political subdivisions, population distributions, climatological data, topographical features in relief, etc. Now when a territory is first explored, large parts of it must originally appear as terra incognita, and the divisions and boundaries that are drawn will appear crude and noticeably different from one cartographer to another. However, certain features of 'the lay of the land' (coastlines, major rivers, etc.) are so apparent that they can be readily identified and compared from one map to another, even though they may not be drawn in exactly the same place. Likewise in concept-formation, the 'constraint of reality' exercises a similar controlling and unifying influence. I suggest that our basic cosmological concepts, such as space, time, and matter, may correspond to these unavoidable geographical features. They were first discovered by the Ionian 'Columbuses of the intellect', and subsequent investigation has been a matter of increasing the accuracy of the representation and adding more details.

The above analogy suggests a cumulative development of science within a framework established long ago, whereas in fact it has been interrupted by a long series of scientific revolutions in Kuhn's sense, which are at once conceptual, methodological, and factual. This fact must therefore be the main disanalogy, and it depends on the fact that we can never have the sort of direct, immediate contact with reality-as-it-is that we can have with a piece of land. It is more like the determination of the geography of a distant planet on which we could never land, but only approach through a cloudy medium, albeit with better and better instruments. Belief in certain features, like belief in the canals of Mars, might pass in and out of intellectual fashion. But returning to the positive analogy, we may say that the conceptual geography as perceived by Plato and Descartes is like that perceived with the far more sophisticated intellectual tools of contemporary geometrodynamics. The details are different; but the outlines of the map are comparable.

The very vagueness of the concepts first proposed to chart the terra incognita is both their strength and their weakness. It is their weakness because their implications cannot be determined unambiguously and thus no precise predictions can be made; but it is their strength because their richness allows for many alternative lines of development. In discussing Boyle's 'definition' of chemical element, which is formally similar to that used today, Kuhn argues that the similarity is misleading since it was introduced into an entirely different context of ideas and was in fact used as the basis of an argument purporting to show that there are no such elements.[2] But as Shapere points out in criticism,[3] this view leads to a general relativism which prevents one from seeing the real continuity in the history of the concept. On this point Kuhn's thought converges with that of the formalists, and both are headed in the wrong direction. The formalists see definition as primarily stipulative, issuing in a set of timeless, analytic 'meaning postulates'; Kuhn sees it as primarily lexical, in large part making explicit the way the concept is actually being used by a scientific community in the context of its relations to other concepts. He further contends that they have little importance since they arise naturally from the way a field of investigation is being viewed. "Given the context, they rarely require invention because they are already at hand."[4] I believe that while both stipulative and lexical definitions have their place in science, the definitions that have actually been given our basic physical and cosmological concepts were neither, but rather, were what I will call 'restrictive definitions'. From the rich realm of possibilities contained in the vague, prescientific 'root concept' a selection is made of certain aspects which seem especially important in the light of presently available knowledge, or especially suitable to precise quantitative treatment. The remaining possibilities may be included in other concepts or left hanging in an intellectual limbo. (It is of course never certain that all these possibilities can be enumerated.) As I have suggested above, it is these selections and restrictions which have the effect of eliminating certain previously legitimate questions and creating the paradigms which make the discipline scientific. These definitions may be explicit proposals, not reflecting any already existing assumptions; but contra the formalists, they are not made in a conceptual vacuum. If the selection proves fruitful, it becomes embedded in scientific thought and may acquire new extensions and possibilities therein. If not, it is possible to return to the root concept and choose a different set. The continued existence of such a root concept depends on the fact that it covers an area in our understanding of the world which must be mapped in some way, but for which we can never be sure what will eventually prove the best way.

Thus within the outlines provided by our root concepts, it is the changing state of our knowledge and the instruments available for extending it that determine how the root concept is to be replaced by a more precise and limited scientific concept and what subquestions which can be answered by scientific methods are taken to follow from the more vague, 'metaphysical' question. Such increasing knowledge may indicate that the concept may be given a broader or narrower scope than was originally expected, or that its range (extension) can be preserved only by changing its formal 'definition', as Putnam has brilliantly shown.[5] Our new (revised) concepts emerge from the old, though this may be a revolutionary, rather than gradual process. It is therefore unreasonable to ask whether a question taken at face value is part of physics or metaphysics. For while remaining essentially the same question, it may be broken down and interpreted in different ways, some of which may be physical—in the sense that they are capable of empirical test—and others not. The mere fact that we do not know how to give a certain question a physical interpretation *now* does not mean that this will be impossible in the future, and it is utter folly to banish such questions forever as metaphysical or meaningless.

With these considerations in mind, let us turn more directly to Plato. In the *Timaeus,* the only dialogue in which he turned away from excursions into the perfect realm of the forms and problems of ethics and epistemology to discuss the physical world as such, he introduced the notion of the 'receptacle' (which later is given other names). This functions both ontologically as a principle of being, and epistemologically as a principle of explanation, or rather as a limitation on the possibility of rational explanation.

The *Timaeus* attempts to provide two distinct things: (1) what may be described as a 'metaphysics of nature'; and (2) the groundwork and outlines of a possible mathematical physics. As Bochner points out,[6] it is highly significant that Plato took them up in this order, and considered the former to be the more important. In this respect he differs sharply from both Aristotle and contemporary physicists, who begin with particular problems and then try to use conclusions drawn from these in the construction of more grandiose and speculative metaphysical hypotheses. Plato would reject this Aristotelian order for several reasons: (1) we *can* have (though perhaps we *do not yet* have) perfect knowledge of one complete and self-contained realm of being, that of the forms; and this can be used as a standard or basis of comparison in terms of which we can locate the deficiencies in any other realm; (2) the physical world of sensible things is by its very nature imperfect and cannot be a source of knowledge, but only of true opinion; and (3) we must have a complete ontological framework, in which no major features of the world are left out, if we are to be able to interpret and locate correctly any particular bit of information within it. Plato presumed (1) and (2) to have been established in the earlier dialogues, and (3) results from his examination of the failures of earlier attempts, which were simplistic in taking a part for the whole, and then finding data that could not conveniently be fitted into their schemes.[7] (Aristotle agreed with (3), but believed he could get this from consideration of a particular, i.e., a single substance with its changes and processes.) Plato certainly believed that 'Nature', as the sum total of sensible things, can be treated as a single whole, and explicitly rejected the possibility of other worlds (55d and 31a,b) which might run parallel to it. But of course it is inferior to the forms; its being depends on that of the forms, but not vice versa.

It might be thought that Plato's point is not really different from that of Scriven, as mentioned above: that physical laws, especially insofar as they are stated in mathematical form, with the use of real numbers and other high-powered idealizations, are only approximations of the actual state of the world as revealed by our measurements and observations. This is certainly part of it, but Plato goes much further. For it has been a regulative ideal in science up to the present that the deficiencies manifested by such approximations are not absolute but primarily technological limitations; that theory and empirical data can be made to come arbitrarily close to each other; and that any residual fluctuations are to be described in terms of the usual physical concepts, i.e., 'forces', whose exact magnitude or effect may as yet be unknown. The uncertainty principle in quantum mechanics has been the first real challenge to this ideal. But at least the uncertainty principle does not lead to a complete lack of rationality in the foundations of the subject. It gives a precise limit to the co-possession of properties by quantum entities, expressed by a definite number, Planck's constant; and it appears in (indeed, derives from) a completely mathematized theory. Furthermore, it plays an essential role only at the quantum level. Although other levels may be grounded in the quantum level, they do not manifest all the features of this level as measurable effects. The basic irrationality which Plato found in physical things is much more pervasive and has no measurable or regular limits. It is not clear whether Plato could have accepted something like my levels hypothesis *within* the physical world. Certainly individual men are not forms, but they seem to be more than just sensible things. But at least in the *Timaeus* men with their bodies are being treated as part of a one-level physical world. The forms are the only other level explicitly recognized.

Thus the 'bastard reasoning' which Plato used in demonstrating the existence of the receptacle seems to be mainly presuppositional and proceeds as follows: (1) we know that the forms exist eternally and are supremely real; (2) these forms account for many features of the sensible world, i.e., they explain the characters or qualities of objects in terms of the mysterious relation of participation or resemblance; (3) but there are certain features of the sensible world which have no counterpart in the world of forms and thus cannot be explained by them; (4) to do so we must postulate some other ground, whose influence works together with that of the forms to produce the world as we see it; (5) this ground is hypostatized and given the name 'receptacle', or later *chōra* or space. Plato of course did mention other factors, such as the Demiurge, the World-Soul, and the Form of the Good, but this still left a host of functions for Plato to assign to the receptacle. Now the question immediately arises, why should all these functions be ascribed to a single factor, rather than to several independent ones? In the first place, Plato would insist that there is no point in introducing several such factors, if all the functions *can* be ascribed to one. But beyond that, it is not simply that he was unaware of the distinctions which Aristotle introduced. Democritus, for example, had already made a sharp distinction between matter and space. Plato believed that such distinctions were not justified, and thus did not introduce them as boundaries in his conceptual geography. Let us consider just what tasks Plato did assign to his chōra.

It first appears (49a) in Plato's account of becoming and substantial change. Since each form is essentially one, perfect, and changeless, it cannot become anything else, nor can it explain why an object can acquire or lose the corresponding character. Yet Plato recognized with Heraclitus the fact of constant change, becoming, process, and instability in the world. Even the Empedoclean doctrine of the four elements of earth, air, fire, and water could not help, since these could also change into each other. There must be something which underlies and persists through such changes and prevents any physical object from being a stable and eternal copy of the form, by giving it the constant possibility of changing into something else. Now the 'four elements' have corresponding forms (51b–e) but their empirical manifestations can only be temporary 'suchnesses', which appear and disappear. The real 'whatness' of the sensible world must thus be something completely neutral, which cannot be identified with any of these local qualities. At this point, its only real property as a whole must be its ability to support change without itself being affected by any such changes.

Plato asserted (50e) that the receptacle has no qualities of its own. Taken at face value, this statement seems to be simply false, since if it were true the receptacle would be pure nonbeing or Aristotle's 'privation', and we could say nothing further about it. But we must distinguish qualities *within* the receptacle from qualities *of* the receptacle as a whole. Each of the 'suchnesses' mentioned above is a local character, something that appears within the receptacle or applies to a certain part of the receptacle at a given time. What Plato was insisting is that the sensible qualities of physical things which come to be in the receptacle cannot be applied to the receptacle as a whole. But this is not intended simply as a point about the logical syntax of predicates, albeit a correct one. For he also has the metaphysical reason that the characters of sensible things must be determined entirely by the forms, not by the third factor. Although the receptacle does somehow oppose the imposition of form, so that these characters can be only imperfect copies of the form, it does so for all forms without discrimination. If, for example, the receptacle were blue, it would reinforce any blues in nature, making them into more perfect copies, while further destroying any reds, yellows, etc. This would prejudice the supremacy of the forms and the power of the Demiurge, as well as being unsupported by observation (unless one is wearing blue glasses). It is essential to recognize here the point that Sellars urges against Hanson,[8] that to be imperfectly one thing is *not* to be perfectly something else. Perfection is a meaningful term for Plato, but the standard resides in the forms, of which there are only a limited number. Since we can perceive only the qualities of the parts of the receptacle, rather than those of its all-inclusive whole, Plato rejects the application of such sensible qualities, particularly when they form a set whose members are mutually incompatible (like colors and temperatures) but is taken to be exhaustive, in that one and only one member applies to every sensible thing. The probabilities of finding any one of these members at a given point at a given moment must be equal a priori. Actual inequalities in distribution must be explained in other ways.

But this very neutrality, this lack of possession of any particular qualities, is what allows Plato's receptacle to be filled with capacities to receive copies of all the forms, and these capacities are not limited by other qualities *already* present in the receptacle. In this case it seems reasonable to look on Plato's receptacle as a source of real potencies just as much as Aristotle's matter. Yet neo-Aristotelian commentators like Eslick have rejected this claim and insisted that the receptacle can be identified only with 'privation' (steresis), a negative concept indicating the denial of qualities, rather than 'matter', a positive concept indicating the possibility of qualities.[9] Why must this distinction be made? Aristotle's concept of matter is basically relative. Rather than postulating a separate world of pure forms, Aristotle remains at the level of physical things and insists that *here* form must always be conjoined with matter in a particular substance. (Plato would of course concede this point.) But Aristotle was concerned with the *appropriate* matter, the matter with the specific potency or capacity to receive a particular form, or one of a limited class of forms. For this purpose the matter may and will have qualities of its own, and these will provide the accidents of the completed substance. Thus Aristotle can speak of bronze or gold as the matter of a statue. But in another context, this bronze or gold might be taken as a kind of form on a more elementary level of matter, and it was by this process that Aristotle reached his notion of primary matter, which is essentially a limit concept. Plato, however, could not make these distinctions of degree and kind of materiality. He would have agreed with Aristotle that matter and form are correlative principles in accounting for the physical world, but would have insisted that the correlation be cosmological, between the whole of the receptacle and the entire world of the forms. Aristotle's correlation is always particular, located in the single substance; indeed, his notion of form reflects the difference in orientation. Whereas Plato's forms refer primarily to universal qualities, in which a wide variety of particulars might participate equally, Aristotle's substantial forms are connected with 'natural kinds' of particulars as their unique essences, and are used in organizing substances into genus-species classifications.

Aside from its connection with potency, the other aspect of matter which would be expected to concern Plato is that of 'stuff', that out of which things are made. Plato discussed this in connection with the analogy of the gold (50a–c). This can be molded and remolded into an indefinite variety of shapes, and the only thing that can truly be said at all times of any of these momentary shapes is that they are gold. If the analogy is to be useful at all, it is clear that Plato must have meant that physical objects are constructed out of the receptacle, which otherwise has no specific determinations. Plato's insistence that the analogy must be imperfect stems from the fact that gold does have many properties of its own, and perhaps can change into something that is not gold, which cannot be allowed for the receptacle. He specifically criticized the Ionian natural philosophers, who had introduced a whole variety of more or less determinate things as the underlying stuff of the whole universe. (Anaximander probably came closest to Plato's conception.) Likewise Aristotle's word for matter, 'hyle', originally meant wood, which has as many specific properties as does gold. However, Aristotle specifically introduced it as a technical term, a gambit with which Plato was unfamiliar. It is certainly reasonable to ask Plato, what then are things made out of? and there is nothing in his system except the receptacle which could be a reasonable candidate for the job. To be sure, this construction proceeds through the intermediary of the geometric solids and their component triangles, but this will be discussed below.

Thus it seems clear that the main functions which Aristotle's notion of matter includes, those of potency and stuff, are contained in Plato's notion of the receptacle. It is true that these functions appear in different ways, which reflect the strikingly different methodological orientations of the two philosophers. But they are undeniably analogs, appearing in the same place in their conceptual geographies. Insofar as Plato recognized the various aspects of Aristotelian matter to be real and legitimate, he could assign them to the receptacle. Others he rejected outright as unnecessary distinctions.

If this can be said for the material functions, what about the spatial functions? Aristotle here introduced a further new concept, 'topos' or place, and Plato did not. However, he does often refer to the receptacle, or 'third factor', as 'chōra', beginning at 52a, and this is usually translated as space. In his introduction to Jammer's book Einstein distinguished between two traditional conceptions of space which have influenced much subsequent thought and analysis:[10]

These two concepts of space may be contrasted as follows: (*a*) space as positional quality of the world of material objects; (*b*) space as container of all material objects. In case (*a*), space without a material object is inconceivable. In case (*b*), a material object can only be conceived as existing in space; space then appears as a reality which is in a certain sense superior to the material world. Both space concepts are free creations of the human imagination, means devised for easier comprehension of our sense experience.

How does Plato fit into this scheme? Aristotle's notion of place clearly seems to fall on the side of (*a*), and his conception of time is even more certainly relational.[11] Plato's cosmological orientation might seem enough to incline him toward (*b*), but we must take care to see just what is involved here. The traditional notion of absolute space, the Newtonian form of (*b*), was bound up with that of a void, and understood primarily as a passive arena for the activity of matter in motion. This certainly is not what Plato had in mind. Plato was familiar with at least one theory in which matter was sharply distinguished from the void, the atomism of Democritus. As Jammer points out,[12] in Democritean atomism space was identified with the void, i.e., "where things (atoms) are not." Only in Epicurus and the later atomists is space conceived as continuous and pervading even the atoms which move through it, and even here it is quite distinct from them. Such distinction is clearly foreign to Plato.

As we shall see, the space of classical physics was infinite, continuous, homogeneous, and isotropic. Its structure was perfectly described by Euclidean geometry, with the accidental feature that it happened to be the Euclidean geometry of three dimensions. It was completely independent of any bodies placed in it and contributed nothing to them except a framework in which distances could be measured and coordinate systems set up objectively. Plato's chōra is related far more intimately to the things in it than is the space of classical physics. If it were merely the pure possibility of extension it would be perfectly passive, and the forms could be perfectly realized in it. But Plato has explicitly denied this. Plato's use of such words as 'receptacle', 'recipient', 'nurse', and even 'mother' suggests that an essential function is to hold and embrace physical bodies, to give them a home as it were, rather than merely allowing them to be arranged. Whitehead saw the superiority of Plato's concept of space to that of classical physics in its provision of a real communion or relatedness for everything placed in it.[13] This is certainly not true of the void, or Newton's absolute space, which required a notion of action at a distance. Plato may have action at a distance, but the distance is not across a completely foreign and passive entity. Chōra is more a medium than an arena.

Let us now consider which features of classical absolute space are indeed present in Plato's chōra. In the first place, it is clearly three-dimensional. It is undeniable that bodies have three dimensions, and insofar as the receptacle is to receive and hold them, it must have at least as many. (Higher dimensionality would have seemed inconceivable and unnecessary.) Plato further emphasized this three-dimensionality by choosing geometrical solids as the material correlates of the four elements.

Continuity presents somewhat more of a problem. Material bodies cannot be subdivided indefinitely, since eventually we must come to the actual solid particles and the triangles of which their faces are composed. But even the two-dimensional space interior to one of these triangles must be continuous; otherwise it would not be a triangle at all, in the sense that many theorems of plane geometry would no longer apply to it. And taken as a whole, the receptacle itself can have no holes or discontinuities, since this would mean either that it is possible for bodies to move into these holes and thus no longer be in chōra, or else the holes would be forbidden to bodies, and there is then no reason why any such regions should exist.

Plato introduces a distinction between the state of chaos, before the Demiurge has introduced form into the receptacle, and the present state, in which there is at least a degree of order and stability (52d–53c). It may be asked here whether 'before' is to be taken as a temporal or logical relation. Plato did say that chōra is eternal, quite independent of the forms and the Demiurge (52b), while time was created by the Demiurge as "the moving image of eternity" (37d). Time is here taken as an element of order and regularity, as measured by planetary motion. It appears that chaos may be taken as a state which actually did exist before time and order, as well as logically underlying temporal reality. In such a state of random, chaotic activity, it is hard to see how regions could be distinguished in such a way as to make a notion of discontinuity meaningful.

Is such a space homogeneous? I think the answer must be no. In his description of chaos, Plato developed the idea that it is a source of energy and irrational motion, with unbalanced tensions and forces distributed throughout its extent. These were purely random and thus unable to maintain themselves in any sort of stable configuration which could serve to define or measure a region with definite boundaries. But there could be no such motion at all without such imbalances in different parts of the space, even if these could be distinguished only momentarily. Since there are as yet no physical objects which may be treated as moving bodies, motions must here be taken as analogous to random convection currents in a fluid. They persist even after the action of the Demiurge, and thus constantly tend to undo his work. In his simile of the winnowing-basket (52e), Plato spoke of these powers as both shaking and being shaken by the receptacle. He may have meant that the unbalanced powers, creating what we would call regions of different potential energy, provide a sort of constant internal earthquake, and the affected parts of the receptacle react in opposite directions, tending to break up these concentrations of power while at the same time creating new ones. Plato's notion of 'vestiges' is even less clear, since it suggests something that should remain *after* the forms or their copies have departed, whereas in fact they appear before this. They must be more like strengthened potencies, with the material conditions for a particular element satisfied more strongly in one place than another. But insofar as chōra is filled with these fleeting potencies and constantly changing tensions and motions, it can hardly be called homogeneous through its spatio-temporal extent.

As for its euclidicity, we must again distinguish chaos from the present state of the world. If chaos does involve such a complete lack of form and measure, it would be ridiculous to consider it a Euclidean space, since the latter contains a very definite and precise geometric structure. But this does not mean that it was non-Euclidean, in the sense of having some other Riemannian structure. It was rather a completely "metrically amorphous continuum" with no "intrinsic metric," to use Grünbaum's favorite phrases. But of course the actual space of physical objects, as determined by any measurements involving them, is assumed to be Euclidean. This is also reflected in Plato's choice of the five solids, perfect only in Euclidean space. Of course Plato knew of no alternative structures; there was only one physical space, for him as for anyone else, and geometry as he knew it was the science of its structure. Confronted with the possibility of other geometries, Plato could very well argue teleologically that Euclidean space was chosen by the Demiurge as being the best or most rational. Since the structure of space does appear as the work of the Demiurge, and he was supposedly guided by the form of the good, such an argument would seem quite reasonable.

A problem arises about whether Plato's chōra is infinite in extent. (I have already argued that it is infinite in divisibility.) Through its function as matter, it must be coextensive with the physical world. Nothing can take place outside of it; but on the other hand it cannot contain potencies which can never be actualized by reason of lying outside the 'body of the world'. This would require cutting the receptacle with an artificial boundary, and requiring a sharp distinction between regions inside and outside the world's body. Since Plato said explicitly that the world's body is a perfectly self-contained sphere with no outside (33b–34a) this would seem to require that space be correspondingly finite. But this seems inconsistent with what was claimed above, that chōra must provide a model for Euclidean geometry, since this requires the notion of at least potentially infinite or indefinite extendability. To resolve this dilemma, we must remember that Plato lived considerably before Euclid's final codification of geometry into an axiomatic system. Up to Plato's time geometry, even when granted the certainty of logical deduction, was taken to apply to local problems in the physical world, to triangles, circles, angles, and line segments which could only be extended a finite (though perhaps very great) distance. I find no indication in Plato, or in his contemporaries, that this local physical geometry might have global or cosmological implications—in particular, that it might require that physical space be infinite. The idea that the validity of a particular system of geometrical relations should reflect, depend upon, and be a key to some central properties of the structure of the universe as a whole, and select one structure out of many otherwise possible, would seem quite extraordinary to Plato. It is only by the time of Descartes, as we will see in the next chapter, that the global implications of geometry take precedence over the local in the relation of matter to space. Only with the development of differential geometry could the local and the global be analytically separated. The central problem in relativistic cosmology today is how to bring them together and understand what implications the one must have for the other. We will consider the local-global problem in more detail below. It is possible that Plato might have a way out. He certainly claimed that there are forms of the mathematical entities, and like other forms they are only imperfectly manifested in the world. He also seemed to want to deny spatiality even to these pure mathematical forms. Plato of course meant here the spatiality of the receptacle, with all its other accoutrements, and he might instead have postulated a pure mathematical space, in which the forms can be organized in a system of 'pure geometry', of which physical space is only an imperfect copy. Such a mathematical space could indeed be infinite, but only as a formal construction, quite independent of the extent of physical space which depends on "the body of the world." But Plato never suggested this possibility directly. I believe we must therefore conclude that chōra is finite even though Euclidean, in the sense of not being a completed infinity in extent; however, since there can be nothing in the physical world outside it, it is unlimited.

Finally, Plato's space must be considered isotropic. Because of its inhomogeneity, it may appear quite different locally in different directions (if one doesn't look very far), but on the average (globally) these discrepancies cancel out, and no direction can be singled out over its full extent as having any special priority. In this respect he contrasts sharply with Aristotle, whose place (topos) was homogeneous but quite anisotropic, with directions singled out as the lines of natural motions leading to natural places. Plato's chōra does not have predetermined natural places for the four elements; instead there are motions governed by the attraction of like to like (53a and 57c), and an internal weight-like property of bodies (63a–e).

Now since all the questions which could be asked about the classical concept of space can also be asked about Plato's receptacle, and reasonable answers can be given, it is fair to conclude that Plato's receptacle is also his 'space', as well as his 'matter'. It also provides a mode of sensible representation of the forms, though a disembodied intelligence can grasp their true nature without such assistance, and a way of allowing copies of different forms to be conjoined together in a single body. It provides a ground for multiplicity in representations, accounting for how a form, while remaining one and self-identical, distinct from all other forms, can nevertheless be manifested in an indefinite number of different things. They can at least be distinguished by spatial position. (This of course does not explain why there should be so many copies.) Indeed Eslick sees this as the main role for the receptacle, as an attempt to bridge the gap between the one and the many,[14] though I think his claim is far too restrictive and extreme.

We must conclude that Plato's notion of the receptacle was deliberately intended to include the functions that might be assigned to separate concepts of space and matter, and thus represents one possible way of unifying the two. (In calling it possible, I do not mean that it is still a "live option" today, at least in the exact form given by Plato.) It has been the bias or assumption that they *must be* separate concepts that has led commentators to misunderstand the role of the receptacle. Thus Shorey,[15] Ross,[16] Jammer,[17] and especially Taylor,[18] noting that the one necessary and invariant property given to bodies by chōra is three-dimensional extension, emphasize the spatial aspects to the point of thinking that chōra is nothing but spatial extension. On the other hand, Zeller,[19] Whittaker,[20] Claghorn,[21] and especially Collingwood[22] emphasize the material aspect and its similarities to Aristotle's hyle. All of these attempts to categorize the receptacle are one-sided because they assume too readily that certain categorial distinctions must be made in a locatable region of a philosopher's conceptual geometry, whereas in fact Plato denied the validity of such a distinction in this case.

Having completed his metaphysics of nature, Plato attempted to develop a 'physics' consistent with it. He began by postulating four of the five perfect geometrical solids as correlates for the four Empedoclean elements. Since Plato was trying to account for the composition of different substances and the possibility of transformation from one to the other, it might be said that he was trying to reduce chemistry, rather than physics, to geometry. This is especially true if physics is taken to deal with motions independent of composition. But such a distinction is not really important in this context. What is important is that Plato was not willing to stop with these tiny three-dimensional figures as elements, although as the smallest possible units of earth, air, fire, and water they must be three-dimensional since the latter are all bodies (53c). They were first analyzed into their bounding surfaces, and then the latter were broken into their component triangles, of which only two kinds were allowed. These triangles are the real elements.

This approach immediately raises a serious problem. For apparently it is not the whole solid, including its interior, but only its surface or boundary that is a real building block for material things. The interior is not an essential part of the elementary solids, but is only that which the combination of planes happens to enclose. Since this interpretation removes the last possibility of having any other substance-specific matter, the 'that which' can refer only to chōra. This further strengthens my claim that chōra is continuous and all-pervasive, and that it is the stuff out of which things are made. Cornford's attempt to restrict chōra to being an 'in which' medium requires him to treat the forms, functioning here as powers, as the material 'out of which.'[23] But this is simply a confusion of form with matter. We will see that these little prepositions again cause great trouble for geometrodynamics.

But if both the inside and the outside of the elementary solids are equally chōra, what is the nature of the boundary dividing them? In Euclidean space, a triangle is a two-dimensional entity, and can be uniquely determined by a set of three noncollinear points. It is also a theorem of geometry that a plane divides a three-space into two distinct regions. But all this lies in the realm of pure geometry, where the division is formal or ideal. As discussed above, Plato's chōra is quite different from such mathematical space. Insofar as it is matter, with unlimited potencies everywhere, the triangles must effect a material division of physical space. If they cannot do so, what then is Plato's point in introducing the triangles and elementary solids? Given his (more or less) mathematical theory of transformations, it would not even do as a model for this theory, since it would be inconsistent with the underlying cosmological doctrines.

Can Plato's triangles do this job? There are in fact severe difficulties in any interpretation, as I have discussed elsewhere at greater length.[24] If they are taken to be two-dimensional, it is not clear what mode of existence they can have within a three-dimensional medium. Yet they are taken to have an independent existence. Plato held that, once a solid has been broken up, the triangles drift about at random until they can recombine in a new figure (56c–57c). They are also able to maintain their identity and perfection of form despite the 'buffeting' to which they must be subjected by the unbalanced forces in the state of chaos. Plato further complicated matters in describing how the solids are broken up. For example, he spoke of the fire tetrahedrons cutting because of their sharpness, but such properties are appropriate only to three-dimensional entities.

But if we say that the triangles have depth as well as length and width, they must enclose a certain volume. But a volume of what? With no other factor available, it can only be chōra and we are back where we started. On the other hand, it might be possible to think of the Demiurge as 'squeezing' some of this chōra into thin little laminae, which would then have a 'density' greater than that of the rest of the space. This would do the job, but at the price of introducing a parameter (density) which would have stable and orderly, though different, values throughout the space, instead of merely irregular fluctuations on a medium that is homogeneous on the average. In any case, we could ask where the Demiurge could get them, if not by such squeezing. They cannot lie outside of the receptacle like a huge pile of shingles, since there is no outside. The basic function of the Demiurge is indeed to impose form on recalcitrant matter, so it might be claimed that the triangles are simply a minimal level of form, and thus a special case of his activity. But this leaves us with the problem of how pure form can perform a material function. Plato's mention of the cutting of the solids again reflects this confusion between the formal (mathematical) and material aspects of a physical process.

The same problem arises when one asks how the triangles are connected together to form a solid. If they are material, one would expect them to have some sort of hooks or adhesive on the edges, and in any case there is no guarantee that they will join at the correct angle. If they are formal, what is involved besides our merely *envisaging* two triangles with a common edge? When such an elementary solid moves, one may ask whether the chōra inside is carried along to the new place. This might be expected if the faces are made of material which cannot let anything pass through them. But since chōra is continuous, it becomes a problem how it can be pushed aside to make way for the approaching solid, and how it can fill in behind it. If they are formal, the form may be propagated as a wave, without creating any motion in the underlying medium. Chōra as a whole must thus be stationary, like the nineteenth-century electromagnetic ether, despite all the random currentlike motions which are supposed to take place in it. We will see this problem become acute in the case of Descartes.

It is clear that the totality of physical objects, as composed of the solids, cannot be continuous, since the perfect solids cannot be fitted together in such a pattern, and in any case there must be room for motion of the solids and the free drifting of temporarily disconnected triangles. But Plato's material atomism does not allow a possible void, since the receptacle underlying them is still continuous. If it were argued that this is a mere dodge, since one can distinguish where physical objects are (within the solids) from where they are not, Plato could counter that the places not enclosed by solids at the moment have equal potential for receiving them and becoming particular substances at some later time, and such positive potencies are a far cry from the absolute emptiness of 'the void'. Plato's richer notion of space made it easier for him to hold on to a belief that the world is a plenum than Descartes' notion. For Plato's space is the totality of material objects, possible as well as actual.

Plato did not fall into the Pythagorean trap of identifying the four elements, with their qualitative aspects, with the solids as geometrical entities. He insisted that a little tetrahedron is not by itself an element of fire, and must be 'informed' by the form of fire before it can achieve this status (51b–e, 55d–56b). The Pythagorean identification here may be considered a primitive form of the logicist error. It identifies an entity which has unique qualitative aspects at a particular level with (possibly) necessary and sufficient conditions for its existence, which can be specified on a different level. Plato was certainly not a reductionist of this stamp. His solids are only correlates, and do not determine the actual qualities of sensible things, which, as he repeatedly insisted, derive wholly from the forms. Pythagoras, like some contemporary formalists, looked only at the quantitative aspects of things and processes in the world; Plato brought in the forms to account for the qualitative aspects, which were also needed in a complete explanation; and Aristotle took the final step in indicating how all the aspects could be brought and held together through his notion of substance.

I have not introduced the above discussion of Plato's 'physics' as a digression motivated by antiquarian curiosity. Its importance lies in the fact that any attempt to identify space and matter (as well as any attempt to distinguish them sharply) runs into serious conceptual difficulties. We will see that the nature of the bounding surfaces and the division they effect is also one of the main weaknesses in the Cartesian system. I believe that geometrodynamics, with its space-concept based on a far richer set of possibilities, is able to avoid this difficulty; but only by giving up a notion of material body which not only Plato and Descartes, but also present-day common sense, want to preserve if at all possible. In doing this we can see some of the philosophical difficulties which geometrodynamics must overcome.

I have been emphasizing the positive features of Plato's receptacle, those which allow it to be considered a precursor of the conceptions of Descartes and Einstein. But there is one essential feature in which the receptacle is strikingly different. I have mentioned above that Plato believes the physical world cannot be fully rationalized or made intelligible; such perfect intelligibility resides only in the forms. But it is precisely the receptacle which places these limits on rationality, because it is the seat of *ananke,* or necessity. As Grote and Cornford point out,[25] necessity

is now usually understood as denoting what is fixed, permanent, unalterable, knowable beforehand. In the *Timaeus* it means the very reverse: the indeterminate, the inconstant, the anomalous, that which can be neither understood or predicted. It is Force, Movement, or Change, with the negative attribute of not being regular, or intelligible, or determined by any knowable antecedent or condition.

This is a fundamental limitation, part of the essential structure of the world. It appears as a force, not active in initiating generation or imposition of form, but always opposing and resisting to prevent a complete realization of form. For this reason any physics can be no more than a 'likely story' (29d), and Plato felt justified in using myths rather than exact concepts. This is certainly a further argument for considering chōra as matter, since philosophers who have separated space and matter but insisted that there was a fundamental irrationality in the world have always located such irrationality in matter. Space, on the other hand, has always been presumed accessible to reason, and indeed the Greeks rightly considered their rationalization of space in (Euclidean) geometry to be one of their greatest accomplishments. These considerations must serve as a warning to epigoni like Friedländer who want to regard Plato as the father of mathematical physics.[26] Despite his use of geometry, Plato did not believe that pure mathematics was applicable to the sensible world, except as an approximation. Sambursky has emphasized this point,[27] which should refute commentators like Taylor who see Plato as an apostle of Pythagoras. On the one hand, this viewpoint enables Plato to sidestep many objections about the details of his system, whereas Descartes was forced into inconsistencies in trying to answer them; but on the other hand, it must appear to us as a counsel of despair and a retreat into other-worldliness. It would be hard to find a sharper contrast with Descartes' confident optimism. Whereas Plato saw in chōra a limitation on possible rationality, Descartes identified space with matter for the very purpose of rationalizing matter. Despite the many objections to the details of his system, Descartes is in this respect far closer to the spirit of modern science.

Let us then turn directly to Descartes' theory of matter. For Descartes, as for Plato, physics was not something which could be developed independently and later perused for possible ontological implications; instead general metaphysical principles were laid down and a physics fitted into the framework outlined by them. Such a procedure is not wholly unreasonable, at least if there is no attempt to deduce every physical law from them in an a priori fashion. As Blake points out,[1] Descartes fully recognized the importance of collecting empirical data, and indeed of performing experiments rather than observing Nature passively, as had the Greeks. However, in contrast to Bacon, he realized that such data could not speak for themselves but needed principles of interpretation under which they could be subsumed and fitted into a coherent picture of the physical world. Such an approach is typical of what Kuhn calls the pre-paradigmatic stage of a science. Without a concrete scientific achievement involving a particular interpretation of the phenomena, there is no indication of which interpretation will work best. Since some starting point is needed, it is reasonable to use one which has been given a putative initial plausibility by metaphysical arguments. Once a paradigm has been achieved, its conceptual system is likely to form the basis of a new metaphysics, which in turn will be used as a first attempt to interpret some other class of phenomena. I will argue below that Newton was the first to make such a conceptual breakthrough and create a paradigm for classical mechanics on the basis of physical reasoning alone. But Newton's philosophical writings indicate that even he did not grasp the significance of this methodological revolution, which had the effect of establishing the value of the hypothetico-deductive method. Furthermore, it is only because of the fact that a certain simple kind of mathematical equation happened to be the correct one, a circumstance that can only be described as fortuitous, that Newton was able to lay down a method for giving exact answers to certain questions and ruling out others as illegitimate *for physics*. Insofar as this sharp distinction between perfectly answerable scientific questions and meaningless pseudoquestions is central to positivism, I will argue that Newton should be considered the 'father of positivism'; but that the very reasons which led Newton to positivism are reasons against its uncritical acceptance today.

Descartes is often referred to as the father of modern philosophy. Despite his conscious efforts to break with his scholastic predecessors, he takes over a substantial portion of their terminology and distinctions, such as that between essence and accident. Indeed, many of his more obscure passages are best understood as attempts to fit his ideas into this framework. However, his real 'modernism' stems from his concern with methodology, and he represents the real beginning of what Rorty calls the 'epistemological

Descartes' Geometrization of Physics 6

turn' in the history of philosophy. Not only is metaphysics prior to physics, but epistemology is prior to metaphysics. Where Plato could blithely refer to the grounds for his assertions about the existence and nature of the receptacle as a sort of 'bastard reasoning', Descartes demands rigor. As I have mentioned above, matter is no longer a limit to knowledge for Descartes; rather it is itself an object of perfect knowledge. But this knowledge is guaranteed only if the reasoning follows the correct methodology. In a nutshell, Descartes' method involves stripping away all unwarranted assumptions about the world by subjecting phenomena to the method of systematic doubt; arriving eventually at 'clear and distinct ideas', either of axiomatic propositions or 'simple natures' of things, whose indubitability may either be a matter of pure logic or a necessary reflection of the goodness of God; and deducing consequences from these ideas. The first step in this method is essentially abstractive from sensory experience. Applied to the physical world, it works as follows: We look at the world of material objects as they present themselves to our senses and inquire what they must have in common in order to be classified together as material objects and distinguished from everything else; or we look at the single body and consider what feature cannot be stripped away from it or replaced by something else without making it something other than a body. The answer in any case turns out to be spatial extension.

But we may well ask, what is the source of this necessity? It is true that every physical object in our experience does appear to us as extended, and that we cannot imagine an unextended body. However, as Kant was to point out, this may reflect a property of our perceptual apparatus rather than a necessary attribute of things in themselves. The assumption that the universal extension of all bodies as perceived is equivalent to their extension in any ontological sense represents the triumph of the epistemological turn. Before this, it would be reasonable to say that a body *is* where it acts, i.e., where it engages in causal relationships with other entities; but with the epistemological turn, a body is where it is perceived to be, where it is located in the spatial-temporal coordinate system of the perceptual world. The latter criterion suggests a relation to a knowing mind in a way that the former does not.

Three comments are in order here. In the first place, it might indeed be possible to develop a theory according to which the only essential attribute of a physical body was extension—and a successful physics based on this notion—without appealing to the method of systematic doubt and abstraction of all dubitable qualities to reach clear and distinct ideas. However, the fact remains that this is Descartes' basic method. The only real argument that he gives for the position is that *le bon Dieu* would never deceive us if we really do have a clear and distinct idea.

In the second place, in any method designed to discover The Essence of all material objects, we have no a priori guarantee that there will be one and only one such attribute. There may be nothing at all which is possessed by every physical object, though there may be many Wittgensteinian 'family resemblances' among them. On the other hand, there may be more than one such attribute, in which case one must decide whether they are necessarily linked in some way or really separate. In fact, we will see that Newton's concept of mass, which Descartes never developed, was at least equally reasonable as an alternative criterion for matter. If a Cartesian objected that one can conceive of massless bodies (but not unextended bodies), a Newtonian could retort that the Cartesian notion of conceiving here requires staying too close to perceptions. The important thing is that one can do classical physics with unextended (but not massless) bodies. The concept of mass is not derived by abstraction from sensible bodies (though weight might be); it is instead required by the exigencies of physics.

It might be argued that the Cartesian and Newtonian criteria for matter are not coextensive, and that they define different classes of things. For instance, Descartes regards fire, which is extended but presumably massless (I neglect mass equivalents of its radiation energy), as a body and Newton does not. However, such disputes illustrate the important points that the class of material objects is not well defined, so that the criterion serves merely as an explication of its boundaries; and that any proposed definition or criterion has some stipulative force, and depends both on what use is to be made of the term 'matter' or 'material' and what other sorts of recognized entities are being contrasted with material ones. We shall see the importance of this fact below. In any case, it must be recognized that all ordinary physical objects seem to be both extended and massive, and it is only the borderline cases which have one of these properties but not the other.

In the third place, Descartes' contention that all material bodies are and must be extended is taken over rather blandly as a basic postulate by the neo-Thomists and Aristotelians, even when they reject the mechanistic conclusions which he derives from it.[2] But their purpose is to give a 'metaphysics of Nature', where 'Nature' is taken to refer to the totality of sensible things. Descartes, on the other hand, wanted to develop a mathematical physics, and there is no assurance that the sensible body, and whatever essence can be abstracted from it, is the ideal basis for constructing such a system. Indeed, the success of Newton shows that a radically different basis proved far more useful. Descartes must have therefore denied one of the fundamental principles of scientific realism, as I have presented them above: that the real physical world, with which physics should be conversant, may turn out to be composed of entities radically different from those in our everyday experience. The latter are still real, but they form just one level of reality, and one which may be grounded in some other revealed by physical theories and their models. For Renoirte and his neo-Scholastic cohorts share with the ordinary language philosophers the assumption that the world of everyday, macroscopic physical objects has an ontological priority[3]—an assumption which I have argued is incompatible with an attempt to create science or take it seriously.

Descartes' matter still plays its traditional roles of being the stuff out of which things are made and being the ground for the observed multiplicity of bodies. But it now acquires a new dignity. Plato spoke of his receptacle as being coeternal with the forms, and described a state of chaos before the imposition of form, but this description may be metaphorical and not really refer to some previously existent state of affairs. In any case there is no uninformed matter existing now. Aristotle and the Scholastics go even further. For them it is logically impossible that prime matter should exist by itself, for it is simply the correlate of form. Every substance (except God, who is pure form) must include both matter and form as contrasting principles. But Descartes elevates matter from the junior partner to the rank of substance itself. Since Descartes is still thinking in terms of the Scholastic definitions, as a substance it must have an essence, and extension is just this essence.

This elevation of matter has the effect of degrading form. One thinks no longer of the matter appropriate to a given form, but rather of the form appropriate to matter whose essence is already known to be extension. Thus form is reduced to geometrical shape, which of course is still its common notion. This was hardly an innovation; as Boodin points out,[4] the original meaning of *eidos* was shape, and this continued through Pythagoras. Plato extended it to include all ideal entities, including qualitative ones, as forming a transcendant realm, and Aristotle made it an internal principle of activity in substances, which in human beings takes the form of soul.

Rorty has pointed out[5] that one of the two main Aristotelian criteria for substantiality was logical independence or self-sufficiency. This is the criterion which Descartes used.[6] It is of course true that Descartes' matter is not absolutely self-sufficient. It does depend on God, especially for its continued existence through time. However, it is completely independent of mind.* It is just this criterion of independence which Spinoza was to carry to its logical conclusion, the monistic result that there can be only one substance (God), and that both matter and mind are simply attributes of this substance. But extension is not only the essence of every material body; it is also the essence of the space in which these bodies could be located. The latter point can hardly be denied; space was even defined as three-dimensional extension. But from this two very important conclusions follow: (1) space and matter are one and the same thing; and (2) there is only one material substance. The latter result does not hold in the case of minds. There appear to be indefinitely many mental substances, each capable of engaging in its essential activity (thought) independent of any others. But material bodies are related much more intimately in that they share the same essence, the one universal extension.

Descartes did not accept this result consistently, since the distinct material body was his starting point and was an appearance to be saved by his theory of matter, but it was implied by his principles and many of his other conclusions depended on it. For example, he argued[7] that two substances are really distinct if and only if we can conceive the one clearly and distinctly without the other, and that any portion of corporeal substance that we can demarcate in our thought must be distinct from the rest. But on the other hand, the possibility of motion is essential for distinguishing such portions, and Descartes could define motion only by referring it to other bodies. Descartes contrasted real, modal, and 'of reason' distinctions,[8] and it would seem more consistent with his definition to make the distinction between separate bodies modal, a move which would bring him close to Spinoza. On the other hand, Descartes claimed that although any one attribute is sufficient to give us a knowledge of substance, there is always an essential attribute on which the others depend, and for matter this is extension.[9] Therefore, although it is possible to abstract out extension from matter and thus distinguish them, the distinction can only be one of thought or reason.

Descartes did take over most of the Scholastic distinctions between space (internal place), external place, body, and extension,[10] so that he could account for such statements as "extension occupies place, body possesses extension, and extension is not body." He even agreed with the Aristotelian definition of external place as the superficies of the surrounding body, which has the advantage of referring only to other bodies rather than an additional

* This does not contradict my contention above that Descartes' conception of matter seems to involve a relation to a perceiving mind. For I was there concerned with the concept of matter, an epistemological notion: I am here considering matter itself, ontologically.

absolute space. However, he showed that in each case the distinction is made by our categorizing reason, and does not indicate an ontological separation. When statements involving these terms are true at all, they are analytically so.

One important ambiguity lies in the term 'extension'. This may be taken to refer either to (1) indefinite extendedness, which is akin to spatiality in general; or (2) definite extension, i.e., occupancy of a bounded region of space. This is significant because (1) seems to be intended when one speaks of extension as the essence of space or matter in general; while (2) is intended when extension is offered as the essence of a particular body. Now Descartes recognized that space must be infinite or at least indefinitely extended if it is to be susceptible of the mathematical treatment that he gave, while any particular body is clearly finite. In this sense a material body can be distinguished from matter itself.

Descartes certainly believed that the intellect alone can arrive at a clear and distinct idea of extension. To be sure, his starting point is indeed the individual material body which is perceived to be extended (and to have various other attributes), but Descartes could argue that this is only part of the 'context of discovery', and that once this is completed the mind has an intellectual intuition of pure extension in all its glory. In similar fashion, Plato had argued in the *Phaedo* (74a) that perceptions of sensible things whose qualities only approximate the forms could lead us to a recollection of that perfect knowledge of the forms possible to a disembodied intellect. No appeal is made to a quasi-perceptual faculty, like Aristotle's active intellect. But in view of the ambiguity above, we may well ask just how much is in this intuition of extension. Does it include a notion of definite interior boundaries, i.e., figure, or mere voluminousness? And in order to be material, must a body occupy a definite portion of this extension, or is it enough that its activity can be described in an arena characterized by a spatio-temporal coordinate system? As we will see, these are recurring questions for any attempted identification of space and matter. Descartes must bridge the gap between the finite and the infinite senses of extension if his subsequent arguments on the nature of matter are to be valid. His most reasonable approach is the idea that the definite extension of any particular body necessarily includes the possibility of its indefinite continuation in all directions beyond its boundary.

One supporting reason for this claim is that Descartes' space (and thus matter) is perfectly homogeneous. Plato's receptacle is full of potencies to receive different qualities from the forms, but within any body the Cartesian extension is the same throughout, as is required for the possibility of mathematical treatment. If one invokes a principle of sufficient reason, this could be used to throw the burden of proof onto anyone who claimed that there were limits to the indefinite continuation, for these limits would disrupt the homogeneity and isotropy of the space at such limit points. Such an argument would be far more plausible in Descartes, whose space-matter was perfectly rational, than in Plato, who saw it as the seat of irrationality. However, it forces Descartes to pay the price of admitting no potencies or inhomogeneities whatsoever into this matter. Except for the boundaries of the bodies which constitute it, it must be an undifferentiated continuum. No new attribute which might vary from place to place may be added to the essence of matter.

Descartes was well aware of the fact that other criteria might be proposed as the essence of matter. He explicitly rejected hardness, weight, color, etc., on the grounds that we are aware of them only because they differ from place to place. If they were homogeneous, we could never notice them; but in this case we could not recognize them as being the essence of bodies.[11] The features which do individuate bodies, and enable us to distinguish them from each other in sense perception, are thus only accidents of matter, secondary qualities whose existence depends on a perceiving mind.

I have suggested that Plato accepted as an irreducible fact the notion that all matter belonged to one of four distinct kinds, and believed that a theory of matter should preserve this distinction. Although the basic units for each of the four kinds are geometrical entities, Plato works out in mathematical detail only laws for the combination and transformation of these basic elements. And the mathematics used here is not geometry, which deals with continuous magnitudes, but the arithmetic of integral multiples. Furthermore, Plato gave no mathematical description of the motions of his elements. Therefore I think it is reasonable to consider Plato's theory of matter to be more like a rudimentary chemistry than a physics. For physics is characterized by its use of a notion of matter as such, regardless of the qualitative kind of physical object being considered. In addition, it deals primarily with the motions of such matter, giving them a complete description in mathematical terms. Since Descartes makes no such distinctions within matter, and since he gives motion a vital role as the source of differentiation and tries to analyze it mathematically, Descartes' theory can definitely be considered the basis of a physics, albeit not an adequate one.

I have said that Descartes' goal was nothing less than the complete rationalization of matter and the physical world. Now Descartes' Scholastic predecessors undoubtedly believed that their natural philosophies were also rational, but Descartes took the additional step of taking mathematics as his ideal of rationality. This was quite novel; it had the effect of eliminating all reference not only to such occult entities as substantial forms but even to the perceptual qualities of everyday experience, except insofar as they were susceptible to mathematical treatment. But the whole tenor of the *Discourse on Method* is to contrast the clarity, certainty, order, systematization, and potential for progress of mathematics with the haphazard and obscure notions of Scholastic natural philosophy, even if the latter did seem to touch on ordinary experience more directly. As Burtt points out,[12] extension has two great advantages as a universal criterion for matter: not only does it appear to underlie all physical things, regardless of their other differences, but it is also capable of exhaustive mathematical analysis. Without the second it is unlikely that Descartes would have emphasized the first so strongly. If physics is nothing but the science of the various modes of extension, it can presumably attain to the same degree of exactness and certainty as can geometry. To be sure, Descartes sometimes hesitated to admit that he was allowing a human contrivance to dictate in ontological matters. For example, Descartes offered permanence as a criterion for distinguishing primary from secondary qualities. But this will not do the job, and Descartes knew it. "His real criterion is not permanence, but the possibility of mathematical handling."[13]

One good reason for Descartes' fascination with mathematics is his own discovery of a powerful new tool, analytic or algebraic geometry. Gillispie has emphasized the revolutionary character of analytic geometry in unifying two previously separate domains.[14] Until then geometry had been the science of the spatial continuum; algebra and arithmetic, of the discrete and enumerable. The Cartesian geometry, which set up a one-to-one relation between (real) numbers and points in space, put these two disciplines under the same rubric and enabled them to enrich each other. Only after this had been achieved could Newton give an adequate mathematical representation of atomistic matter in continuous space. More directly, it gave Descartes a new freedom in his handling of figure. No longer are we restricted to the perfect solids of Plato, whose perfection stems largely from the fact that only they are capable of convenient analysis in nonalgebraic geometry. In fact, the notion of perfection in this context loses most of its importance, for we can now write an equation for any curve or surface, and once we have this equation, we can apply any algebraic methods to it that we wish in order to derive further conclusions. Perfection is soon dismissed as a vague and undesirable teleological notion.

Having laid down the requirement that everything in physics must be describable in mathematical terms, Descartes soon found himself in the difficulty of trying to carry it out consistently. Blake argues that Descartes recognized that there are some things which seem rightly to belong to the physical world, but for which no mathematical treatment is readily available.[15] Most of his detailed analyses of phenomena are in fact sketches or outlines, with the mathematical details to be filled in later, assuming that this is possible. Perhaps Descartes considered this task to be no more than filling in initial and boundary conditions in his general laws, but even these are not put in a definite quantitative form. He felt a need for finding some measure for density and weight, no matter how they may ultimately be represented, but never did develop one. Nevertheless, Descartes' optimism prevailed. In a letter to Mersenne in 1638,[16] he asserted that all his physics was nothing but geometry, and certainly expected that such a physics would prove adequate. He was not always perfectly true to this vision, but given the conception of geometry with which he had to work (three-dimensional Euclidean geometry), he probably went as far in this direction as anyone could. And he bequeathed this vision to geometrodynamics; for there too the goal is to produce a physics that is nothing but geometry.

The famous mechanism of Descartes, which has been so widely praised and condemned in the subsequent history of philosophy, is simply the hypothesis that everything in the physical world can be explained by reference to matter (= extension) and motion alone. The important feature is not that some things can be given one level of explanation in this way, but that this level is complete: "all phenomena of nature" can be subsumed under this rubric, and nothing further is needed.[17] Thus Descartes is certainly a reductionist, and it is this reductionism that is the basis of his mechanism. He allows only one level of explanation as having ontological priority, and insists that it take the same form for each phenomenon. In this respect he contrasts sharply with Scholastics, who insisted on bringing all of the four causes into a complete explanation and introduced separate principles for each, often specific to a particular kind of occurrence.[18] Descartes and his successors often used analogies, but the analogies were always devices of matter in motion.

Dijksterhuis sees four significant conclusions derivable from the Cartesian form of identification of matter and space: (1) a vacuum is impossible; (2) the world consists of the same matter throughout; (3) matter is infinitely divisible; and (4) matter is infinitely extended.[19] Let us take up these points in more detail.

For the first, other philosophers had denied the physical possibility of the void, but Descartes goes further and argues that it is logically impossible, since the notion of void space is meaningless. The mere fact of three-dimensional extension is a sufficient condition for the presence of matter.[20] Whittaker sees the nonexistence of the void as a special case of the impossibility of having any attribute, either essential or accidental, without the substance of which it is an attribute.[21] Since extension is the essence of matter, there can be no extension without matter. Since spatial extension is continuous and admits no gaps, there can be no void, and the material world is a plenum.

The problem here is how to make sense of the possibility of motion in a plenum. Čapek argues that the only possible plenum is motionless Parmenidean being, and that the notion of a perfect fluid, which underlay the fluid theories of matter of Descartes' successors, is nothing but an unrealizable mathematical abstraction. In fact, there must be interstices, however small, into which fluid can flow.[22] Čapek criticizes Russell for claiming that the difficulty is overcome by the 'equation of continuity' in hydrodynamics. He contends that in Descartes' example of the swimming fish[23] there must be an instantaneous transmission over a finite distance if fluid is to flow in at the back to replace that which must be pushed aside at the front by the moving fish. A true plenum, he claims, cannot allow the variation in density which would take place otherwise. In Descartes' case, this much seems correct. If Descartes' extension is homogeneous, there can be no variation in density. (This, however, need not be true of all plenum theories, if the fluid is inhomogeneous although perfectly continuous to start with.)

The other variables in the equation of continuity are velocity and pressure. Descartes can certainly use velocity, but it is not clear whether he can make sense of pressure. On the one hand, it would be hard to represent in geometrical terms; but on the other hand he speaks of light as a pressure in the ether propagated with infinite velocity.[24] Nevertheless, Čapek's criticisms do not really apply to Descartes. In the first place, Descartes insists that all motion is essentially circular. If something moves in one direction, something else must move in the opposite direction to take its place. In this way a circular whirllike motion can be set up, and Čapek points out that in 1867 William Thomson did propose just such 'vortex-atoms' in a homogeneous, uniformly dense plenum as the basic constituents of matter.[25] Furthermore, Čapek admits that such vortex-rings are uncreatable and indestructible, mutually impenetrable, and constant in volume, even though they can vibrate and change in shape, and thus can function like the atoms of classical physics provided that a concept of mass can be represented in terms of them. But in the case of the swimming fish, Čapek's primary mistake is to assume that the motion begins or has its source at the moving head of the fish, and must be propagated, either gradually or instantaneously, back to the tail. But in fact Descartes' insistence that all motion is circular and is really many rather than one[26] can only mean that all parts of the total motion from head to tail begin and continue together. The notion of propagation is quite out of place. For Čapek is thinking of the pushing of the head as being the cause of the motion, while Descartes always traces motion back to God as its first cause.[27] Motion requires the passage of time, and Descartes' theory of motion depends on his theory of time and its relation to God, which we will see is one of the great inadequacies of his system. But within that theory of time Descartes is perfectly consistent.

As for the second point, I have already argued that Descartes' criterion for matter must lead to a material monism, that there can be only one material substance. The matter of all parts of the world is united, in that each shares the common essence of extension. Thus heaven and earth must be composed of the same matter, and united as one, essentially undifferentiated world.[28] The effects of this conclusion were radical and far-reaching. The Medieval philosophers had assumed a radical separation between the celestial and sublunary realms, arguing that the constitution of matter and the laws of motion were different in the two. The planets and fixed stars were taken to be composed of a more perfect 'fifth element' not found on the earth, and the planets could engage in an 'inertial' motion in perfect circles. Descartes' reasons for his unification of the two realms are entirely logical and metaphysical, rather than physical, but they opened up the possibility for a unified physics and allowed Newton to place planetary and terrestrial motion under the same general laws. However, they did leave the serious problem of how this uniform world can be differentiated to 'save' the appearance of diversity.

A related conclusion is that there can be no plurality of worlds. Descartes' argument again is logical. He claims that the matter of any other possible world would have extension as its essence. But the universal extension of *this* world is complete and gapless, and leaves no room for any other. One cannot have, in effect, two worlds sitting on top of each other, with their extensions in the same place. From this it follows that one need not postulate impenetrability or 'solidity' as an independent property of bodies. If a body is in one place, it is logically impossible that anything else could be there, since in that case it would have the same extension. Observable differences in the impenetrability of solids and fluids are to be explained only by the presence or absence of movement, not by some mysterious cohesive force.

The infinite divisibility of matter also follows from its definition as extended substance. No matter how small we envisage a body, it must have a finite length, breadth, and depth, and when given these dimensions we can always divide them in thought into something still smaller. This conclusion is again related to the possibility of mathematical treatment, and the fact that mathematical division of any given magnitude can be repeated indefinitely many times. Whittaker points out that the Scholastic concept of extension, on which Descartes was raised, made infinite divisibility seem even more analytic, for it meant mutual externality of different parts into which the extended thing could be divided.[29] As we will see, Descartes does postulate small elementary particles in working out the details of his physics, and admits that there may be limits to the possibility of actual (physical) division by human beings, but concludes that such particles cannot be real atoms in the sense of being absolutely indivisible. For if we can divide them in thought, so too can God, and if He could then not divide them in fact it would be an unwarranted limitation on His power.[30]

Perhaps the most striking result, at least for the breakdown of the Medieval notion of a finite, orderly, purposeful cosmos congenial to man, was the infinite extension of space and matter. Koyré discusses at length the fact that Descartes insisted on calling his space indefinite rather than infinite.[31] For Descartes infinity was a positive attribute of God, about which our finite minds could not reason, while indefiniteness merely implied the absence of any limits and the possibility of continuation.[32] But theology aside, the effect is clearly the same. We can envisage no possible boundaries of the world, for at any such it will always be legitimate to ask what lies beyond the limit. The answer can only be more space or continued extension, and therefore it must be part of the material world. We cannot grasp an actually completed infinity, but the Cartesian world is at least potentially infinite, and this is all that is needed.

I have suggested that this question, perhaps surprisingly, does not seem to arise in Plato, but it comes to the forefront with Descartes. For we must remember that Descartes took his mathematics, and all its implications, very seriously. By his time it was recognized that Euclidean geometry could not be absolutely valid unless it referred to an *infinite* Euclidean space. The Medieval philosophers did not take geometry this seriously. They regarded it as a tool for solving particular local problems of place, whereas Descartes saw it as the key for unlocking the cosmological secrets of the world. Since its axioms were clear and distinct, any structure underlying them must be accepted. Disputes about the local application of geometry had long since ceased, and interest now centered on its global or cosmological implications. It became taken for granted that once the local validity was accepted, global results would follow, and the two could not be proposed separately. I have suggested (p. 72) that Plato's spatial continuum is metrically amorphous, at least in the state of chaos, and measure has to be introduced by the Demiurge from outside; Descartes' space is essentially and intrinsically metrical, and geometry simply reflects this inherent order. Indeed, it is the guarantee of any other order prevailing in physics.

Gillispie has shown that Descartes was responsible for the idea of rectilinear inertia, which was taken over by Newton: that any inertial motion takes place along straight lines in this infinite Euclidean space.[33] Rectilinear inertia requires infinite space; otherwise, a body could not continue to move indefinitely in a straight line, but must eventually approach the boundary and bend backward in some way. Circular inertia, of course, can continue indefinitely in either a finite or an infinite space. But on the other hand, it is impossible in a space that is perfectly homogeneous and isotropic, like Descartes'; for it requires distinguishing one point in the space from others as the center of the circle, and some finite value as the radius, and these cannot be justified by the principle of sufficient reason as rectilinear inertia can.

Yet Descartes did not really arrive at Newton's first law, though he recognized that uniform motion is as 'natural' as rest and can be continued indefinitely.[34] For without a concept of inertial mass or even mass density as a basic nongeometrical property of matter, he could not use Newton's momentum as the quantity of motion which is preserved for all time in the universe by the action of God.[35] The only definition which he could give for 'quantity of motion' would be volume times velocity, and this will not work. Furthermore, he failed to recognize the vector character of velocity or motion, i.e., that its algebraic sign must change when its direction changes. Thus when he tried to develop more detailed laws of motion his results were not only wrong but not even close to observational data.[36]

We recall that the identification of matter and space has two facets: (1) everything material must be extended, and (2) everything extended must be material. I have also claimed that the ordinary notion of matter does allow borderline cases in which it is hard to decide whether the entity in question is or is not material, so that any proposed criterion has a stipulative component. The decision is usually made in terms of what other things in one's ontology are to be contrasted with matter. For Descartes, the main contrast was with minds (and of course God), which he believed were clearly unextended as well as being immaterial by definition. But of course there are other possibilities: form, spirit, etc. Van Laer illustrates the attempt to contrast the material with the spiritual, which he praises as having a higher grade of being.[37] He then rejects such properties as weight, mass, etc. as possible candidates for a universal criterion for matter, before arriving at extension. But his primary argument against mass is the fact that physics introduces such entities as photons and neutrinos, which have no rest mass. But are they extended? If we consider the spatio-temporal spread of their quantum wave functions (which Van Laer does not), one might be able to insist that they are extended; but this is far from the definite extension of the sensible body, which Van Laer, like Descartes, takes as his starting point, and far from being essential it is functionally far less important than some of the other properties ascribed to it. To retain extension as the primary attribute here seems to require stretching its definition far beyond its elastic limits. Van Laer actually seems to mean by 'extended' no more than capable of activity describable in spatio-temporal terms, or capable of interaction with more obviously extended bodies, but this is hopelessly vague. In particular, it would include Descartes' minds as extended, since they can certainly interact with bodies.

Most physicists, if they wanted such a contrast between the material and the immaterial at all, would probably make it within physics, so that something like radiation, rather than spirit, would be ruled out. I consider it a more reasonable approach to recognize that matter is a cross-level term in the sense given above. Photons and sensible bodies are on different levels, and it is unreasonable to expect any criterion to cover all these levels, though it may work for a particular one. The mistake here again lies in assuming that there is only one legitimate level of explanation.

We will consider below the possibility that, at least on one level, matter may not be extended, even though it is capable of producing observable effects which will therefore be perceived as extended. The opposite possibility is that immaterial things might be extended. This was the argument of More, who used it against Descartes.[38] More claimed that a whole variety of spirits were extended, and indeed, so was God Himself, though His extension was of course infinite. Two things follow from this: (1) We can have parts of space with no matter, although they are filled with spirit. Thus we avoid the need for a material plenum without having some extension predicated of a pure nothing. (2) Two different extended things can occupy the same place, one of which may be material and the other spiritual. Koyré raises the interesting suggestion that More's position is important, despite the quaint-sounding reference to spirit. For some of the things which More includes as spirits are very close to what would today be considered fields of force, which do extend through regions of space although they are not readily assimilated to the concept of the material body.[39] Clearly more than one such field (i.e., electromagnetic and gravitational) may be present at a point, whether or not there is a sensible body there as well. We will see that in geometrodynamics it is just such fields whose extension is to be identified with matter. At the level of geometrodynamics they are indeed material, though when they are used at other levels to calculate the force of sensible bodies they may still be considered immaterial.

The great problem for Descartes was how to differentiate this essentially homogeneous matter, in order to account for the sharp boundaries of the particular bodies as actually observed. Descartes answered that the differentiation is accomplished by figure and motion. Descartes conceived of both of these as a second class of attributes, modes of the primary attribute of matter, which of course is extension.[40] They differ in this way: figure may be adequately represented at any given moment, while motion requires time for its realization. However, aside from this Descartes treated them similarly.

The first problem that arises is what is the nature of figure, the geometrical shape that serves as the boundary of a particular object? We see that Descartes immediately ran into a problem similar to Plato's. The two problems are not precisely the same: Plato was not talking about the boundaries of sensible objects, but rather those of their elements. In addition, he did not require the little solids to constitute a plenum. The plenum was rather the underlying chōra, the medium in which and out of which they were built and which was potentially any kind of thing in all of its parts. Thus Plato's task was to make a stable distinction between the outside and the inside of his solids. Descartes had no outside-inside distinction, since his plenum requires that to be outside one body is to be inside another. Thanks to analytic geometry, Descartes could also use a more general notion of figure than Plato's. Any two-dimensional surface will do, no matter how irregular its curvature, while Plato was restricted to perfect planes and the solids formed out of them. Nevertheless it becomes even more important that such figures effect a real division of matter. There is no free mobility, and it is impossible to treat the boundaries, which are quite tangible, as purely formal elements. And yet if figure is more than a two-dimensional geometrical abstraction, if it is more like the skin of an apple, it is hard to see how the space can still be considered homogeneous. One does not have to 'cut through potencies' and work against the uncalculable forces of ananke, but more is required than merely envisaging two bodies with a common border as distinct and abstracting out that boundary in thought. Like Plato, Descartes did not give us any other material out of which they might be constructed, nor did he allow variations in density.

If understood as a mere separation or cut, figure seems to belong equally to the bodies on both sides of it, which was certainly not true according to Plato. But in fact this happens only in the static case where there is no motion. When a solid body moves, it normally preserves its shape, even though it becomes surrounded by bodies of different shapes from those in the original configuration. This suggests that figure somehow belongs to the moving body. But then each boundary must contain two figures, one on each side, for the bodies that it separates. And since figure is what encloses the particular extension (extension$_2$) for each body, we would then have no assurance that the various extensions$_2$ and their corresponding figures could combine together into the one universal extension$_1$ necessary for Descartes' use of geometry. These remarks should make it clear that the problem of figure and boundary will be vital for any proposed identification of matter and space, no matter how it is carried out.

Descartes then made the striking suggestion that motion somehow generates figure, as well as all the other properties which provide the observed variety in nature.[41] Hall claims that motion is especially important in the division of matter into the little particles which Descartes uses in presenting the details of his physics, and he uses motion much more often than the shape or nature of these particles in carrying out particular explanations.[42]

The net effect seems to be that motion becomes itself a substance interacting with matter but independent of it as a principle of explanation. This is the conclusion of Kemp Smith, who contends that the idea of motion cannot be derived from that of extension and must be introduced independently, despite Descartes' insistence that it is only a mode of extension and that matter and mind are the only kinds of substance.[43] Čapek reaches the same conclusion, though in the context of the development of classical physics in general.[44] But their criterion for substantiality in this case is that it obeys a conservation law (essentially conservation of momentum and Newton's first law). Now in classical physics matter obeys a conservation law, as do energy, charge, etc., but this is certainly not the reason why Descartes considered matter to be a substance. He used logical independence instead. Now it is true that Descartes did have a law of conservation of motion, but because it requires time it is even more directly dependent on God than is matter. Besides, motion must be the motion of something, viz., extended substance. Nevertheless, Descartes did move in the direction of classical physics, which Čapek claims involves the absoluteness and mutual independence of space, time, matter, and motion.

Descartes was also responsible for the doctrine of the relativity of all motions. Bodies are not 'in space', but among other bodies. Therefore motion, 'properly speaking',[45]

> is the transference of one part of matter or one body from the vicinity of those bodies that are in immediate contact with it, and which we regard as in repose, into the vicinity of others.

In view of the material plenum, it is impossible for one body to move by itself against an absolute static framework; there is no place for it to go that is not already occupied, nor can it leave a void behind. Once a motion and its compensatory countermotion begin (simultaneously, as they must) it can continue indefinitely, but at no time can we point to either one, or any other body in the background, as the one that is really moving or at rest. Our frame of reference is composed of bodies, not points in absolute space; and we can choose any selection of these bodies to be the static frame for the purposes at hand. It likewise follows that nothing new is added to a body by its being in motion. One major difficulty in the Cartesian doctrine of relativity is that Descartes did not distinguish between inertial and accelerated motions here. For Descartes did not have the precise concepts of velocity and acceleration that Galileo was developing and that would achieve their complete mathematical expression only in Newton. Thus he did not give us a real mathematical theory of motion but rather a description of its effects in general, qualitative terms. While changes in the speed and direction of a motion were as perceptible to him as to anyone else, he saw no reason to single them out for special treatment.

I have mentioned (p. 89) that the main weakness in Descartes' theory of motion is his inadequate concept of time. Plato also kept time quite distinct from his chōra and discussed it only in poetic and mythical terms. However, Plato was not interested in a theory of motion, or the quantitative aspects of the actual *processes* of breakdown and recombination, so this defect is not so glaring. Descartes appears to have been bothered by the fact that time or duration seems to pertain equally to mental and material substance; they both endure. Kant seized on this same fact as the crucial link between the pure a priori concepts of the understanding and our perceptions of the external world, and introduced the schematized concepts, in which their operations were represented as taking place in time. But Descartes would have found this unfortunate. If mind and matter are to be completely independent of each other as substances, no feature common to both can be part of the essence of either. Therefore he refers time back to the only remaining source, namely God.

Descartes asserted that neither mind nor matter has the power to maintain its own existence. The concept of neither contains the ground for its perpetuation. Yet both bodies and minds do endure. Descartes therefore took this fact as a proof of the existence of God.[46] The whole world must be recreated anew at each instant, and this happens only through the power and goodness of God. Furthermore, it is only by His will that quantities of motion or anything else remain constant through time. Spinoza's God is immanent in the world, and matter and mind are among His attributes. Descartes' God is transcendent; matter and mind are not part of Him, but He creates them by His will.

But this has the unfortunate effect of making time atomistic. The world lasts (statically) for an instant, disappears, and is re-created in the next instant. On the one hand this gives us a universal and objective time, since all bodies (i.e., all parts of matter) are re-created together; but on the other hand it interprets duration as an enumerable sequence of instants, rather than as a real continuum. Now, there is no reason to believe that motion can take place *during* one of these instants; quite the contrary, since there is no 'during' in them and no way of separating temporal parts. Evidence of motion appears only when the world is reconstituted in the next instant. The process is thus more like a reshuffling of parts than a continuous flow. There is no reason why God should not put back the pieces (bodies) in different spatial relationships to each other when He re-creates them. The possibility of motion is thus not inherent in matter itself, which has only a static, momentary existence, but derives directly from God. The laws of motion do not derive from the nature of matter, but are imposed by the goodness and will of God. Thus the cause of being and the cause of coming-to-be of material things are conflated. Matter itself is causally inert and has no power of independent action.

This theory did have certain advantages for Descartes. It enabled him to do without a notion of physical causality, and thus made it more reasonable for him to identify matter and space. For classical space was always taken to be causally inert, allowing the transmission of actions while being itself unaffected by them. Matter was distinguished from it on the grounds that it was active, and that bits of matter could exert influences on each other, through gravitational forces, etc. In addition, it resolved the problem of motion in a plenum. The cause did not lie in the front of the moving object and have to be transmitted to the rear; the parts just reappeared in their new positions at the next instant.

But these advantages are far outweighed by the disadvantages for geometrizing physics. For geometry deals with the continuum, and if duration is not continuous, it cannot be treated geometrically. Since physics includes motion, and motion requires time, Descartes' physics is not completely geometrized. Insofar as it must make reference to time, it can always be distinguished from pure mathematics. Since time must be a fourth dimension, it of course cannot be represented graphically along with the three spatial coordinates, but Descartes' own discovery of analytic geometry had obviated this necessity in any case. Once he had developed the differential calculus, Newton was able to recognize that the notion of derivative was equally applicable to change in one spatial coordinate with respect to another (spatial slope), and change in any spatial coordinate with respect to time (velocity). Since Descartes never achieved this, he tended to downgrade the concept of time, defining it as only a mode of thinking of that mysterious duration, or a measure of movement, rather than an entity on a par with space and matter.[47]

Without the central notion of derivatives for continuous change in time, Descartes could not reach an adequate mathematical kinematics. Had he been able to do so, one might be able to describe his physics as a 'geometro-kinematics', since it would describe motions in purely geometrical terms without reference to their causes. But it could not be a 'geometrodynamics' unless an active causal principle could be defined and represented in the matter so described. Before this could be done, a whole host of special dynamical notions had to be introduced, and then given an interpretation in terms of a far more complex geometry.

If motion can divide the spatio-material continuum in such a way as to reproduce the perceptible boundaries of sensible things, it should be capable of subdividing them into finer particles. It is this assumption on which Descartes' particle theory of matter is based, and this theory then governs his physics. It must be emphasized that this is not an atomism, although it performs many of the same functions and Descartes can speak about his particles in most of the ways that a Newtonian can speak about his. For the Cartesian particles must have a definite extension, however small, and thus can be further subdivided. If in fact they never are cut below this size, it is only because God does not choose to do so. Thus the particle hypothesis is no more incompatible with the plenum theory than is the assumption of figure in general (although we have seen that the latter causes difficulties).

However, it is strictly a physical hypothesis, and is not derivable from the notion of matter in general. But there is a further problem. Renoirte insists that we must give an account of the fact that sensible bodies have sharp boundaries, and that these boundaries can remain stable for long periods of time.[48] Descartes would certainly accept this requirement. But each such body contains a great number of these little particles, some of which will touch the boundaries while others remain in the interior. Then if the boundaries of the particles in the surface layer are to constitute the boundary of the body (its figure), we must assume either that they have a clarity, hardness, toughness, or cohesion greater than that of the interior boundaries; or that they differ in some property from those on the other side of the figure. But the only way in which they could differ, consistent with his geometrical approach, is in size. Neither alternative is very satisfactory. The first merely aggravates the difficulties mentioned above concerning the nature of figure or boundary, for it now allows different kinds of boundary; the second requires that there be particles of different sizes for any pair of things that do not flow together into a mixture upon being brought into contact, and introduces volume quanta and discontinuous variations in particle size in an ad hoc way. The crux of the problem is Descartes' assumption of a single level and single principle of explanation. As in the case of Eddington's two tables, I must insist that the perceived table and the particles of which it is composed, according to some physical theory, are on two different levels. Unless we are trying to prove a reduction, the latter level need not reproduce all the distinctions of the former. But Descartes' insistence that there is only one material world rules out this move for him. The principle of the particle level and that of the sensible level are one and the same: division of the one material plenum into figure by motion. Thus the particles are properly parts of the larger bodies and must reproduce their outlines.

Descartes does in fact divide his particles into three size classes. In *Le Monde* these corresponded to the traditional fire, air, and earth (no water!) in order of increasing size; in the *Principles* they refer to the sun and fixed stars, the intervening ether, and the earth.[49] But this is hardly an answer to the problem above. In the first place, one would need a far greater number. Especially in the case of earth, there are too many different kinds of things for them to be lumped together as having particles of the same size, and even the possibility of compounds will not resolve this difficulty. Secondly, the fact that these do not fit together compelled Descartes to postulate additional interstitial particles, whose function is primarily to allow the others to move. It is such particles which Čapek thinks are really bits of vacuum.[50] Thirdly, Dijksterhuis claims that the particles increase gradually in size as their distance from the sun increases, although Descartes did not give any equation governing this increase.[51]

The particle hypothesis has two main roles: it is used to explain rarefaction and condensation,[52] and it is the basis of the famous vortex theory of force and motion. The particles move together continuously in a complex series of circular whirllike motions which can interact with each other. Bodies are caught up in these vortices and carried along in motions governed by their size and relative position. There are no natural places or motions, although the cumulative effect is to separate the three kinds of elements. Thus vortices of the second (ethereal) kind drive the small spherical particles of the first (luminous) kind toward the center of the vortex, and the large earth particles cluster together in the areas of minimum turbulence.

Since in this scheme motion is a natural consequence of the interaction of vortices, and since Descartes' law of conservation of the arithmetical quantity of motion is more inclusive than Newton's, Descartes saw no need to introduce a concept of force to account for change of motion, as Newton was to do in his second law.[53] If we consider the quantity of motion for a whole system, rather than for each separate part, the essential circularity of motion in a plenum may require that its vector magnitude remain zero. This might be one reason for Descartes' not interpreting it in vectorial terms. Since there is no change in the totality of motion, thanks to God, there can be no real force. To be sure, all motion is not inertial (rectilinear), but the deviations from uniformity can all be explained in terms of the influence of vortices in the vicinity. We will see that general relativity is also able to abolish the distinction between natural and forced motions and thus eliminate force except as a local result of choosing an inappropriate coordinate system for the problem at hand. One further objection to force, which occurs not only in Descartes but which is especially important for him, lies in its connection with action at a distance. A particle can respond to something that is going on somewhere else only if it can 'know' about this remote action; but such 'knowledge' by particles would break down the rigid matter-mind distinction. This seems to be particularly true of attraction. Since it was usually assumed to take place over longer distances than repulsion, and to involve 'pulls' rather than 'pushes' which could be understood in terms of local impenetrability, it was more suspect than repulsion.

It is hard to give a complete account of what the Cartesian vortices are. Insofar as they direct motions through regions of space that may not be filled with tangible, solid bodies, they function like the force fields which were introduced in the nineteenth century to eliminate action at a distance. But they are purely geometrical entities, which do not introduce any new kind of element. Furthermore, there is only one field of vortices, rather than a different one for each kind of force; and the vortices must affect every particle in a given place in the same way, except insofar as the interaction is governed by particle size. In any case, their essence seems to be simply motion. They do not generate a curvature of the space, nor do they affect it in any other way. Although they raise fascinating possibilities and anticipations of contemporary ideas, they have one great weakness: Descartes never defined any precise measure for them or subjected them to full mathematical analysis. Their activity is always described in qualitative terms.

Now, among other things, the vortex theory is supposed to be able to account for what we would call gravitational phenomena. For Newton this required assigning a new parameter, gravitational mass, to all material bodies. Descartes could certainly not use this. If it were proportional to volume for all bodies, it would be gratuitous as a separate concept; and in fact it is clearly not proportional, as Descartes must have recognized.

But we may well ask, granting that Descartes wants to treat gravitation geometrically through the vortex theory, could he construct a concept comparable to gravitational mass? He could deal only with the extension, figure, and motion of the particles comprising the vortex. Let us grant that figure is real, measurable, and a useful concept in physics. Then, since the particles come in different sizes, it is possible to construct a variable 'density of figure' as the amount of figure in a given region. More precisely, it could be the ratio of the total surface area of all the figures or particle boundaries in a region to the volume of that region. To be sure, this is only an average and cannot be made into a point function. But if the particles are small enough and the total volume is large enough, it could be useful, even though it is not clear how one might measure it. One could also define total mass, or quantity of matter, as the total volume of these constituent particles, which may differ from the volume of the macroscopic figure.

We can also use the average velocity of the particles in that region. Descartes did seem to indicate that both particle speed and size of the vortices are important factors in determining motions. Then perhaps some combination of the two, whether their product or something more complicated, might work as an accurate measure. Given the observable effects, especially a comparison of the different motions of two bodies having what we would call different masses in the same place, Descartes might even have been able to work backward and calculate the magnitude of these two geometrical parameters and then use the results for further tests. It is certainly regrettable that Descartes never attempted such a calculation. However, we may be demanding too much in the way of twentieth-century methodological sophistication from him. In any case, there is no guarantee that any combination of the geometrical parameters available to Descartes could have functioned as an equivalent to the Newtonian mass or mass density and led to a physics observationally equivalent to Newton's.

As an alternative interpretation, Toulmin and Goodfield suggest that only the third class of large earth particles have weight.[54] Differences in density are explained by the presence of the smaller luminous and ethereal particles in the interstices of the different varieties of 'earth.' But this is hardly reasonable. In the first place, it makes weight a purely terrestrial phenomenon, which cannot be linked to planetary motion by universal gravitation. Descartes' insistence that heaven and earth have the same kind of matter and his desire for a unified science would make him unsympathetic to such a move. Second and more important is the fact that this would introduce a nongeometric differentiation of matter. Matter would be of two kinds: earth particles, which have this mysterious property of weight; and everything else, which would be weightless. This is quite contrary to all Descartes' ideas. More exciting is Koyré's claim (unfortunately undocumented) that [55]

> vortices which surround fixed stars limit each other and prevent each other from spreading and dissolving under the influence of centrifugal force; if they were limited in number, and therefore in extension, then, first the outermost ones and then all the others would be dispersed and dissipated.

This passage is pregnant with striking implications. First, it suggests that universal gravitation, as represented by the vortex action, is what unifies matter and holds it together, though it may possibly prevent the realization of any perfectly inertial motion in the world. In addition, it suggests that the magnitude of such a 'force' depends on the amount and distribution of massive (vortical) bodies in the world. As we shall see below, such ideas are central to the thought of Mach, Sciama, and Dicke. If Descartes really believed this, he may be considered a precursor of Mach.

Burtt points out one further line of thought which Descartes never really followed up. In the *Regulae* he suggested the possibility of treating weight, velocity, and other physical magnitudes as dimensions, on a par with length, breadth, and depth.[56] By dimension he seems to mean nothing more than something that is amenable to mathematical treatment and that is needed for a complete description of a physical situation, which of course applies to the three spatial dimensions. It is "nothing but the mode and aspect according to which a subject is considered to be measurable." Burtt goes on to see all sorts of exciting possibilities for physics if this had been carried out, with the elimination of all sorts of distinctly physical concepts in favor of mathematical dimensions. But as we will see in discussing geometrodynamics, which tries to do this very thing, the matter is not nearly so simple. Certain features of dimensionality are very closely bound up with the notions of space and time, and their universality and continuity. In any case, Descartes never came up with the exact measures that would be needed before he could count physical magnitudes as dimensions, even by his own criterion. Had he done so, he would have needed to enrich his conception of geometry, and thereby would have needed to come much closer to contemporary geometrodynamics.

7 Newtonianism and Atomism

During the preceding lengthy discussion of Descartes, I have referred on several occasions to Newton for purposes of contrast. It is not my intention to treat Newton's natural philosophy in similar detail here, since my effort is rather to show how far one could proceed in an identification of space and matter, given the conceptual and experimental material available at a period of history. Newton did indeed give a philosophical commentary on his physics, replete with orthodox theological references, which both made his work more palatable to the church and the general public, and helped form the basis of a new philosophy of matter. But the important thing is that Newton developed his physics quite independently of his philosophy. He laid down his definitions and axioms in precise mathematical terms, and once he had the laws and techniques he simply began to calculate, applying them to a variety of problems. The commentary is more like those "Reflections of a Physicist," which most great twentieth-century physicists have felt themselves obliged to make during their later years. Descartes had assumed that before one could give any physical law one must postulate an intelligible mechanism to represent the sort of effect being analyzed, and too often never did get beyond the description of the mechanism to the mathematics. Newton contented himself with laying down the equations describing how the motion proceeded; why it should necessarily proceed in this way, i.e., how it could be made intelligible, usually lay outside of the scope of physics. If this were so, Newton simply came to a stop, guided by the positivistic slogan "Hypotheses non fingo."

Newton's abilities as a philosopher were certainly less than his abilities as a physicist, even by his own admission. Since his own philosophical commentaries are neither especially original nor derived from the exigencies of his physical formalisms, I will spend very little time discussing them. My main purpose is rather to elucidate the model (more precisely, the model_1 in terms of the distinction in Chapter 4) inherent in the formalism which Newton developed for mechanics and gravitational theory. On such important questions as the meaning of 'mass' and the proper character of 'atoms', Newton's own views were really not very different from those of Descartes. However, his model had a far more revolutionary character, even if this was not fully appreciated until after about 1770. Thus, I will be primarily concerned with this later 'Newtonianism'. Since it does derive from Newton's *physical* theories, I will find it convenient to attach Newton's name to these new ideas. However, the reader should understand that I am not reporting on Newton's own metaphysical pronouncements, unless I explicitly state this.

The most striking new concept which Newton introduces is mass, especially since he uses it as a criterion for matter. I have claimed that Descartes did not have a concept of mass (or mass density) as an essential attribute of matter independent of its geometrical properties, although he might possibly have constructed something equivalent from the other concepts in his system. The motions of all bodies in response to a given force is governed by the second law of motion, $\mathbf{F} = m\mathbf{a}$, and the magnitude of the acceleration thus depends on the inertial mass of the body. Likewise all bits of matter are united together by the law of universal gravitation, where the magnitude of the attraction depends solely on the gravitational masses of the two bodies and their spatial separation. No other parameter which might otherwise be used to describe the body enters into these equations. In particular, the size (spatial volume) of the bodies is quite irrelevant. *For the purpose of doing physics,* mass rather than extension has become the universal essence of matter.

Thus the model of Newtonian physics came to include the unextended mass-particle as the basic building block out of which sensible bodies could be constructed. Since it seems undeniable that the essence of space must be extension, it is of course impossible that matter so conceived could be identified with space; in fact, it must be contrasted with it in the sharpest possible terms. Space is the universal, perfectly passive arena through which matter moves and acts; matter is the essentially foreign entity that moves through it. Neither is ever affected by the other, and they are logically quite independent. To this degree it was quite reasonable for Newton to give space an absolute status: it has no communication with any of the bodies in it but merely provides an absolute and unambiguous location for them. Since for Descartes the essence of any bit of matter, even the smallest of his elementary particles, was extension, the notion of unextended matter was a logical contradiction and a conceptual impossibility. The contrast between Descartes and Newton (or rather Newtonianism, the model inherent in Newton's physics) could not be sharper, as Jammer has recognized.[1]

How can such a conceptual dispute be mediated? In the first place, Newton's particles are not really unextended in an ultimate sense. In discussing planetary motion, for example, he treats the sun and planets as particles, and they are surely very extended bodies. But as Törnebohm has brilliantly demonstrated, the point is that their actual extension is irrelevant to the problem.[2] The concept of 'particle' is implicitly defined by the laws of classical mechanics. When applied to concrete physical problems, these laws require that some features of the total situation must be ignored as unnecessary complications in order to make reasonable approximations. 'Particle', in the sense of 'unextended mass-particle', belongs only to the model of Newtonian mechanics. Thus it is possible to assent to all of the following three statements, which are on the face of it mutually incompatible: (1) this is a particle; (2) a particle has no extension; and (3) this has extension, where 'this' refers to some particular physical object. Statement (2) is analytic; (3) is synthetic but perhaps a synthetic a priori truth if one stays in the realm of perceptible objects with Descartes. But (1) simply reflects a conviction that the extension of the body does not influence the motions to be computed, so that it may be treated as if it were a particle. 'Particle' is thus an idealization which may be more or less closely approximated by various bodies in different contexts. It belongs to a different level from 'material (or sensible) body', and an attempt to represent such a body as a particle involves trying to ground the phenomenon in another level, the level of classical physics.

But why must particles be considered unextended, even in the model of classical physics? The reason is the same as that used by Descartes in arguing that all material bodies must be extended, namely the exigencies of the mathematics used. Consider the law of gravitation, $F = Gm_1m_2/r^2$. All of the quantities on the right side of the equation—the universal constant G, the gravitational masses m_1 and m_2, and the distance r—are assumed to be measurable with indefinitely high accuracy. To be sure, we may not be able to calculate many significant figures with the instruments presently available, but there are no theoretical quantum limitations. Since r is subject to the operations of the calculus in calculating potentials from forces, etc., it is taken to have all positive real numbers as possible values. Now, given any two points in a Euclidean space, the distance r between them can always be determined as a real number.

Suppose now that we want to measure the gravitational force between two bodies which are essentially extended. Then the distance between them is ambiguous: we may take any point in one body and compute its distance from any point in the other, and for each choice we will get a different answer. Which value of r shall we use? Three plausible answers might be: (1) the minimum possible value of r, measured from the points on the surfaces closest to each other; (2) the maximum value, between the points furthest apart; and (3) the distance between their centers. Choices (1) and (2) might seem reasonable for attractions and repulsions respectively, if we assume that the bodies want to get as close or far away as possible, and "choose" the most convenient distance to measure such an accomplishment. Choice (3) may be more symmetrical for both cases, but we may ask, centers of what? The Cartesian can understand them only as geometrical centers, and if one has the exact equation of the boundary this point can be computed. But if the mass density within the extended body is not uniform, the use of geometric centers for r will in general give the wrong result. The correct answer turns out to be the distance between their centers of mass. But 'center of mass' is not a geometrical concept. It is developed only after a painstaking analysis, which I will discuss below, that begins with the notion of a point mass, for which distance is certainly unambiguous. (It might be noted that (1) and (2) correspond to the assumption that the density is located on the surface and can move freely to concentrate at a point, a situation that in fact occurs with charge density in electrostatics.)

It might be argued that the correct value of r could be calculated indirectly, as follows: the equation can be written in the form $r = (Gm_1m_2/F)^{1/2}$. If we can measure F to the desired degree of accuracy, using Newton's second law, we can then calculate r from this as the 'equivalent distance'. But there are two difficulties in this move. In the first place, r is the magnitude of a vector. Even if we assume that the direction of the r-vector is parallel to that of the force vector, it can be moved by parallel transport. Any pair of points in or on the two bodies for which the r-vector has the same length and direction as that between the pair first chosen will be equally qualified as candidates for 'centers of force'. It is possible that we might cut down on the number of such possibilities if we consider the gravitational interaction between a great number of pairs of bodies, and assume that the center of force must be the same in each case, but such a procedure would be arduous and time-consuming, and could not guarantee that one single point would emerge for each body. A more serious objection is this: it has often been pointed out that $\mathbf{F} = m\mathbf{a}$ could be no more than a definition of force and an instruction for measuring it. In order to make the statement synthetic, we need an independent law for a particular force whose magnitude could be determined by other means and then compared with that obtained by inertial mass times acceleration. The law of gravitation was designed for just such a purpose, and the fact that it gives the same result suggests that the second law was at least a very wise choice of definitions. But if the law of gravitation is used to measure distance rather than force, this advantage disappears, and we are back in a closed circle of analytic statements.

But the great value of the Newtonian theory of gravitation is that it is not restricted to determining the effects of unextended mass-particles; indeed, it is applicable to any sort of distribution of mass, whether discrete or continuous, and however irregular its shape may be. Furthermore, it can do this without ever introducing a notion of 'center of mass', though this concept can be precisely defined and is convenient for some purposes. This extraordinary result depends on (1) a *mathematical* fact about the form of the equation of the gravitation law, and (2) a mathematical technique developed by Newton. The mathematical fact is just this: the law of gravitation is *linear*; therefore it is possible to calculate the effects of each source of gravitation independently and then simply *add* them together (vectorially) to determine the total effect at any point in space. This can be brought out more clearly if we write the law in the form of a differential equation, of which $F = Gm_1m_2/r^2$ is an integral form. The differential equation is simply Poisson's equation, $\nabla^2\phi = 4\pi\rho$ where ϕ is the gravitational potential and ρ is the mass density. In the special case of $\rho = 0$, which holds in space outside of matter, we have Laplace's equation, $\nabla^2\phi = 0$. It is a theorem of pure mathematics that any linear combination of solutions to Laplace's equation is also a solution. It is also a theorem that to any solution of Poisson's equation we may add a linear combination of solutions to Laplace's equation (the 'corresponding homogeneous equation') to get other solutions. Thus we can take each pair of particles separately and calculate the mutual force between them, without worrying about the possibility that it might be altered by the presence of other particles in the vicinity, which would of course also interact with the pair under consideration.

The mathematical technique is simply that of the integral calculus, which is what allows us to make the move from discrete to continuous sources represented by a mass-density function. We begin with the single unextended mass-particle as a source and calculate its effect. Then we proceed to a group of such point sources scattered through space, adding their effects together in accordance with the principle of superposition. But the integral calculus allows us to represent a continuous distribution as follows: we divide the space (represented by the Cartesian coordinates, x, y, and z) into an indefinitely fine mesh of elements dx, dy, and dz. In each of these 'infinitesimal boxes' we select a point and associate with it a mass dm, where $dm = \rho(x,y,z)\,dxdydz$. Then we simply add the effects of the various dm's together as in the second step, although here the 'addition' is integration. At every step the elements added or integrated together are taken to be located at a particular point and thus have no real extension of their own. But such a procedure is applicable no matter what the form of $\rho(x,y,z)$ may be. This is true even if ρ has a finite value at every point in the entire infinite Euclidean space, i.e., if matter (as characterized by mass) forms a gapless plenum coextensive with space, as Descartes required. The scope and power of such a method is unlimited, despite its conceptually unpromising starting point. Once we have calculated the force on a test body at any point (i.e., the gravitational *field*) by using it, we can if we like define a center of mass and 'equivalent distance'. But such a concept is useful primarily when the distribution is spherically symmetric, since otherwise there will be variations with direction at the same distance from the center of mass of the gravitating body.

The importance of this result for Newtonian positivism cannot be overemphasized. For it enables one to avoid completely the ultimate question, what is the nature of matter? Is it continuous or discrete? and if discrete, are the particles really unextended or do they occupy small but finite regions of space, with a vacuum in between? The mathematics is equally applicable to all of these possibilities, and any one may be used as an approximation for the others. The ultimate nature of matter may indeed be a legitimate and important scientific question, but it is quite independent of the classical theory of gravitation, which does not need to wait for an answer to it. In particular, it is unreasonable to look for an answer by metaphysical shortcuts, as Kant pointed out in the second antinomy. Thus it would have been quite reasonable to Newton to consider the ultimate nature of matter to be a metaphysical question of the kind which he wished to eliminate, since there was no scientific means of answering it available at the time.

But if the law of gravitation had turned out to have a form different from the simple r^{-2} dependence, and the corresponding differential equation had been nonlinear by involving some power or product of the derivatives, none of these blessings would accrue. Superposition would not hold; therefore one could not break the source down into elements of some sort and compute their effects separately. Instead one would have to do the whole thing at once, which for a complex distribution might be incredibly difficult. In any case, one would have to know the exact nature of the source, and if separate sources could not be isolated as having negligible influences on each other's gravitational activity, the only legitimate 'source' would be the mass distribution of the entire universe. We shall see that this is precisely the situation in general relativity, which cries out for a characterization of sources that can be used in its equations and for a possible integral form of them.

The moral for contemporary positivism should be obvious (though it will probably not be taken to heart). It is incorrect to make a distinction between physical and metaphysical questions in terms of what seems answerable (even if still unanswered) *now*. In the first place, as I have mentioned above, we cannot predict how a question will be interpreted in the future in order to make it resolvable by means of techniques that will be available then. But a second—and new—reason is that the legitimacy and necessity of answering certain questions is intimately bound up with the mathematical form of the prevailing physical theory. If the mathematics used for expressing the laws of nature takes one form, a certain question may be quite gratuitous; but if it takes another form, the same question may demand an answer before further progress can be made, whether or not anyone can think of a reasonable way of finding such an answer. Newton was thus very fortunate. If his law had not fitted the physical situation exactly (as apparently now it does not, given our greater experimental accuracy), he might have modified his equations to achieve a better fit, but paid the price of bringing such questions to the fore; or he might have remained with the more convenient and positivistically advantageous mathematics, but had to settle for approximations. As long as we cannot dictate the true laws of nature a priori, we must always be prepared for such a dilemma.

Having established that Newton's mass-particles are extensionless, let us consider what would happen if extension were taken as an essential attribute, along with mass. We would then need at least one additional parameter characterizing the size of each of these particles, and this parameter would enter essentially into some of the basic laws of physics. It might also introduce a quantum limitation on length. But it is quite clear that no such minimum length, or particle size parameter, appears anywhere in classical physics. Indeed, it is just this lack of such a definite magnitude that allows the term particle to be applied equally to planets and to putative atoms, albeit in different contexts of measurement. Any appeal to the observed extension of bodies would be quite out of place in considering whether extension is essential here, since we are on a different level.

Extension certainly does not lose its importance in physics, but it is now referred entirely to space, which thus appears as an independent coprinciple to matter. Between them they exhaust the 'primary qualities' in the physical world-picture. One can say nothing about space except what refers to its three-dimensional extension, eliminating all potencies or other propensities to receive matter or act on it; one can say nothing about material particles except their mass, with no reference to possible sizes, shapes, or such tactual qualities as hardness and pointedness. This represents a considerable abstraction and simplification over the classical atomism of Leucippus, Democritus, Epicurus, and Lucretius. These philosophers had used a primary-secondary quality distinction, and made the latter qualities depend on a perceiving mind. However, they believed that their atoms must be imaginable and thus thinglike, rather than mathematical abstractions—which could be modeled only at another level less rich in properties—growing out of a physical theory. As we have seen, even Plato made this mistake when, for example, he interpreted the action of fire as a real cutting by his tetrahedrons.[3]

Newton himself did not fully recognize these implications. It is quite clear that he interpreted his particles as 'really' being small but finite corpuscles with an extension which could be divided in thought, though probably not in fact.[4] Since such a magnitude had no importance for his physics, he saw no reason to speculate what it might be. But in any case they must have mass, and exert attractive and repulsive forces on other similar particles, whatever 'accidental' properties they might also have. Such particles could move and collide with each other, but nothing else was needed. From the standpoint of classical mechanics and the theory of gravitation, the 'extended' interpretation of the term 'particle' is physically unwarranted and gratuitous. It is rather a $model_2$ in Hesse's sense, where the properties other than mass belong to the negative analogy.[5] The unextended particle is the $model_1$, the sense which I have been using. But such an interpretation had definite advantages in unifying science. The unextended particle might well have remained solely in the model of classical mechanics, but when interpreted as a real corpuscle it introduced an idea which could be fruitfully applied in other areas of science. It suggested the very important kinetic theory of gases (which did introduce a size parameter in van der Waal's equation!), and permitted the beginning of a truly quantitative chemistry. As each of these other sciences also found the corpuscular hypothesis useful, it came to acquire a metaphysical status as a universal $model_2$ for all of science, or at least a regulative ideal for it.

Extended particles might have one additional advantage. It has often been noted that Newton did not offer mass as a primitive concept, but instead defined it as the product of density and volume. But density, despite its obvious importance for any theory of matter as extended, had not itself been defined.[6] This approach led to much subsequent criticism, especially when the notion of unextended particle required that mass be prior to mass density. But there is an interesting possibility. Čapek has argued that the constancy of volume as well as mass for atoms, and the proportionality of mass and volume, is the very core of classical atomism.[7] This applies to Newton's own 'philosophical' interpretation of mass, though not to the notion as it appears in the model of Newtonian physics discussed above. But if this is true, a concept of mass is not needed at all, at least as an additional primitive. If all particles are the same size (both mass and volume) we can simply define density as the *number* of such particles per unit volume. Even if the particles are of different sizes, we could still define density as volume of particulate matter per unit volume of space, assuming there is some independent way of determining such a percentage. (We may note that this would lead to a maximum limit on possible density, when all the region became filled with material particles.) All these definitions refer only to arithmetical and geometrical magnitudes, and thus may be more intelligible than the mysterious 'mass', which seems to be related to quantity of matter as 'stuff'. However, they require an un-Cartesian dichotomy of matter and space without giving a criterion for distinguishing the two. Furthermore, no justification is given for the universal proportionality of mass and volume. Without any way of measuring the percentage of a region of space that was actually filled by material particles, this proportionality could be little more than a mere definition parading as geometrical representation of 'mass'.

It should be clear by now that even when a philosopher does not try to identify space and matter, his concept of the one will be greatly influenced by his concept of the other. Thus Newton makes mass, including both the inertial mass (which governs a particle's response to any sort of external force) and the gravitational mass (through which it exerts an influence on other particles), the only intrinsic property of matter. But before a particle can be fully characterized and its motion analyzed, it must be possible to assign a position to it unequivocally at every moment of time. It is primarily for this purpose that Newton was led to introduce his static, unchanging, absolute space. Absolute space and mass-points are correlative principles in his system. External criticism of any one concept in such a system is usually beside the point, unless it takes into account the role of the concept in filling the gaps in the system and making it a unified whole consistent with the empirical knowledge available at the time of its construction. Kuhn has pointed out that the criticisms offered by Leibniz, Berkeley, Huyghens, and others of the Newtonian concept of absolute space were logical, metaphysical, and even esthetic. They did not maintain either that absolute space was the wrong sort of correlate of the mass-particle, or that a more relativistic system (which was not developed at the time) might have empirical consequences more closely in agreement with observations.[8] Thus I will not discuss these criticisms here.

But if we grant this duality of space and matter, yet a third concept seems to require an equally prominent status, that of force. For their common possession of gravitational mass entailed that each particle was capable of exerting an instantaneous attraction on other particles which might be separated from it by considerable distances; indeed, it attracted every other particle in the universe in this way, though in most cases the force, as determined by the inverse-square law, was negligibly weak. The nature of force, and the intelligibility of action at a distance, rather than the meaningfulness of absolute space, became the central problem for those who accepted the general principles of the Newtonian system and assumed that it led to the correct empirical results. For the minimal requirements for a model for Newtonian mechanics did not include a unique representation of force, and Newton himself took a positivistic attitude toward the nature of attraction. Thus it was possible to suggest alternative ways of completing the model in order to include gravitational attraction.

One approach was simply to deny that there is any such thing as action at a distance and postulate instead that all interaction between particles takes place by contact, according to the laws of kinematics. Long-range attraction thus appears as a macroscopic manifestation of the motions of microscopic particles rebounding off each other. This hypothesis assumes the existence of tiny ethereal particles, which are set in motion by contact with an (actively) gravitating body. Further contacts transmit this motion through space until a presumably different set of ethereal particles impinges upon the (passively) attracted body. Insofar as most subsequent theorists went along with Newton in postulating the existence of some sort of ether permeating space, they interpreted its activity in this way.[9] But in assigning a corpuscular structure to the ether, they were following Newton's gratuitous model_2 rather than the more basic model_1. The former assumed that the particles did have some small but definite extension as well as mass. Now the ether, by definition, consisted of 'imponderable' (massless) matter, in that it had no inertia and could not generate gravitational forces directly, even though it might transmit them. Thus ethereal particles were like those of ponderable matter only in being extended, the nonessential property of the latter. But if the ether did not have the very property (mass) which gave matter its essence, why should it be assumed to have any other properties in common with matter, and why should any reasoning by analogy to determine its structure be appropriate?

Such a hypothesis also has many difficulties of detail. Presumably the ethereal particles must move very rapidly, since there is no obvious time-lapse in gravitational attraction. But how rapidly? If they can move at infinite speed, do we not have a notion even less intelligible than action at a distance? And if there is a limit, what is its value and how can this magnitude be explained? Besides, how can such motion be incorporated into the law of conservation of momentum, if this is taken to the basis of kinematics? The notions of impenetrability and action by contact, which are taken as primitive and fully intelligible, depend on the surface areas of the interacting particles, and thus require that we determine this parameter and make it an essential property. Such a theory, which probably received its fullest expression with Lesage, may well be considered the culmination of classical Lucretian atomism, since it takes as its elements impenetrable material atoms and the infinite spatial void—and nothing else.* But since

*Lesage recognized this fact in entitling his magnum opus *Lucrèce Newtonien.* See Milič Čapek, *The Philosophical Impact of Contemporary Physics*, pp. 85f., Van Nostrand, Princeton, 1961; Max Jammer, *Concepts of Force*, pp. 192f., Harper Torchbooks, Harper & Row, New York, 1962.

it gives a mechanistic explanation of gravitation by interpreting it in terms of matter in motion, Lesage's theory is also an extension of Cartesianism, and might be called 'Cartesianism without continuity (i.e., the plenum)'. Although it separates matter from space, it does not appreciate the importance of mass and its connection with force, as the classical atomists had also failed to do.

The dynamist school, on the other hand, accepted the unextendedness of the primary mass-particles, and therefore gave a central role to the new concept of force as a third irreducible element. In doing so, I believe that they gave a better explication of the true $model_1$ for Newtonian mechanics, although they were forced to give up the requirement of making their model correspond closely to perception and to previous philosophic tradition. For they recognized that action by contact was neither capable of explaining all other apparent forces nor necessarily more intelligible in itself. In the first place, as mentioned above, it depends on the surface area of the particles, so that as their size decreases to zero, the probability of such actual contacts becomes vanishingly small. In the second, the rebound which takes place upon contact with an impenetrable particle involves an acceleration; and by Newton's second law, this requires a force. The suggestion thus arose that the impenetrability and internal cohesion which gave macroscopic bodies the appearance of stable extendedness might be an effect of forces not essentially different from those of gravitational attraction.

This was the approach of Boscovich. He follows Leibniz in regarding force as the essence of matter, but does so in the context of Newton's absolute space. Matter is given in the form of unextended points which possess inertial mass; but each such point is surrounded by a sphere of force extending indefinitely far outward. This force has only a radial component and is a function only of the distance from its point center.[10] At great distances this function is (at least approximately) identical to Newton's inverse-square gravitational attraction; as one approaches the central mass-point it becomes infinitely repulsive. Thus it is physically impossible for there to be any contact between mass-points at all, though Boscovich admits the logical possibility of compenetration. This repulsion accounts for impenetrability. But the force function also changes sign from attraction to repulsion at other points near the center. By introducing this complication Boscovich hopes to include the three forces recognized by Newton (gravitation, cohesion, and fermentation) in one function.

The reader may be tempted to say that Boscovich is talking about a *field* of force. To be sure, it lacks some of the features later associated with fields: (1) it is propagated instantaneously; (2) it requires a central point source; and (3) it does not allow nonradial components. However, it has important similarities, especially the fact that it is taken to exist in its own right as something extending through all space and persisting through time, rather than being a mere name for particular actions between localized particles. It is this 'substantialization of force' which is one essential requirement for the notion of a field. In field theory a particle interacts directly (i.e. by spatio-temporal contiguity) with the field at the point where it is located, and only indirectly with any possible sources of that field. For Boscovich seems in fact to be creating a trichotomy of space, matter (identified with the point inertial masses), and force. While it is true that mass and force appear to be proportional, they are different sorts of entities, and Boscovich would certainly want to keep inertial and gravitational masses as separate concepts only accidentally related. Inertial mass is localized at the center of force, but gravitational mass really extends throughout space. But most important of all, insofar as Boscovich may be said to have a field theory, it is a *unified* field theory. There is no multiplicity of forces surrounding the central mass and exerting independent influences on any test particles elsewhere in space, but only one. Although the total force-function may include many terms, they are all functions of r which may be simply added together, i.e., $\mathbf{F} = \Sigma_i f_i(r)\mathbf{r}$. This force $\mathbf{F}$ will then affect all bodies in the same way, depending (presumably only) on their respective inertial masses. Boscovich's vision is certainly admirable. Its main weakness is that he never gave an analytic (algebraic) expression for the total force; the most he achieved was a graphical representation of it. A reasonable expression might be a sum of increasing negative powers of r, so that $f_i(r) = \alpha_i r^{-i}$, where the first term would be $i = 2$, $\alpha_2 = -Gm$ (the gravitational term). The terms would alternate in sign, with the last term being opposite to that of the gravitational. (We will see that general relativity introduces correction terms of just this sort into the law for the gravitational field of a single mass-particle.) But there is no indication of what the magnitude of these other terms might be, what physical interpretation could be given to each of them, or whether the α's would require introducing any new parameters which might have to refer to other essential properties of matter.*

Kant very definitely belongs to the dynamist school, even in his postcritical phase when he had given a philosophic rationalization of Newton's absolute space. Kant is perhaps the first to suggest that specific physical laws should

*Boscovich might in fact have been able to resolve Olbers' paradox that the night sky would be infinitely bright in an infinite universe with uniform average density of matter under Newton's law; he could simply have introduced an additional term (proportional to $1/r$, say) effective at great distances.

be bound up with specific physical geometries, and claims to have proved that the three-dimensionality of space and the inverse-square law imply each other, so that when either one is established, the other follows.[11] Kant also opposed the distinction of three-dimensional regions of space occupied by cohesive and impenetrable matter and those that were empty, as introducing radical discontinuities into nature at the boundaries. (Presumably point singularities gradually approached would not have this defect.) But Kant did not make Boscovich's distinction between unextended matter and space-filling force. Matter is extended, but only because it is essentially force. As repulsive force, it fills a region of space and repels bodies which try to penetrate it, though this impenetrability is only relative; as attractive force, it draws other bodies toward its own region.[12] This theory eliminates the singular center; makes sense of the fact that bodies have definite, if sometimes irregular, sizes and shapes; and recognizes that impenetrability or lack of compressibility is only relative to the strength of the applied force. But it makes an ontological distinction between attractive and repulsive force, while Boscovich's forces differ only in algebraic sign. Repulsive force, which acts only at the surface of the region of space that it fills, seems no more than a new name for action by contact of really extended particles. Nevertheless, the theory is able to represent such actions as stemming from an essential activity of matter, rather than its supposed passiveness, and has the additional advantage of making the void physically though not logically impossible, since the repulsive forces of different bodies are in dynamic tension and will fill an otherwise unoccupied space, although with decreasing magnitudes as one goes further from the center.

I have introduced this brief discussion of dynamism for two reasons: (1) in order to show that the interpretation of the Newtonian particles which became orthodox and which thereby provided many of the conceptual difficulties in developing the plan of general relativity was by no means the only possible one; that indeed dynamism may have remained closer to the *physical* requirements; and (2) because the concepts of the dynamists may prove useful in giving an adequate interpretation of geometrodynamics. It is fascinating to speculate whether some of the paradoxes of present-day quantum mechanics would seem so paradoxical if physicists had interpreted the notion of atoms from the start in some sort of dynamist terms, instead of as hard little bodies with sharply defined boundaries and positions in a causally inert absolute space. If such a conception had been more effective, it would be legitimate to criticize Newton and his philosophic followers on the same grounds as Descartes: that they stayed too close to the apparent disclosures of perception rather than the exigencies of the underlying physical level which his theory was trying to explicate.

Although the more orthodox notion of extended particles came to dominate most philosophic interpretations of classical physics, physicists in their practice did incorporate some elements of dynamism, if only as idealizations. The total conception of matter which emerged by the end of the nineteenth century is beautifully summarized by Vigier in the following five propositions:[13]

A. The world exists objectively, independently of any observer.
B. Any motion in nature can be described in the frame of the space-time "Arena." This implies that we can describe such motions by successive positions occupied at different times.
C. One can consider any physical system as an assembly of "material points" without dimensions; points endowed with a finite number of properties (mass, etc.).
D. The motion of these points results from "forces" independent of these points; forces that are governed by differential equations which represent the laws of nature. In other words these "laws" exist independently of any object, and if we know all initial positions, velocities, and forces of a given material system, we can predict its evolution for all subsequent time.
E. This system of laws is "complete": meaning the laws of nature are finite in number and well determined. This implies that if we knew all initial positions and velocities in nature, we could predict the future with absolute certainty.

We note that (A) indicates the acceptance of realism as opposed to any idealism. Proposition (B) accepts the independence and substantiality of space, whether or not it is taken to be absolute in the Newtonian sense. Proposition (C) indicates that physicists recognized that particles were to be taken (for the purpose of physics) as having no definite extension or other properties that were not essential in the context of a particular physical theory. Proposition (D) accepts the irreducibility of force, and especially the emphasis on representing it in differential equations which emphasize continuous action in the spatio-temporal framework, as well as the causal aspect of such laws. Finally, (E) represents the belief, more pronounced in Descartes than in Newton, that there is only one level needed for explanation; and when all its laws are completely known (as they can be) one need not appeal to any other level and can reduce everything else to this one. As Vigier points out, it is (D) and (E) in particular that constitute the basis of Laplacian determinism.

Čapek also tries to summarize the classical world-picture in five propositions.[14] However, he believes too readily in the necessity of the corpuscular hypothesis, and its conjoined emphasis on action by contact over action at a distance. As a manifesto for post-Newtonian philosophic mechanism, Čapek's theses are correct, and rightly emphasize that all qualitative features of nature were to be reduced to configurations of matter in motion, whose appearances depended on the psychic additions of the human mind. But they are not the basic tenets of classical physics, but rather those of a philosophy which drew its support from classical physics, yet never was an accurate $model_1$ for the theory, and became increasingly inadequate by the end of the nineteenth century. In the next part we will consider how these were used to interpret general relativity, why they failed, how they created resistance to new ways of thought, what finally replaced them, and how the new concepts in the model for geometrodynamics are related to the old.

PHYSICAL THEORY: CLASSICAL GENERAL RELATIVITY 3

Having completed our discussion of the metaphysical, methodological, and historical background of the theory, we are finally ready to embark upon an analysis of general relativity itself, or at least of those features which should interest philosophers. It should be mentioned again that my approach will be primarily ontological, rather than epistemological; that is, I will be concerned with the theory as a whole, as an intellectual enterprise aimed at achieving an adequate understanding of a certain level of reality, rather than with the procedures by which its concepts can be applied to observational data and its particular hypotheses tested, except where this is essential for an understanding of the nature and function of the theory itself.

A good starting point for an analysis will be the statement, "General relativity is essentially a field theory of gravitation." For despite its innocuous appearance, this statement makes two important points: (1) the primary *subject matter* of general relativity (hereafter to be abbreviated GR) is the phenomenon of gravitation, about whose real nature Newton dared not hypothesize beyond simply giving his law for the magnitude of gravitational interaction; and (2) the *concepts*, *principles* and *procedural techniques* of GR are all taken from field theory. It might be suggested that this characterization would not suffice to differentiate it from Newton's theory of gravitation, which can also be formulated as a field theory. However, although Newton's theory *can* be expressed in this way, it *need not* be, and historically, it was first presented in the very different form of action at a distance. (3) Unlike Newton's theory, GR *must* be expressed in this way. I will argue that it is impossible, in the sense of logically contradictory, to use the original Newtonian model and try to interpret GR in terms of mass particles acting instantaneously at a distance.

What justification can there be for a theory of this sort? In the first place, we have no guarantee that the phenomena of gravitation are ontologically distinct from those of, e.g., electromagnetism, so that we may not be able to give an adequate account of them by themselves. If they are never really separate in Nature, it is likely that the only reasonable theory will be one encompassing all these phenomena together. Einstein was indeed worried about this very point. Special relativity (SR) seemed quite adequate for electromagnetism, but because of the relativistic increase of mass with velocity and the fact that the field energy now had a mass-equivalent which could serve as a source of gravitation, it seemed inconsistent with the facts of gravitation. Einstein, reviewing his own intellectual development, asserted,[1] "This convinced me that, within the frame of special relativity, there is no room for a satisfactory theory of gravitation." Einstein's lifetime goal was a theory of the total physical field, but since such an attempt seemed hopelessly difficult during the period 1908–1915 he decided to proceed in two

Field Theories 8

steps, first setting up a new kind of formal framework and applying it to the (possibly ideal) case of a pure gravitational field; and then, if it seemed to work there, trying to extend it to a more general field.[2] In fact, nongravitational fields were not completely excluded, since they could be considered as possible *sources* of the gravitational fields being treated; but they were represented in a way which Einstein himself found unsatisfactory, as we shall see below. In this treatise, I will make the same separation: Thus, I will first take up "classical GR," also called "Einstein's theory of gravitation"; then later I will consider that development by Wheeler and others known as "geometrodynamics" (GMD). The latter is, among other things, a unified field theory, albeit within the framework of Einstein's original equations, and indeed, it seems to be the most successful so far developed. Ironically, it has been developed largely since Einstein's death, and Einstein was apparently not fully aware of its possibilities.

I will argue below that GR gives an ontological priority to the field as the basic concept at its proper level. Before I can make any sense of this notion, however, it will be necessary to explain the meaning and importance of fields in physics. The area in which it first came to prominence was nineteenth-century electromagnetism, culminating in the discovery of Maxwell's equations. The notion of a field entered the particle ontology of classical physics in a very innocent way, as potential force. We note, for example, Coulomb's law in electrostatics, $F = qq'/r^2$, where q' is the charge on a particle considered as source, and q the charge on another test particle. This relation was taken to hold whatever the relative positions of q and q'. Thus if q had been in a different place from that where it actually was, there would still have been a force acting on it, although if r had been different the corresponding F would also have been different. As q moved around (relative to q'), F would still remain proportional to q. Thus it became reasonable to think of q' as exerting a "potential force" at all points in space. If q were at one of these points, the force would become actual. One could define a quantity $E = F/q = q'/r^2$ at all points in space as representing the contribution of q' to the "potential force." Then for any test charge q, the actual force on it would be qE. The quantity E received the name "electric field," or more precisely, the *magnitude* of the field at the point where the charge was located. I am neglecting the vector character of the force and the field here. A more important imprecision arises from the fact that any finite q contributes to the field, so that we should really take as a definition, $E = \lim_{q \to 0}(F/q)$, even though we now know from quantum considerations that we cannot make q come indefinitely close to 0. It is assumed that F can be measured independently by $F = ma$ for the test particle.

We may note first that such a definition is quite in keeping with Carnap's method of "reduction sentences".[3] Secondly, an identical procedure could also be used to define a gravitational field, $g = Gm'/r^2$, which also involves an inverse-square law. Thirdly, if this were all there is in the notion of a field, it would be an insignificant mathematical convenience. In Newton's theory of gravitation this seemed to be the case. One could either add up all the individual forces $Gm_i'm/r^2$ on a test particle m, or add the gravitational fields g_i and multiply the sum by m, always getting the same result since actions between particles were taken to be the only source of gravitational fields. All gravitational phenomena resulted from such individual forces: any total g at a point could be broken down into its g_i components, and there were no other kinds of gravitational fields. For these reasons fields were not part of the *essential* conceptual apparatus of Newtonian gravitational theory.

However, electromagnetism proved to be a far more complicated domain. It will be worthwhile simply to list some of the differences between electric and magnetic fields on the one hand, and gravitational fields on the other. (1) Gravitational forces in Newton's theory always act along the line joining the centers of the two particles involved; so do electrostatic forces, but magnetic forces do not. (2) For a given field, only one parameter is needed to specify the total force on a particle in gravitation, its (passive gravitational) mass, and before SR this was taken to be a constant intrinsic property of the particle; in electromagnetism one must know both the charge and its state of motion or velocity, according to the Lorentz force law

$$\frac{d\mathbf{p}}{dt} = q\left[\mathbf{E} + \frac{1}{c}\,(\mathbf{v} \times \mathbf{B})\right].$$

(3) The effect is always on single particles in gravitation, but it might be on whole current loops in electromagnetism. Only by an act of abstraction guided by the particle model did Ampère succeed in writing down an equation for the magnetic force between two such loops, taking line integrals over the differential elements $I_i ds$ which appear in the original force law.[4] (4) Electromagnetic fields are greatly affected by various shielding devices, conductors, dielectrics, surface charges and currents, electromagnetic induction, etc., which make it virtually impossible to treat problems in simple two-body fashion and thus require the intermediary of the field; gravitation has none of these difficulties. (5) The total electromagnetic field comes from a variety of sources—static charges, currents, multipole terms, inductions by other fields, etc. Thus it cannot be conveniently broken down into sources of the same kind, whereas gravitational source-particles differ only in mass. (6) In practical problems, only a few sources needed to be considered in gravitation—the sun for the solar system (except, e.g., when invoking other planets

to account for perturbation) or the earth for terrestrial phenomena; a commonplace electromagnetic field might have many sources present together. (7) This ability to consider only very few sources at a time arises from the extreme weakness of the gravitational field, which nevertheless can build up since every additional piece of matter must increase it; in electromagnetism various effects of destructive interference are possible. (8) Gravity is always attractive, while electromagnetic forces may be either attractive or repulsive (bipolar).

Along with these everyday effects, which are observable without the aid of a sophisticated theory, we may add (9) according to the last two of Maxwell's equations governing the electromagnetic-field variables, a change in the magnetic field always sets up a change in the electric field, and conversely. For this reason it is important to recognize that Maxwell's theory is, at its own level, a unified field theory. It indicates that electric and magnetic fields cannot be treated adequately in isolation from each other, as would be the case if only his first two equations held. Furthermore, it reveals the striking fact that fields can be set up, or existing fields changed, without any sources, particle or otherwise, being present. In effect, the changing fields function as source terms themselves.[5] Maxwell's equations, as partial differential equations, are structural equations giving the structure of the total (unified) field at all points in space and time, whether or not either source or test particles are present.[6] SR simply carries the unification a step further, by showing that the division of the total field into electric and magnetic parts depends on the reference system chosen, and in going from one inertial system to another one can make electric fields appear as magnetic and vice versa. Clearly there seems to be only one indivisible electromagnetic field. Furthermore, (10) according to Maxwell's theory electromagnetic influences are propagated with the speed of light c, so that one could not have the instantaneous action at a distance assumed in classical gravitation. This was indeed recognized to be a crucial difference; Faraday used it to argue that radiation fields were real, while gravitational fields were not.[7] For the form of Newton's theory presupposes instantaneous transmission in the same way that it presupposes point masses. Suppose we have our two gravitating masses m and m' moving in an arbitrary manner with respect to each other, so that their relative distance $\mathbf{r}(t)$ may be any continuous function of the time t. In this case the law still holds $\mathbf{F} = Gmm'\mathbf{r}(t)/(r(t))^3$. But unless the transmission is instantaneous, t is ambiguous. Should we use the time when the test particle feels the force? Should we use the time when the force is first transmitted? Should we use some intermediate time or try to combine the extreme times using some sort of retarded potentials? Our simple equation does not provide a definite answer, so that it seems indeed incompatible with a field that takes time to propagate from one point in space to another.

Having put forward a claim that the approaches to gravity of Newton and Einstein were essentially different in that the latter required the notion of a field while the former did not, I must explain more clearly what is meant by a field in general. First of all, one requires some set of background elements. Normally these are taken to be the points in space and time, or (after the growth of relativistic ideas) in space-time. These are assumed to constitute a continuum, so that the mathematics of partial differential equations will hold. At each one of these points the appropriate field magnitude is defined. These magnitudes may be of three kinds: (1) scalars, characterized by a single number whose value at different points constitutes a scalar function of those points; (2) vectors, characterized by a set of numbers with the same dimensionality as that of the space; and (3) tensors, characterized by m-dimensional square arrays of numbers.

The mere existence of such a set of numbers at each point, or such a set of functions over the space, is necessary but not by itself sufficient to guarantee the existence of a vector or tensor field. The problem arises from the multidimensionality of the background space. If the space has n dimensions, we can locate points in it only by some set of n independent *coordinates*, whose intersections form a mesh or grid. We have required that vectors must be characterized by exactly n quantities, and tensors by n^m quantities, where m is some integer. These are then called the *components* of the vector or tensor. But clearly the numerical values of the components will depend on the coordinate system chosen, and this choice is quite arbitrary. Therefore an additional requirement is made that if we pass from one coordinate system to another (in the same space) by means of a transformation, the components of the vector or tensor in the new system must be related to those in the old in a definite way, depending only on the partial derivatives of the new coordinates with respect to the old. It is this restriction that allows us to associate invariant magnitudes with vectors and tensors, such as the magnitude ("absolute value") of a vector in Euclidean space.

But such a characterization is still entirely mathematical, rather than physical or metaphysical. In mathematics the "background space" may be any set of points that we like; in physics it is the physical space in which we have our being. This raises the possibility that genuine field magnitudes should be of a certain kind, with some sort of natural or intrinsic relation to the points of the background space. In a Riemannian space, for example, curvature seems to be a natural (because geometric) *property* of the points of the space in a way that a magnetic field, considered as a sort of foreign entity, is not. I believe that it was precisely this worry that underlay most of the nineteenth-century attempts to construct various "ethers" as the media in which fields might inhere, especially the very complex electromagnetic field. As mentioned above, the Euclidean space (and time) of nineteenth-century physics was completely homogeneous, isotropic, and passive. Thus it was difficult to see how it could function as an active medium transmitting causal influences, or how it could have properties which varied from one point to another as fields do. Geometry and physics were entirely separate: the former dealt with the static, uniform background as an arena, the latter with the dynamic, varying changes which took place in this arena. All that geometry could provide for physics was a system of fixed locations. On the other hand, it was recognized that various kinds of space-filling fluid and elastic media could and did have properties which varied from place to place in the medium and thus allowed it to be causally active. This fact was combined with three important though implicit ontological assumptions: (1) everything real must be either a substance or an attribute of a substance; (2) fields cannot be substances, but only attributes, i.e., a field cannot exist in its own right, but can only inhere in something else; and (3) a given substance cannot have any arbitrary set of attributes. Some attributes are essential, derivable from its "real definition;" others are accidental, but none of these can contradict the essence. It should be clear that all these assumptions stem from Aristotle. For these reasons, ethers had to be introduced, despite the difficulty of giving a consistent account of them. They were the intermediary substances, continuously extended through the otherwise empty Euclidean space, of which fields could be the natural attributes. Epistemologically, one worked backward, of course, starting with the attributes and using them to find the substance.

The ether had appeared to be a necessary division in the conceptual geography of physics, but unfortunately it was impossible to find a consistent model for it. But it could not be simply dismissed. Some other assumption had to be given up. A positivist might immediately vote to deny one of the metaphysical assumptions, i.e., that fields must be either substances or attributes or that Euclidean space by itself can support only certain field properties. But in fact this did not happen, largely because such moves would leave a conceptual void, with no clear-cut alternatives. Instead we seem to have a combination of the following two positions: (1) fields exist in their own right; they have an absolute and irreducible character;[8] and (2) "a field is characterized as *any property of the points of empty space*"[9] (empty in the sense of not containing ponderable matter). As Misner points out, the effect of (2) is to deny the nineteenth-century interpretation of space. Empty, matterless space has been allowed a more general structure just in order that fields, especially those fields considered in GR and GMD, may plausibly be considered properties of it. What GR claims is that it is impossible to accept (2) if one restricts oneself to classical Euclidean space, and thus GR takes as its only proper field magnitudes the natural geometric properties of a general Riemannian space-time. We may then use Ockham's razor to rule out the existence of additional ethers as being only apparent divisions of this level.

It might be thought that (1) contradicts (2), by suggesting that a field should be considered a substance, rather than an attribute. But in the context in which Einstein uses it, (1) has a rather different meaning. In order to understand it, let us review the ontological progress of the field concept. I have argued above that the basic entity in the ontology of classical physics was the particle, treated as unextended though usually represented as extended. Since fields were first introduced as an auxiliary concept ("potential forces") they clearly had a lower priority. But by the nineteenth century it was becoming more and more apparent that one could not give an adequate account of all the phenomena discussed by field theories in terms of particles and mechanical interactions. Consequently a new dualistic ontology arose, in which particles and fields entered on an equal footing. By the time of Lorentz this was well established for electromagnetism. Particles (electrons) were needed as the seat of charges; but elsewhere space was filled with the field, and the interaction of the two gave rise to such mechanical phenomena as mass.[10] Relativity (both SR and GR) goes at least this far, and makes this assertion: fields are not just properties or relations of particles; they are at least as real, and as entitled to be considered independent substances, as are particles. The question of whether they are still only properties of space (or space-time) may be partly semantic. If one has enough absolutist leanings to consider points or the whole of space as a substance, it is reasonable to think of fields as its properties; if one insists that space is nothing but a system of relations one should consider the field itself as a substance. I will return to these issues later, especially in connection with the support that GR and GMD give to one of these points of view.

But might one be able to carry out a reduction for this level in the opposite direction? Might one set up an adequate ontology in which there were no particles, but only fields? But then the question would arise, what is the source or origin of these fields? How could they exist if there were no source particles anywhere in the space? Matter was still identified with particles. Physics in fact seemed to require a trichotomy of source particles, fields, and response or test particles. Törnebohm, accepting this interpretation, properly concludes that a field theory must then have two distinct kinds of equations: (1) *field equations*, which enable one to calculate the fields at all points in space generated by a given distribution of sources; and (2) *ponderomotive equations*, which describe the behavior of a test particle in a given field (regardless of its sources).[11] It is a vitally important *empirical* fact about the world that there is no ontological distinction between source and test particles. This is part of the content of Newton's third law, that every interaction between two particles is symmetric. One can easily imagine possible worlds in which this did not hold. For example, some particles might be capable of setting up fields but could never be affected by them, while others might be affected but could never generate them. In fact, the definition of a field as the ratio of the force to the response parameter as the latter approaches 0 suggests a difference; if the source parameter went to 0, the field itself would vanish. Particles treated as sources are normally "larger" than particles treated as test particles. Indeed, such a discrepancy in size in the relevant parameters is the only justification for using one-body approximations in genuine two-body problems. This fact will be important in our later discussion of the different concepts of mass, and it becomes a crucial problem in GR.

Our question about the dependence of the field on its sources may be broken into two subquestions: (1) Can a field be generated by anything other than sources ontologically distinct from itself, like particles? In other words, can another field, or the same field at a different point in space and time, be a source of itself? (2) If (1) is answered affirmatively, can the entire field be thus self-generated? Or can such fields only modify a basic contribution from distinct sources, so that if all the sources vanished (if there were no ponderable matter in the universe) all fields would necessarily vanish? The possibility of sourceless fields was established by Maxwell's last two equations, where a change in the electric field sets up a change in the magnetic field, and vice versa. The status of (2) is less clear, and must be discussed below. In the electromagnetic case, one might argue that there must be some charges somewhere to get the whole cycle started.

In any case, the elimination of distinct sources, and the corresponding reduction of everything at a given level to fields, is one of the goals of a *unified field theory*, towards which Einstein considered GR the first step. A unified field theory is, a fortiori, a field theory, but it is unified in two distinct ways, which should not be confused. Let us call a theory in which there is only one overall field, with gravitational, electromagnetic and other "fields" simply aspects of this one field, a *total field theory*. On the other hand, a theory of any field, total or partial, which does not admit any entities other than the field and does not allow the field equations to break down at any "singular points" in the spatial manifold may be called a *complete field theory*. Now it is certainly possible to have one without the other. A theory might introduce magnitudes purporting to describe the total field but still need singularities or extended bits of ponderable matter; on the other hand, we might reach a complete theory of, say, gravitation, but still need to postulate irreducible point charges as a source of the electromagnetic field. Only if it is both total and complete have we achieved a genuine unified field theory; without one of these there remains an "esthetic imperfection" which leads us to seek the further unification. If we find that GR and GMD seem to require singularities—of whatever dimensionality—we know that we have not yet reached our goal.

One might interrupt here that such a theory must in any case contradict everyday experience. We "know" that in fact there are distinct bits of matter which generate fields, and that we must have some sort of test particles to measure them. The answer to this problem again lies in the levels hypothesis. What I will argue is that GR, in attempting to give a unified field theory, does so on only one level. What may appear as a certain complex field configuration on that level may appear as ponderable matter on another.

In considering the problem of motion, for example, it is important to distinguish three questions: (1) *Within* the level described by the theory, is there a phenomenon analogous to motion? (2) Can this phenomenon be coordinated with the motions of possible objects in *another* level which is taken to be observable? And what are the relevant properties or parameters that such objects must have? (3) Do such objects exist and move in these ways *within* this other level? The first question is ontological, and discoverable only by investigating the structure of the theory and its proper model; the second is epistemological, since I take the primary task of epistemology to be that of establishing coordination relations between different levels; the third is also ontological, concerned with the actual existence of objects or phenomena of a certain sort. However, since its level is taken to be observable, it is also empirical, discovered by observation and experiment. This schema is very general: in certain cases, especially the so-called phenomenological theories, entities within the original theory may be considered observable, so that we need not go to another level in (2). More importantly for GR and GMD, (2) is likely to require intermediate steps. We may be able to coordinate the entities of the first level plausibly only with those of another still unobservable level, and repeat these coordinations several times before reaching an observable level. It is also possible that different coordination chains could lead to different sets of observables. Thus 'test particle' as such is an epistemological notion and belongs properly to (2) rather than (1), though the ontology of (1) may contain as basic entities 'particles' which may be readily coordinated with them. The use of actual objects in making measurements belongs to (3). Coordination relations are necessary for the possibility of testing a theory, but not for its internal development and ontology, and we will be primarily concerned with explicating the latter.

Thus any objections of this sort to the possibility of eliminating particles within a field theory rest on a confusion of levels. And in this connection it should be recalled that in Newtonian physics 'particle' is as much a "theoretical term" as 'field' is here.

We can now turn directly to GR. In his *Autobiography* Einstein claimed that his development of the more general theory was motivated partly by his natural desire for theoretical unification and partly by the difficulty of explaining a crucial experimental fact: the universal proportionality of inertial and gravitational mass. We should in fact distinguish between three concepts of mass in classical physics: inertial mass, active gravitational mass, and passive gravitational mass.[1] Active gravitational mass (m_a) is the mass that serves as the source of gravitational fields. At any distance, the field is proportional to m_a, as seen by $g = Gm_a/r^2$. Passive gravitational mass (m_p) is a response parameter for a test particle, which determines the gravitational force on that particle in a given field, $m_p = F_g/g$. Inertial mass (m_i) is also a response parameter of test particles, but of a very different sort: it measures the ratio of the total force on a body, whether caused by gravity, electromagnetism, or anything else, to the kinematical property of acceleration, $m_i = F_{\text{total}}/a$. Thus m_a and m_p are relevant only in gravitation, while m_i has a much more general significance.

In view of these very different definitions, it is clearly not necessary that any pair of these three concepts should be equal. In order to bring out the discrepancy more fully, let us compare gravitation with electromagnetism. Formally, the law $F_{\text{grav}} = m_p g$ is exactly analogous with the electrostatic law $F_{\text{elec}} = qE$. It might indeed be more appropriate to refer to m_p as "gravitational charge." Likewise it would be appropriate to use a subscript, referring to q_p instead of q, and call it "passive electric charge" to distinquish it from the "active electric charge" q_a which appears in the law $E = q_a/r^2$. Now even if the contention above were valid, that every particle is capable of functioning both as a source and test particle, i.e., that it possesses both source and response parameters, it would not follow that such parameters must be equal, either in gravitation or electrostatics. The ratio of the two might vary from body to body, or even depend on the body's position and orientation in space and time. What *does* guarantee their equality is Newton's third law. With any pair of particles we may look on either one as the source and the other as the test. Since r is the same in both cases, and G is a constant, if the forces are equal in magnitude, $m_{1a}m_{2p} = m_{2a}m_{1p}$, or $m_{1a}/m_{1p} = m_{2a}/m_{2p}$. Since Newton's third law is taken to hold for *all* forces in nature (not just gravitation), a similar argument proves that q_a/q_p is also the same for all particles subject to electric forces. The ratios may then be set equal to 1, and we may drop subscripts.

Inertial mass presents a quite different problem. Presumably no one would think of identifying inertial mass with (active or passive) charge. Not only does their ratio obviously differ from one body to another, but they seem to be conceptually different entities. But from the above argument by analogy, we should no more expect m_p/m_i to be a constant (and use the same name 'mass' for both) than q/m_i. In each case we would be equating a response parameter for a single kind of field to one that responds to all fields. (Classically, m_p was as irrelevant in electromagnetism as q in gravitation, but m_i was needed in each case.) And yet, as a matter of empirical fact, m_i and m_p did seem to be equal for all bodies. This had first been recognized, if only as a curious fact, by Newton, and by Einstein's time the experiments of Eötvös and others[2] had established this to an accuracy of 1 part in 10^8. No theoretical explanation was available.

Now if this fact were strictly true, it would effect a considerable simplification. For since $\mathbf{F} = m_p\mathbf{g} = m_i\mathbf{a}$, we could then drop the m's and write $\mathbf{g} = \mathbf{a}$ in a situation where the only forces acting were gravitational. Every object, whatever its size or internal constitution, would receive exactly the same acceleration $\mathbf{a}$ when placed in a gravitational field of strength $\mathbf{g}$. Such a force would be universal in Reichenbach's sense, since it would affect all bodies in the same way. But then an interesting phenomenon arises. Suppose we have a set of objects in uniform (unaccelerated) motion, as measured in a so-called inertial system. If we now go to a coordinate system which itself has an acceleration $-\mathbf{a}$ with respect to the original, all our objects will appear to have an acceleration $+\mathbf{a}$, as if there were forces acting on them. In classical physics, these were called inertial forces, such as the "centrifugal force" experienced when a framework in which the observer was at rest was actually rotating. Classical physics considered inertial forces unreal, unlike gravitational forces. But clearly the same effect could be achieved if there were a real gravitational force $\mathbf{g} = \mathbf{a}$. (Other forces which affect bodies differently, depending on such ratios as q/m_i, would not do for this purpose.) Likewise, suppose there were a real gravitational force $\mathbf{g}$ acting, but we transformed to a system with acceleration $\mathbf{a} = \mathbf{g}$. In this system the net acceleration would be zero, and the system would appear inertial, as if there were no forces. How then could we possibly tell if there were a real gravitational force present in a given case or if our system were simply accelerated, if its dynamical behavior in the two cases would be identical?

This was the problem Einstein faced in trying to explain the results of the Eötvös experiment. Two possible ways of distinguishing the two cases seemed to be available. First, we could decide to restrict ourselves to an inertial system. But how could we know whether or not a given system was really inertial or accelerated—and with respect to what? Newton had postulated an absolute space, and this at least gave an ontological meaning to inertial systems: those which were at rest or in uniform motion with respect to absolute space. But epistemologically, absolute space in this sense could never be identified. If this were so, it would be better to allow a more general class of coordinate systems, since any one we chose might turn out to be really accelerated. Einstein himself suggested a second Newtonian alternative: we should look for source masses. If they were present, the forces were gravitational; if not, they were inertial and the system was simply accelerated.[3] Now in practice, it might of course be difficult to find the sources. They might be nearby but weak, or distant but powerful, and they might be able to account for part but not all of the observed acceleration. Nevertheless Einstein takes the argument seriously, and uses it to show that the field concept is indispensable. Only if we can get away from the source-field dichotomy and allow the possibility of fields not derivable from individual source masses can we overcome this objection. He concludes, "As far as we are able to judge at present, the general theory of relativity can be conceived only as a field theory. It could not have been developed if one had held on to the view that the real world consists of material points which move under the influence of forces acting between them. . . . It is clear that relativity presupposes the independence of the field concept."

This then was the slim amount of empirical data available to Einstein for the construction of a new theory of gravitation. He wanted to develop a field theory which would explain the observed equality of inertial and gravitational mass in a natural and necessary way, without presupposing the existence or distinguishability of inertial systems, the necessity of distinguishing inertial from gravitational forces, or the necessity of uniquely reducing every gravitational field into contributions from individual source masses. It cannot be emphasized too strongly that he was *not* trying to give an explanation, which might have been simply ad hoc, of the three small effects which are often cited as the crucial tests of his theory, as will be discussed below. If he was aware of their existence, he had no assurance that the theory actually developed could predict them. It was far more important that any new theory should account for all the phenomena for which Newton's theory had been so successful. In fact, few theories were better confirmed than Newtonian gravitation, and in such a situation any alternative could only be put forward with a pious hope that it could do at least as well, as Kuhn points out.[4] There was no sense of crisis or inadequacy in the Newtonian theory, quite unlike the situation which led to SR. Thus Einstein could well feel a need for caution, and mull over his ideas for seven years (1908–1915) before publishing them in more or less final form.

As a first step in constructing a new theory of gravitation, Einstein decided that if inertial and gravitational forces were observationally indistinguishable, they should be considered conceptually identical. This postulate was the famous principle of equivalence. But as Dicke has rightly pointed out, there are really two distinct principles involved here.[5] What he calls the *weak principle of equivalence* simply asserts that all bodies do receive the same acceleration in a given gravitational field, so that by a proper choice of accelerated coordinate systems this field may be made to disappear locally in the immediate vicinity of a test particle. This clearly does no more than reassert the result of the Eötvös experiment, though it makes it into a basic principle. The *strong principle of equivalence,* on the other hand, goes much further. It asserts that all the laws of physics, including their numerical content, will be the same, and an observer can expect to make the same observations, whether he is falling in a gravitational field or in empty space in which his laboratory is being accelerated with respect to the coordinate system which he has chosen. The weak principle deals only with gravitational effects; the strong principle includes electromagnetic, interatomic, and all other natural forces. It asserts, for example, that the (dimensionless) ratio of the masses of a proton and electron must be the same in all such systems, and similarly for other "constants of nature." If only the weak principle were true, we could not distinguish gravitational from accelerational phenomena by any gravity experiments; but we might well be able to do so by looking at these other effects. The strong principle is needed to rule out the possibility of making this distinction on the basis of any physical experiments whatsoever.

Dicke and his co-workers have refined the Eötvös experiment and established to an accuracy of about 1 part in 10^{11} that all bodies receive the same acceleration from gravity, regardless of their chemical composition.[6] On this basis the weak principle may be considered pretty well established, although as with other "null experiments" still greater precision is desirable. Dicke contends that the actual constancy through time of "physical constants" is the crucial implication of the strong principle. He concludes that the Eötvös and other recent experiments do give inductive support to the strong principle, especially in the case of electromagnetic and strong interactions, though they say very little about its compatibility with the weak interactions of beta decay.[7] It seems clear that Einstein was making an incompletely warranted inductive leap when he put forward the strong form of the principle. Once more the problem seems to be whether gravitational or any other kind of phenomena can be adequately treated in isolation from the others. We will see that Einstein definitely needs and presupposes the strong principle in deriving GR, even though he is offering only a theory of gravitation and does not presume that the theory should have major consequences in other fields.[8]

Not only does the strong principle seem to be falsifiable even if the weak one were rigorously true, but one can also derive definite ways of distinguishing gravitational and accelerational fields if the strong principle were not true. Weber suggests that possible "radiation reaction" by accelerated charged particles might provide such a test, though the results are still inconclusive.[9] Another possible breakdown might occur if antimatter were used. The sign (positive or negative) of any of the masses of an antiparticle might be reversed. Weak equivalence principles might hold for both matter and antimatter, but both inertial and gravitational mass might be negative for antiparticles, so that the strong principle would not hold in matter-antimatter interactions. Bondi has shown that an uncharged particle-antiparticle pair would move off with uniform acceleration of the center of mass in this case, a possibility which GR does not rule out.[10] But Schiff claims that the Eötvös experiment shows that inertial mass definitely, and passive gravitational mass probably, are positive even for antimatter. In any case particles which are their own antiparticles, like π-mesons, must have both masses positive, and if the others did not, there would be a fundamental asymmetry between matter and antimatter.[11]

Although it does seem to involve an extrapolation, the equivalence principle is based on at least one important piece of experimental evidence. Its significance is clearly physical rather than merely formal, since it can be expressed in a way that makes no reference to high-powered mathematical techniques or notational devices. It is certainly falsifiable, since a variety of experiments were available to overthrow it, though in fact they have given null results. Its status in GR is thus somewhat analogous to that of the constancy of the speed of light in all inertial systems within SR, which reflected the null result of the Michelson-Morley experiment.

At this point one might be tempted to see a difference in that the constancy of c was actually inconsistent with the other assumptions of classical electrodynamics, while classical gravitation theory could live comfortably with the equality of inertial and gravitational masses, even though it could not explain it. But Schild has given an interesting argument to show that, taken just by itself, the equivalence principle has a more revolutionary significance.[12] For we can derive the gravitational time dilation and red shift from this principle directly, without any of the other postulates of GR. The gravitational red shift is usually listed, along with the precession of the perihelion of Mercury and the bending of light rays in a powerful gravitational field (e.g., the sun's) as one of the three novel and verified predictions of GR. Schild then makes two important points: (1) The observed red shift provides inductive support not only for GR, but also for a host of other gravitational theories which may be quite incompatible and radically different in outlook from GR. All they need have in common is the equivalence principle, and we shall see that GR includes many other principles. It is these others which account for the remaining two effects and thereby provide a possibility for deciding experimentally in favor of GR. But (2) the equivalence principle also has a restrictive effect. For Schild claims that it is inconsistent with SR and all other flat space-time theories of gravitation.[13] This assertion seems too strong, for by making suitable ad hoc assumptions within this framework one can introduce devices that will predict such effects as the red shift. If this were not so, flat-space theories would never have received a hearing once the results of the Eötvös experiment and the actual observations of the frequency shift were accepted. But I think Schild is correct in saying that they cannot be incorporated into such theories in any natural or straightforward way. Clearly *something* had to be modified in the theoretical formalism—one could not just add on equivalence as an additional classical principle. Flat space-time was certainly a candidate for the particular classical principle to be rejected; but it was only Einstein's perception of the possible alternatives and advantages of using a more general curved Riemannian manifold that enabled physics to give it up.

A further important immediate consequence of the equivalence principle is the so-called principle of local inertial systems. Suppose that in a given coordinate system, all bodies were found to have a uniform acceleration in, say, the x-direction. Then by transforming to a new coordinate system moving with that same acceleration with respect to the original, one could make the accelerations of the bodies vanish so that the new system appeared to be inertial. This was quite acceptable classically. By the equivalence principle, one could similarly transform away a uniform or homogeneous gravitational field. Unfortunately, most gravitational fields are quite inhomogeneous. Suppose, for example, we drop a ball at some height above the surface of the earth, measure its acceleration towards the center of the earth, and then try to transform this away. We cannot do so. The magnitude of the earth's field depends on the distance from the center, so that in the new system the ball would appear to have an upward acceleration at any greater height and a downward acceleration at any lower. Furthermore, since the direction toward the center changes as the longitude or latitude changes, the ball would appear subject to lateral forces if we moved it to a different position at the same altitude. However, if the field changes continuously, we can introduce an apparently inertial system at every point in space-time, no matter how strong the field in that region, provided that we restrict ourselves to an infinitesimal region about that point. Thus we need worry no longer about whether a system in which a particle is in uniform motion is really inertial or not, as we did in SR and classical mechanics. By lowering our sights from the cosmic to the minuscule, we can create inertial systems by fiat anywhere we find them convenient. As Reichenbach puts it,[14] "Whereas according to Newton the astronomical inertial systems form the natural systems for all phenomena, Einstein maintains that it is the local inertial systems which form the normal systems." As long as we do not try to go beyond these regions, the notion of 'inertial system' is still meaningful and useful and no longer raises epistemological problems.

This is about as far as one could go in trying to build up a new theory of gravitation in a more or less inductive manner, without developing an appropriate formalism. Thus Einstein was led to make a second step and introduce the principle of *general covariance.* This he stated as follows:[15] "The general laws of nature are to be expressed by equations which hold good for all systems of coordinates, that is, are covariant with respect to any substitutions whatever (generally covariant)." Unlike the equivalence principle, that of general covariance is difficult to state precisely in words. For other typical examples of such attempts, we may cite Weber,[16] "the laws of physics should have the same form in all coordinate systems, since all are equally good for the description of nature"; and Tolman,[17] "The laws of physics should be expressed in a form which is independent of the coordinate system." Among the questions which are immediately raised are, what are the *general* laws of nature (or physics)— does that include all or just some of the totality of generalizations which might be called laws? What does 'the *form* of a law' mean? What constitutes a genuine 'description of nature'? How large and variegated is the class of all coordinate systems—are there any ruled out? To what degree is the expression of a law affected by the choice of a particular coordinate system? Can laws be represented in a form which makes no reference at all to possible coordinates, or can we do no more than write down expressions with 'free variables' for which particular coordinate systems may be substituted? All these questions concern the very meaning of the principle, prior to any considerations about its verifiability or its role in the development of GR.

In attempting to understand covariance, we should look first to the simpler case of SR, for Einstein wisely allowed his ideas to develop from analogies which had worked in other areas. The constancy of the speed of light required that all inertial systems be absolutely equivalent for electrodynamics as well as mechanics; we could not single out any one as being at rest in absolute space. The laws of SR had to reflect this fundamental fact, and this proved to be true only if the relation of magnitudes in the two systems was governed by the Lorentz, rather than the Galilean transformation, which brought out the intimate relation of space and time in a single four-dimensional continuum. The use of three-dimensional vector notation had already made it much easier to write down something like Maxwell's equations, since one did not need a separate equation for each component. Einstein had at first been suspicious that Minkowski's introduction of four-vectors—which made apparently distinct quantities such as momentum and energy appear as mere components of a more general quantity[18]—was a mere mathematical trick with no physical significance, but he later found in it the key to his more general principle. For when written in terms of these four-vectors, the laws of electrodynamics and classical mechanics were automatically made to appear the same in all inertial systems, without the nuisance of writing down detailed Lorentz transformations in each case. Here was a new mathematical *language* that provided a natural vehicle for representing the invariance properties of physical laws. I do not mean to imply that the mere introduction of the notation accomplished this by some mathematical hocus-pocus; the crucial problem still remained of deciding which quantities could be put together to form four-vectors, and in what ways. However, it opened a new range of possibilities. Sometimes it was possible to write down four-vectors in the form **P**, which made no reference to any particular components; in this case the laws had an *invariant* character. On other occasions this could not be done. The **B** and **E** fields of electrodynamics (with 6 components altogether) could only be combined by using a second-rank tensor F_{ik}, and sometimes even **P** was more conveniently expressed as P_i, where the subscripts could run from 0 to 3. In the latter case the laws were simply *covariant,* since the components signified by the subscripts would change in magnitude in different coordinate systems, but the subscripted quantity as a whole would appear in the same way.

We may therefore say that at least in Minkowski's version of SR, a fundamental principle was that the laws of mechanics and electrodynamics should be covariant under the group of all Lorentz transformations. This statement is simply an attempt to express in mathematical language the principle that these laws should have the same form in all inertial systems. The form of the law thus means its representation in terms of four-vectors or tensors; the set of all inertial systems is the set that can be derived from a given one by a sequence of Lorentz transformations. In attempting to use such a law to make physical predictions, we must indeed introduce some particular inertial system and make our measurements in terms of it. However, any particular choice is quite arbitrary, and the transformation law enables us to predict what values we would have found if we had started with another such system. This principle of *special covariance,* as it might be called, is certainly not *derivable* from the constancy of the speed of light; nevertheless it was *motivated* by this fact and enables us to deduce it as a central feature of SR.

This was the basic idea that Einstein tried to generalize in arriving at the principle of general covariance. His new theory of gravitation needed an appropriate mathematical formalism in terms of which the physical principles above could be expressed and naturally derived, and which could lead to a new set of fundamental laws for physics. P. Bergmann asserts, "General covariance is the mathematical representation of the principle of equivalence."[19] Such a claim suggests that there is really no difference between the two principles, i.e., that covariance and equivalence are the formal and material aspects of a single principle. But if this were so, then we could derive nothing more from covariance than the effects mentioned above. Let us see if this is really the case.

Now SR had put all inertial systems on the same footing, but had not admitted accelerated systems or those that required curvilinear coordinates. Clearly this was the first restriction that Einstein had to remove. Instead of just Lorentz transformations, GR admitted the whole group of continuous and differentiable transformations from one set of coordinates to another. The only requirement was that the Jacobian, the determinant formed from the partial derivatives of the new coordinates with respect to the old, must not vanish anywhere.[20] Furthermore, since we had to allow accelerated systems corresponding to any kind of gravitational field, we could no longer look at our coordinate system in global or macroscopic terms. Since we were confined to the infinitesimal neighborhoods of points, the only geometry possible was differential geometry. But fortunately this branch of mathematics did provide an absolute means for determining the four-dimensional separation between two infinitely close points, whatever coordinate system was used. If one knew the components of the *metric tensor* g_{ik} at a point P (even though the g_{ik} would themselves vary from point to point) one could calculate the separation ds at P in terms of the metric tensor and the coordinate differentials dx^i by $ds^2 = g_{ik}(P)\,dx^i dx^k$, where we follow the convention of summing over repeated indices. But this is simply a generalization of the Pythagorean theorem. At the point P we can, if we wish, introduce a coordinate system such that the g_{ik}-matrix will be diagonal, with diagonal components $(-1,1,1,1)$—the Minkowski matrix. This is the mathematical expression of the principle of local inertial systems. In general, at any point other than P the g_{ik} will take on different values in this system (they will be functions of the four coordinates x^i) so there are no guaranteed global inertial systems.

The equivalence principle has led us to introduce one tensor, g_{ik}, into physics. Extending this further, general covariance commits physics to the use of tensors (including scalars and vectors as special cases) for the representation of all its fundamental quantities. Furthermore, if we think of tensors as n-dimensional arrays, with the same number of components in each dimension, any tensors used must have exactly four components in each dimension, corresponding to the unification of space and time into a single four-dimensional manifold that SR had begun. Finally, since physics was no longer restricted to flat (Minkowskian) space-time, it could no longer use only Cartesian tensors, where contravariant and covariant components are identical; it was forced to use the full resources of the general tensor calculus, which had been laboriously developed by Riemann, Christoffel, Ricci, Levi-Civita, and others. The mathematical device which Minkowski had introduced as an interesting but inessential way of looking at SR became the central feature of GR. It is still possible to state the principles of SR and give their physical significance without using the formalism of four-vectors and tensors; it is *not possible* to do this in GR. Even without knowing the final form of Einstein's gravitational field equation, we can know that it will *look* different from Newton's, in that the latter makes explicit reference to a quantity r that represents a macroscopic separation and is uniquely defined only in particular coordinate systems, or in flat space-time.[21]

We have still not considered the empirical character of general covariance. Almost as soon as Einstein put forward the principle, Kretschmann argued that it had no physical consequences by itself, since it was logically possible to write down any law in covariant form, provided that one had no qualms about the complexity or lack of naturalness in such a representation.[22] Einstein and most others have concurred in this judgment. Tolman in fact argues that this must be the case, since he believes that any coordinate system is simply a human convenience and nature must continue to be governed by the same laws no matter how men try to represent its behavior. "We thus come to regard the principle of covariance as in any case an inescapable axiom, and to regard it as merely a task—possibly difficult but theoretically possible—for the mathematician to find a form, invariant to coordinate transformation, for the expression of any desired physical law."[23] Tolman appears to be taking an ontological standpoint, and suggesting that if there were no human beings, there might be no coordinate systems at all. In any case, he clearly seems to consider covariance an a priori necessary truth. But at the same time he also regards it as synthetic in that it extends our physical knowledge. "Nevertheless, as further emphasized by Einstein, the explicit use of the principle of covariance does have important actual consequences in our investigations of the axioms of physics."[24] Is it a genuine synthetic a priori principle as Kant would understand it, even though it might rest on a different foundation from his epistemological deduction? Is it falsifiable in any way? Or is it an analytic truth, a mere linguistic recommendation?

I believe that part of the confusion stems from the fact that covariance is being taken in different senses when these different claims are being made. If covariance is taken in the broadest possible sense, with no restriction on the number, form, or complexity of the tensors allowed nor of their individual physical significance, one probably can write any given law in covariant form as a matter of logical possibility. But Einstein clearly meant it in a narrower sense. Above all, it must enable one to express and derive the equivalence principle, as mentioned above. If the latter had to be added through other arbitrary assumptions and unexplained coincidences, like the equality of inertial and gravitational mass in classical physics, covariance would be a pointless metaphysical assumption. But if covariance is taken to restrict the kinds of tensors used, it may enable us to make verifiable predictions with the help of other auxiliary assumptions, even though these would not follow from covariance alone. If one were to object that it is these other assumptions which carry the whole "empirical weight," I would counter that one cannot always separate out the contributions from different principles so easily; they work together as a whole, and no one would carry us very far without the others. Furthermore, a principle of the generality of covariance may have the heuristic value of suggesting the proper auxiliary principles to look for, especially since it provides the linguistic framework in which they are to be expressed.

General covariance has at least a crucial methodological significance in controlling the plan of GR. If we try to write down a law in tensor form, before considering its representation in particular coordinate systems, we can avoid many unwarranted and unsuspected assumptions that merely reflect the choice of a particular system.[25] Thus it has a restrictive effect on the choice of possible physical laws or at least the order in which they are considered and tested. If we start with the covariance requirement, we are likely to take as our first candidates laws quite different from those (like Newton's) that we get by beginning with a particular coordinate system. Thus it affects what I have called above the *plan* of the theory—the expectations of significant results, the procedures for developing and testing it, the mode of combination with other ideas, etc. For this is where considerations of simplicity play a vital role. The point has often been made that given any finite set of plotted points, an infinity of continuous curves can be drawn through them. This is of course true enough as a matter of logic, but utterly irrelevant to the physicist who is trying to use these points to discover a law. He must take his curves one at a time, in order of increasing complexity, and make his tentative predictions of additional points on this basis. But there is no criterion for any choice being *absolutely* the simplest. What he will consider the simplest will depend on the framework principles that he uses to interpret the data. The claim has often been made (and will be discussed below) that Einstein chose his law of gravitation as being the simplest. To anyone who has ever tried to use the tensor calculus, this statement sounds highly ironic—calculations are far more difficult than in the Newtonian theory. It is indeed *possible* to express Newton's laws of mechanics and gravitation in a covariant form, and Cartan and Havas have suggested how such a representation might look.[26] However, any attempt to include in this form all the special features that reflect Newton's assumption of Euclidean geometry would make the law appear so incredibly complex as to be far down on the list of simple laws to be tested.

Such a methodological requirement may also be considered falsifiable in the following sense: it is as difficult to show that a law first expressed in terms of specific coordinates cannot be written in covariant form as to show that it can. Likewise, it is not clear what would constitute discovering a particular coordinate system in which an established law did not hold. In view of these difficulties, one always *could* continue to require that scientists (or their mathematical colleagues!) look for covariant representations. But if, as a matter of fact, there did appear to be systems in the world in which all the laws of physics had a particularly simple form, and which could readily be distinguished from all others, the justification for covariance and the use of tensors would disappear, and it would be more reasonable for physicists to use those preferred systems. Nature could be this way, and it is just an empirical fact that we can find no such justification for any preferred system.

In the opposite direction, one might object that general covariance admits too large a group of coordinate systems, including many in which the space and time coordinates gyrate so wildly as to be quite useless for representing concrete situations. If the only real justification for covariance is the equivalence principle, why shouldn't one just add a small additional group of coordinate systems, corresponding to those encountered in practice, and restrict covariance to that group? This is essentially the point of view of Bridgman, though his criticisms were based on his operationist assumptions.[27] Bridgman considered SR and GR to be very different theories. In SR, according to Bridgman, the entire emphasis was on individual observers and the problem of comparing the measurements each made in his own proper system; and Einstein spoke constantly of the need for rigid rods and clocks in making such measurements and establishing criteria for simultaneity. Indeed, Bridgman considered the emphasis on particularity in SR to be the idea that led him to operationism.[28] But for Bridgman, GR was a backsliding on Einstein's part.[29] Einstein now had no qualms about admitting systems which could never be used by any observer or measured by anything called rods or clocks, and even blithely quantifying over the totality of them. As Bridgman put it,[30]

One is prompted to ask why *should not* different coordinate systems play a different part? In the last analysis the whole scheme of nature must make sense to me, who am a special observer and who employs a special coordinate system. . . . It cannot be too strongly emphasized that there is no getting away from preferred operations and a unique standpoint in physics. . . . Insofar as the general spirit of relativity theory postulates an underlying "reality" from which this aspect of experience is cancelled out, it seems to me to be palpably false, and furthermore devoid of operational meaning.

Bridgman's mistake lay in his interpretation of the special theory, where he too readily identified inertial systems with the proper systems of actual human observers in uniform relative motion. Since the systems are defined and compared on a macroscopic scale, no great harm is done by this move until it is treated as an additional *requirement* and used as a basis for criticizing GR. We will discuss the meaning of measurement in GR in the next chapter. But the very fact that GR is concerned with the micro-regions of differential geometry suggests strongly that 'observer', 'rigid rod', and 'clock' *cannot* mean the same thing here. GR must be given its own kind of epistemological interpretation, not one borrowed wholesale from a particular version of SR. Within the theory proper, 'coordinate system' is equally a theoretical term for both SR and GR, and part of its meaning is given implicitly by the way it is used in the respective formalisms.

Einstein clearly appreciated the force of Bridgman's argument and the difference between SR and GR which it displays. In fact, it was the crucial factor that prevented him from formulating GR sooner. To quote him at some length,[31]

Why were another seven years required for the construction of the general theory of relativity? The main reason lies in the fact that it is not so easy to free oneself from the idea that coordinates must have an immediate metrical meaning. The transformation took place in approximately the following fashion.

We start with an empty, field-free space, as it occurs—related to an inertial system—in the sense of the special theory, as the simplest of all imaginable physical situations. If we now think of a non-inertial system introduced by assuming that the new system is uniformly accelerated against the inertial system (in a 3-dimensional description) in one direction (conveniently defined), then there exists with reference to this system a static parallel gravitational field. The reference system may thereby be chosen as rigid, of Euclidean type, in 3-dimensional metric relations. But the time, in which the field appears as static, is *not* measured by *equally constituted* stationary clocks. From this special example one can already recognize that the immediate metric significance of the coordinates is lost if one admits non-linear transformations of coordinates at all. To do the latter is, however, *obligatory* if one wants to do justice to the equality of inertial and gravitational mass by the basis of the theory, and if one wants to overcome Mach's paradox as concerns the inertial systems.

If, then, one must give up the attempt to give the coordinates an immediate metric meaning (differences of coordinates = measurable lengths, times), one will not be able to avoid treating as equivalent all coordinate systems, which can be created by the continuous transformations of the coordinates.

Particular coordinate systems may be, as Bridgman suggested, a human convenience, a tool for relating individual observers to physical situations. Nevertheless, this does not mean that each person must single out one such system for himself and analyze everything in terms of it. If he is a theoretician with sufficient mathematical ability, he can use an indefinite number of them together for a more complete view, just as in binocular vision one uses the different perspectives of the two eyes simultaneously to see a three-dimensional object. Within the framework of GR it is not appropriate to ask, as Bridgman did, what are a set of coordinates the coordinates *of?*

From his own operationist standpoint, Bridgman's criticisms were perfectly natural and consistent with the rest of his ideas. General covariance does appear as an attempt to get away from particularity and find invariants which may be taken to have a more objective (or at least impersonal) significance, while Bridgman recognized that operationism must eventually lead to a completely subjectivist position.[32] His criticism may thus seem to have been merely external, a matter of holding up GR against his own methodological biases and rejecting GR wherever they did not agree. However, he cannot be dismissed so lightly. For in GR we have the paradoxical situation that although all coordinate systems are taken to be equivalent, and the general laws of physics cannot single out any one, particular systems are important for several reasons: (1) measurements capable of giving numerical results, which are essential for any experimental tests of the theory, must be carried out in some definite system; (2) for any given physical situation, there are particular systems in which the general laws take on an especially simple form, both in terms of manipulatory convenience and emphasis on the central features of the situation; and (3) since we already have another theory of gravitation available (Newton's) and others more or less highly developed (Whitehead, Birkhoff, Nordström,

Brans-Dicke, etc.), it is highly desirable to compare them directly and see which effects they predict in common and which are peculiar to a particular theory and thus possible candidates for a "crucial experiment." For example, the best known exact solution of Einstein's field equations in GR is the Schwarzschild solution for a single, spherically symmetric mass (e.g., the sun for planetary motion). Since the corresponding Newtonian solution is written in a particular set of coordinates, it is important to find the corresponding system for Schwarzschild's so that a point-by-point comparison can be made. In fact, such a system does exist, though it has other undesirable features which may be avoided (and important new characteristics discovered) by transforming to other systems.

Fock has used these observations as the basis of an *internal* criticism of the role and importance of general covariance within GR. Fock introduces into GR additional requirements on the metric tensor, namely, that its components satisfy the equation $((-g)^{1/2}g^{ik})_{;k} = 0$, where g is the determinant of the metric tensor and the semicolon stands for covariant differentiation. This has the effect of singling out a particular system (or rather a set related by Lorentz transformations), the only one in which the g^{ik} satisfy this requirement.[33] Fock shows that this restriction singles out the inertial systems in flat Minkowskian space, and considers it the natural generalization for curved space. Such a requirement, which adds four additional equations to Einstein's field equations, is at least mathematically possible, and the systems thus defined are indeed especially convenient for many problems. But, as Anderson points out, it is by no means clear either (1) that the requirement has any *physical*, rather than merely mathematical, significance; or (2) that another set of coordinate conditions might not single out equally interesting, though different, systems.[34] The derivation of Fock's conditions is obscure, and would apparently rule out spherical as opposed to Cartesian coordinates for flat space. Since Fock admits that solutions in other coordinate systems are still equally legitimate,[35] it is not clear how he intends his requirement except as a simplicity principle for trying out coordinate systems in a definite order. As such, it would be acceptable to Einstein for applications of the theory, but would hardly be a fundamental principle.

Fock is among those who believe that SR and GR are markedly different in character, and objects to the use of the term 'relativity' for both, especially if it includes the implication that GR is just a generalization of SR. Fock interprets relativity as uniformity, i.e., as equivalence of spatio-temporal directions.[36] No direction is singled out as being, for example, the direction of motion of an inertial system with respect to a stationary ether. In GR, on the other hand, this uniformity is lost. In the small, space-time is not uniform, homogeneous, or isotropic, though it may still be on a cosmological scale. Since the different points in space-time have different characteristics, relativity or indifference is lost. We can no longer form the various vector-invariant quantities possible in SR, but must be satisfied with covariance. Fock claims that GR is a relativity theory at all only if 'relativity' is here identified with covariance, although it was identified with uniformity in SR; and since other theories can be written in generally covariant form, this feature would not be enough to distinguish Einstein's from other theories.

Insofar as SR is a theory of flat space-time and GR a theory of curved space-time, Fock is right in pointing out a fundamental difference. But his strictures on the use of 'relativity' are unwarranted. For in both SR and GR Einstein was concerned not with the equivalence of points and directions in space-time, but rather with that of the coordinate systems used to locate and describe them. In each case there is an indifference principle which asserts that all systems within a group are relative, i.e., equally capable of representing the physical situation without distorting features. The difference is that GR admits a far more general group, and in this important respect is indeed a generalization of SR.

'Relativity' has many senses which can lead to semantic confusion, and I have not yet touched on the crucial issue of whether the space-time postulated by GR is itself relative or absolute. This problem will be considered in detail below. At least, one should note that both SR and GR are concerned with finding a proper set of *invariant* magnitudes, which are thus *not* relative to a coordinate system or anything else. In this sense they represent a continuing quest for the absolute, though with the recognition that the old absolutes of Newtonian theory can no longer be considered valid.

Törnebohm has attempted to counter Bridgman's operationist objections in the following manner: He agrees that general covariance can have no operational significance by itself, nor can any general law p, written in covariant form using dummy indices instead of actual coordinates. However, if we now substitute actual coordinate systems S_1, S_2, etc. we can derive from p the specific propositions p_1, p_2, etc. which do contain numerical values that can be compared with actual measurements. If we choose S_1 and introduce appropriate operational definitions, we may observe the values O_1. Then according to our transformation laws, these should have the values O_2 within S_2. But as Bridgman had pointed out,[37] it is by no means clear when we can be said to be performing the same (or even an analogous) operation in the two systems, especially if their relative motion is very rapid or changing wildly. Thus without further stipulations we canot say unambiguously whether we have observed O_2 in S_2 or not. Törnebohm suggests that we have two alternatives in this dilemma: (1) We may *stipulate* that two operations in the two systems are to count as the same. In this case the covariance clearly becomes an empirical, testable hypothesis. However, any such stipulations must appear as arbitrary without some further justification, and the latter is not readily available. On the other hand, (2) we may *define* an operation in S_2 to be analogous to that in S_1 if and only if it gives the expected result O_2. This gives us a clear-cut basis for a comparison, but at the cost of making covariance analytic and unfalsifiable.[38] Physics has chosen the latter alternative, partly because of the advantages of keeping covariance as an axiomatic framework principle, and partly because, with the vast number of irregular coordinate systems allowed, it would be virtually impossible to make stipulations covering all such systems in any other way.

These considerations on framework principles which are not directly falsifiable bring us back to what is probably the most crucial function of general covariance: to function as a test or criterion for laws of nature. Wigner has emphasized the importance of having principles which stand in the same relation to statements correlating particular events as these statements do to the events themselves.[39] Invariance of such relations under change of position in space and time has long been held up as a criterion for a general law; anything which held only in some local area would be merely a statement of a complex phenomenon in that region and not a law of physics. Likewise covariance (over whatever group may be chosen) can function as a test. Any genuine law of nature can indeed be written in covariant form, and it is advantageous to do so, since in a particular system one may attach too great a significance to features that merely reflect the peculiarities of that system, rather than genuine aspects of the objective world. This was recognized even in classical mechanics where, however, the Galilean group was considered the only appropriate system of transformations. Any choice of a covariance group has the effect of limiting possible statements correlating events from the category of genuine physical laws: the larger the required group, the smaller the number of laws allowed, with those eliminated having only a more limited range of validity. Insofar as any science is concerned with unification and economy of concepts and independent laws, enlarging the covariance group can contribute to such unification.

Törnebohm argues for the need for such principles in a similar fashion.[40] If we write a putative scientific law in a form that gives one physical magnitude as a numerical function of some others $y = f(x_1, \ldots, x_n)$, we must make at least implicit reference to some particular coordinate system in which these quantities may be measured. Suppose this coordinate system is described by some set of parameters Q. If the coordinate system were changed, and the new system described by Q', the value of the function might also change. Therefore we may regard y as a function of Q as well as the x_i's, $y = f(x_1, \ldots, x_n; Q)$. But in this form the expression is semantically incomplete, since until a particular value of Q is specified y cannot be calculated—it involves a 'free variable'. We may then either restrict ourselves to a particular value of Q, i.e., single out a particular coordinate system for the expression of the physical law, or we may attempt to quantify universally over Q. The rules governing the procedure for such quantification are what Törnebohm calls "closure rules", and unless we have some such rules we can never get beyond particular coordinate systems. The principle of general covariance is thus the proper closure or quantification rule for GR, and the higher-order propositions formed by quantifying over Q are the covariant physical laws. (There are also occasions when we might want to use an existential quantifier; for example, when we assert that in any field there is a coordinate system which is Minkowskian in the immediate vicinity of a given point. However, these are not the physical laws themselves, but deductions from the general principles of the theory.) SR and GR are not unique in having closure rules; however, they are perhaps the first theories to recognize their cardinal importance and state them explicitly.

Let us now attempt to summarize the foregoing conclusions and see what can be said about the principle of general covariance taken by itself. In the first place, it is *not* a particular law of nature, on a par with, e.g., the law for the gravitational field. It is rather a law of laws, an essentially "metanomological" principle.[41] Although it is itself a lawlike statement, its referents are not objects and events, but physical laws of a lower order. Although it can be stated in words (albeit with some difficulty) it cannot be stated within the tensor calculus, the language appropriate to GR, since it is responsible for setting up that language and making it the required mode of expression. It is completely general, since it is taken to include all the laws of physics considered by GR, though of course it is most important in gravitation theory. Therefore it is the only closure rule necessary. Like the framework principles of other theories, it cannot be falsified directly by particular experiments, but could be dropped if it were not fruitful in allowing the derivation of particular hypotheses which agreed with experiment. For it does indeed provide a framework and language, though as I will argue in the next section, it does not yet provide a full model or ontology for GR. And like the choice of other new framework principles, it is essentially a gamble, a shot in the dark, a commitment. The number of additional phenomena which have been directly analyzed and explained by means of GR (and thereby provided inductive evidence for the theory) is far smaller than that explained by Newtonian mechanics, and in view of the much greater mathematical difficulties in handling the equations, this situation is not likely to change greatly in the future. Insofar as it breaks away from the viewpoints of particular observers with their unique coordinate systems, and looks instead for features which are invariant in all possible systems, it represents a commitment to a new criterion of objectivity and reality that, as I will show below, can have ontological significance.

Törnebohm certainly considers covariance crucially important for GR. He asserts (in italics),[42]

> We shall give a short general characteristic of the theory of relativity. The theory of relativity is essentially a reconstruction of classical physics based on new closure rules. The reconstruction has been accompanied by a basically new schematism whose most important features are new concepts about space and time. It has furthermore led to new general field equations. On the basis of these the theory of relativity has been successfully used to derive propositions about the universe at large.

I contend that Törnebohm goes too far here. The new closure rules and new schematism do not *by themselves* lead to all these conclusions. But by its very character as metanomological, covariance cannot, should not, and does not work by itself in setting up GR. When combined with a new conceptual interpretation that I will discuss next, it does provide the criteria of form and simplicity that will give these results, but this conceptual revolution required an additional act of genius on Einstein's part. Thus Bridgman is wrong in claiming that a physicist in using general covariance is trying to get information about the world *just* from the language that he is using.[43] But while it cannot function alone, the other principles in GR cannot function without it either.

I have claimed above that the formation of a new physical theory requires a delicately balanced combination of three different things: (1) availability of experimental or observational data, (2) knowledge of mathematical techniques, and (3) awareness of conceptual possibilities. All three must be present together and synthesized in the mind of the theoretician. If any one of these is lacking, no amount of development of the other two may be sufficient to produce the new theory and thereby lead to a scientific revolution, though in particular cases the material needed for any one of these may be elementary. For example, in many 'phenomenological' theories, and in theories outside of physics such as Darwinian evolution, the mathematics is not difficult (though evolution was strengthened when it was combined with mathematical genetics). On the other hand, Newton had to develop the calculus for an adequate presentation of his own mechanics. Furthermore, the observations may be of an everyday sort, or may require very elaborate and expensive apparatus. In the case of GR, the experimental datum was the observed equivalence of gravitational and accelerational fields, or gravitational and inertial mass. This has really been known since Galileo's famous experiment on the tower of Pisa, and Newton, Eötvös, and Dicke have only refined it. The mathematical technique was the development of the tensor calculus. This had been done within pure mathematics by Riemann, Levi-Civita, and others, but except in certain areas of hydrodynamics it had not been used in physics, at least not in its most general form. At the same time, the mathematics of general Riemannian geometry had been worked out. To a degree, tensor calculus and Riemannian geometry go together; however, many interesting theorems in Riemannian geometry do not require this tensor representation, and tensors may be used in much more general and abstract problems quite remote from geometry.

The crucial conceptual innovation in GR, the one that allowed the possibility of a scientific revolution and a new synthesis, was the interpretation of magnitudes that had previously been considered physical in terms of geometrical magnitudes; and at the same time, the rejection of physical geometry as something given a priori, replaced by its treatment as wholly determined by the physical situation. Physics and geometry became so intertwined that any assertion about the one necessarily implied an assertion about the other. General covariance had simply restricted physics to the use of tensor quantities, without specifying how they might be derived or what they should signify. The identification of physics with geometry added the new requirement that the tensors in terms of which the fundamental physical laws were to be expressed must be derived from or equated to tensors which have a purely geometric significance. But the fundamental tensor used to characterize a space is the metric tensor g_{ik}, and all other geometrical tensors are derived from this by various mathematical operations. Thus if the metric tensor were given at all points in physical space-time (and the coordinate system in which it was being represented) we would know not only all the geometrical properties of space-time, but all the physics there as well. This at least is the *ideal* of GR. It was not perfectly realized in the classical version of GR, a fact which Einstein clearly recognized and which led him to suggest several modifications in the original formalism of the theory. On the other hand, it was such a radical ideal that until about 1955 physicists were unwilling or unable to try to use it consistently. The attempt to realize this ideal while operating within the original formalism has been the primary task of geometrodynamics as it has developed since 1955. In this connection GMD has made explicit and ramified the model appropriate to GR, and developed a new plan, or procedure for use, to provide a more fruitful internal development of the theory and ultimately to generate more testable consequences.

Now the idea of interpreting physics in geometrical terms was by no means new. The whole point of the earlier chapters was to show that this very thing had been said by Plato and Descartes in their different ways. One aspect of this (the aspect of substance) was the identification of matter and space. We have seen that Descartes had the best conception of space and geometry available in his day, but his work was vitiated by an inadequate conception of matter and the actual laws governing its behavior. Plato avoided some of these mistakes by refusing to identify his chōra, as physical space or matter, with the perfect Euclidean space of geometry. But the result was that his chōra was so hopelessly vague and unstructured that it could not be used for making precise, testable predictions, a conclusion which suited Plato's metaphysical assumption that one could not have genuine knowledge of the physical world.

If the space of Cartesian geometry was too undifferentiated to explain the different features of matter known in Descartes' day, it became increasingly out of date as new properties of matter and its interaction with fields were discovered. Without the development of a more general notion of space Descartes' identification of space and matter would undoubtedly have been forgotten except by scholars, and even with Riemannian geometry the complexity required for such an identification would have been hopeless without some general guiding principle. Nevertheless, the ideal remained sufficiently attractive so that abortive approaches to it were still made before GR. Dugas points out that in 1889 Darboux attempted to give a geometrical representation to all of classical mechanics.[44] However, Darboux did not attach any *conceptual* significance to his work. He was concerned rather with formal analogies between geometrical and mechanical theorems and states of affairs that might have mutual heuristic value in the development of the two sciences.

The more daring conceptual leap was made in 1870 by Clifford, when he proposed,[45]

(1) that small portions of space *are* in fact of a nature analogous to little hills on a surface which is on the average flat; namely, that the ordinary laws of geometry are not valid in them. (2) That this property of being curved or distorted is continually being passed on from one portion of space to another after the manner of a wave. (3) That this variation of the curvature of space is what really happens in that phenomenon which

we call the *motion of matter,* whether ponderable or ethereal. (4) That in the physical world nothing else takes place but this variation, subject (possibly) to the law of continuity.

Read today, against the background of GR and its successes, this dramatic and forthright statement sounds like an extraordinary piece of insight and prophecy, far ahead of its time. Insofar as he recognized that the space required would have curvature, and indeed curvature that varied in space and time, Clifford was certainly far ahead of Descartes. However, there are still marked differences between Clifford's conception and that of GR. For Clifford still requires that space be Euclidean, or at least uniformly curved in the large, in order to preserve the possibility of free mobility of extended objects without distortion. Only on a very small scale do we find these local variations in curvature, which eventually cancel each other out. Nevertheless, while requiring that the overall curvature must be uniform at any particular time, Clifford allows that this average value might change in time.[46] This gives space and time a quite different significance, and further indicates that Clifford was thinking in terms of a three-dimensional space quite independent of time. In GR the identification of matter with space is achieved only by using the full four-dimensional space-time manifold. In the second place, although Clifford's conceptual idea was fundamentally sound, he did not attempt to work out in mathematical detail how or why it might be so. Strangely enough, although he was himself a mathematician and had translated Riemann's original paper, he did not see or use it as the crucial mathematical tool for expressing his ideas precisely. And there is no indication that he had anything like the notion of a general tensor. In the third place, neither he, nor Darboux, nor anyone else at the time attempted to connect these ideas with the observed equality of inertial and gravitational mass, or any other empirical fact. Clifford did consider the possibility that he might be able to explain double refraction using this hypothesis, but never came up with definite results. In fact, he suspected that we could never distinguish the effects of space curvature from those due to real physical forces, and that both models would always lead to the same predictions.[47] There is no real indication that Clifford believed that his interpretation could explain some observed phenomena that more orthodox theories could not, even if in some metaphysical sense it was the "true" one.

We must now make clear what 'geometry' is taken to be within GR, and what assumptions and restrictions are included. In the first place, it is four-dimensional. Secondly, there is a fundamental invariant for any space considered, the differential form $\Phi = g_{ik}dx^i dx^k$, where the i and k (and Latin indices generally) take on independently the four values 0,1,2,3, no matter how the four independent coordinates are chosen. This form Φ is then given a geomterical significance by being identified with ds^2, the square of the *interval* between two infinitesimally separated space-time points. If $ds^2 < 0$, the interval is called timelike; if 0, it is lightlike (or null); if > 0, it is spacelike. Thirdly, the metric tensor g_{ik} is symmetrical, no matter what system of coordinates is chosen.

The priority of the four-dimensional space-time continuum over any separation into individual spaces and times is of course taken over from SR, and for the same reason: it is impossible to find a scalar invariant comparable to ds^2 in either of these lower-dimensional manifolds. Reichenbach raises the possibility that a quadratic form in the dx^i might not be available as an invariant; we might, for example, have to use a quartic form like $ds^4 = g_{ikmn}dx^i dx^k dx^m dx^n$. Such a form would be vastly more difficult to manipulate, but it might allow more degrees of freedom, and in the simpler cases the majority of the g_{ikmn} would vanish anyway.[48] However, we know that in flat space the quadratic form is sufficient, and it is therefore reasonable to try this for the most general case, going to more complex forms only if the quadratic proves inadequate. As a matter of empirical fact, the quadratic does suffice. Since $dx^i dx^k = dx^k dx^i$ for all i and k, it might seem that g_{ik} must be symmetrical, or at least that it could always be made so by choosing $g'_{ik} = g'_{ki} = \frac{1}{2}(g_{ik} + g_{ki})$. But this depends on our choosing the g_{ik} as the basic tensor for investigating the geometry. It is possible to begin with other geometrical quantities like the so-called affine connection[49] so that the g_{ik} must be derived and may not be symmetrical. In such a non-Riemannian geometry, two rods which have the same length when juxtaposed at a point may differ when brought together later if they have travelled over different paths.[50] This seems to be a hypothesis which could easily be tested, but in fact there have been no precise or rigorous tests. Marzke and Wheeler have suggested, however, that according to the Pauli exclusion principle, if such lengths as the Bohr radius of an atom were affected by its path of motion, there could be many more low-energy quantum states near the nucleus into which the orbital electrons could fall, and all atoms (and thus the whole world) would ultimately contract in volume. Since this has not been observed to happen at all, Marzke and Wheeler claim that any path-dependence of lengths must be far too small to measure.[51] Since there is at least no evidence that lengths are path-dependent, we may take the use of Riemannian geometry, rather than a more general non-Riemannian geometry that uses something other than the symmetric g_{ik} as its basic magnitude, to be empirically grounded.

Thus in 'geometry' as we are using the term, the metric tensor is the fundamental tensor magnitude, in that every other piece of information about the geometry of the space can by appropriate mathematical operations be derived from a knowledge of the metric and the coordinates in which its components are expressed. For this reason it has epistemological priority. However, it is much less fundamental ontologically. It does not describe a purely geometrical and invariant property of the manifold, such as its curvature at a point; rather it gives the relation between a particular set of coordinates and the basic invariant form, Φ or ds^2. Fortunately, there are other derived tensors, such as the Riemann tensor R_{ikmn} and the Ricci tensor R_{ik}, which do refer directly to geometrical properties and may thus be taken to have ontological priority. Now if we consider four-dimensional space-time to have any sort of independent existence, we must assume that (1) it has definite properties at its various points; (2) these must be geometrical or spatial (in the general sense, not as opposed to temporal), such as the various curvatures; and (3) they must be quite independent of any coordinate system, since the latter is a mere human convenience, what Wheeler calls a "bookkeeping device," and cannot affect the geometry.

It is important to note that (3) is quite independent of general covariance. It follows only from the conceptual possibility of identifying physics with geometry. This in turn follows from the possibility of identifying matter with space, since geometry is the science of spaces and spatial objects in the same way that physics is the science of matter and material objects. If the properties and relations of physical objects are taken to be real, objective, invariant, or independent of the observer, the same must hold for the geometrical properties and relations with which they are to be coordinated. Now it is certainly true that this assumption of objective physical properties was a central postulate of classical physics, even though it is now being challenged in quantum physics. Classical GR was developed before quantum mechanics, and insofar as it attempts to reconstruct classical physics, it shares these assumptions. It may in fact turn out that we can make no such complete correspondence between geometrical and physical magnitudes, just as Descartes was unable to do this. If so, this might be a good (at least heuristic) argument against the objectivity of geometry. However, any epistemological arguments which do not depend on the success or failure of GMD in making these identifications are quite beside the point. For they *begin* with assumptions identifying the geometrical quantities with physical objects such as rigid rods and paths of light rays, while GR develops the mathematics and geometrical model of the theory *first*, and only *later* considers the 'second interpretation' of this model in terms of actual physical objects. The expression ds (even for spatial intervals) does not *mean* 'the length of an infinitesimal rigid rod', whether or not it may eventually be correlated with such a thing or an approximation of it.

Wheeler points out that the requirement that GR must have a purely geometrical content goes beyond general covariance in placing an additional physical restriction on the form of the laws allowed. As mentioned above, covariance in the broad sense allows any law that can be written in tensor form. But Einstein realized that if such a law enabled us to find *unique* solutions for the ten independent components of the metric tensor g_{ik}, it would give us information not only about the geometry but also about the coordinates to be used, since these particular values will give the invariant ds^2 only when combined with a particular set of coordinates x^i. But if these coordinates have no ontological or geometric significance, as postulated above, we have violated our requirement that the theory have a purely geometric content. Thus the field equations *must not* determine the g_{ik} uniquely. They must give us information about geometry but not about coordinates. In satisfying covariance, they rule out no systems of coordinates, but many possible geometries. For the same reason, it is clearly desirable to write the equations in a form that makes no references to any coordinates, but little has been accomplished along these lines so far.[52]

It is vitally important to recognize that this identification of space with matter and geometry with physics *is* the central conceptual feature of GR. Such philosophers as Reichenbach and Grünbaum have misinterpreted the theory by attempting to keep them separate. They have been concerned with such questions as: Once we have laid down a set of "coordinative definitions" (even if by arbitrary convention), to what degree is the geometry uniquely determined, and once this is done, what is the determinate form of the physical laws in this geometry? Are universal forces (considered part of physics but not geometry) required? They recognize the interdependence of physics and geometry in that physical laws which involve geometric magnitudes such as lengths must change their mathematical form when the geometry changes. However, they think in terms of adjustments by successive approximations, leading ultimately to the correct geometry and correct physics for that set of coordinative definitions.[53] While these questions are interesting in their own right, the whole approach is antithetical to that of GR. They also run the danger of forcing a dichotomous distinction between 'standards', which are physical objects and processes and thus subject to physical laws, but which also have the function of determining the geometry of space-time; and, on the other hand, 'ordinary' physical objects which are governed by the physics but do not influence the geometry. Such a distinction seems ontologically unwarranted. The net result of such a procedure is an unfounded worry about whether physics influences geometry or geometry influences physics the more. Reichenbach reflects this when he claims[54]

We are therefore reversing the actual relationship if we speak of a reduction of mechanics to geometry: *it is not the theory of gravitation that becomes geometry, but it is geometry that becomes an expression of the gravitational field.* (Reichenbach's italics)

From the standpoint of GR there is no difference between the positions that he wishes to distinguish. However, it would be natural, if geometry were nothing but the articulation of relationships among standard objects, to expect that somehow physics would be lost if it were reduced to or identified with geometry. It is his basic point of view that is mistaken. Neither the conventional character of the choice of coordinative definitions *nor* the possibility of using a particular stipulation to make an empirical determination of the geometry is really relevant for GR, though it may be important for an understanding of space in general (apart from any particular theory in which the term 'space' appears). For a given physical configuration, GR does single out a unique geometry, but this uniqueness results from the identification of physical with geometrical entities, *not* from a choice of coordinative definitions. Rods, clocks, and other possible measuring devices may indeed be considered implicitly defined by the equations of the theory and given an ideal representation in its model. Bridgman was thus correct when he claimed that a 'clock' within GR is simply anything that behaves in a certain specified way.[55] Ironically, this is not the criticism that he intended it to be unless there are no such objects (or reasonable approximations of them) within the world. There is no reason to assume that we have a priori (pretheoretic) knowledge of what a genuine clock must be.*

Anderson has argued that in characterizing theories it is desirable to distinguish between their absolute and dynamic elements.[56] The former are fixed and invariant under whatever group of motions is admitted by the theory; the latter are governed by the physical characteristics of the system to which the theory is being applied. In both classical physics and SR the geometry was such an absolute element. The metric tensor was fixed and constant, and the only problem remaining was to determine the relative values of the components of physical magnitudes in systems displaced, rotated, or translated at uniform velocity with respect to each other. Fock has argued for the need for such fixed points in any theory if definite results are to be obtained,[57] and Whitehead made it a cardinal feature of his system to distinguish the particular physical relations from the absolute geometry that provides a background for the possibility of their uni-

*Arguments similar to mine here have been advanced by Hilary Putnam in "An Examination of Grünbaum's Philosophy of Geometry," in Bernard Baumrin, *Philosophy of Science—The Delaware Seminar,* vol. 2, pp. 205–255, Interscience, New York, 1963.

versal relatedness.[58] However, Anderson points out that this distinction introduces a fundamental asymmetry. The absolute elements certainly influence the rest of the physics (in the example above, the form of the physical laws depends on which geometry is chosen), although they themselves are not affected by them. The dynamic elements, on the other hand, both affect and are affected by each other. If we accept what Anderson calls a general principle of reciprocity, that each element of a physical theory should be influenced by every other element, it is highly desirable to eliminate absolute elements if at all possible.[59] Anderson's reciprocity is clearly a regulative principle: it need not be accepted as a methodological ideal, and presumably Fock, Whitehead, and many other physicists and philosophers would not. However, Einstein did so, and thus was led to make the geometry a dynamic element of the theory, something which is both fully determined by the distribution of matter in the physical system and which fully determines the motion of such matter. But once we have reached this stage of complete reciprocity of dynamic geometry and matter, there is no conceptual reason for requiring a sharp distinction to be made between the two.

However, there is an additional problem, which Wigner has rightly emphasized.[60] In solving a system of differential equations, we cannot get definite numerical results unless we add initial and boundary conditions to the field equations. This data is normally specified 'from outside', and is taken to be unaffected by other elements, so that it functions like an absolute element. For example, if we specify that a field is to vanish along a certain boundary, this is to hold regardless of the distribution of sources inside it, although such distributions may be restricted by this boundary condition. One may ask, what sorts of boundary conditions are required by GR, what physical situations do they represent, how is their introduction justified, and what is their position in the absolute-dynamic dichotomy? In practice, they have taken the form of global properties of the whole of space-time, i.e., that the metric should be Minkowskian at infinity or that space-time should be closed in a certain way.

No satisfactory general answer has yet been found to the questions raised in this chapter. It may in fact be the case that GR cannot have any arbitrarily specifiable boundary conditions (which would make it unique among physical theories), which would in turn require that we must always deal with the universe as a whole, as the only truly legitimate physical *system*. This possibility will be considered further in the final chapter.

Measurement in General Relativity 10

I have argued in Chapter 9 that geometry is not simply an empirical description of the behavior of certain kinds of physical objects (rigid rods and clocks), that geometrical entities and structures are taken to exist in their own right. However, this makes even more acute the problem of how to measure or determine this geometry. How are we to get the numbers to be substituted for the geometrical variables in the equations of GR? We must therefore turn to the problem of measurement in GR.

Bridgman, Reichenbach, Grünbaum, and others with their general viewpoint have apparently assumed that measurements in GR *must* be made by rigid rods and clocks, just as they are in classical physics. Now a rigid rod, considered as a material object with definite boundaries which cannot be deformed except by perturbing forces whose influence can be 'corrected out', is indeed an appropriate measure of distance or three-dimensional spatial separation, though if the space has a variable curvature it must be considerably smaller than the radius of any curvature in its vicinity. Likewise a clock—any device with regular periodicity or 'ticks'—is an appropriate measure of time, the one-dimensional temporal separation. In classical physics nothing further was necessary, since space and time were entirely independent of each other and measurements in either continuum had an absolute status. But both in SR and GR we have only one four-dimensional continuum, space-time, and the only absolute or invariant is the four-dimensional interval ds. If we must still use rods and clocks, we have the awkward situation that we must always use a mathematical combination of their readings to get an invariant, and it would certainly be more desirable if we could measure this invariant directly.

Bohr and Rosenfeld have put forward the principle that every proper theory should define in and by itself the means to measure the quantities with which that theory deals.[1] This principle seems not only correct and relevant, but should be extended further. Such general concepts as mass, energy, and motion, which pervade the whole of physics, can appear quite differently in different physical theories. Any attempt to require 'absolute' definitions of them which do not take account of these differences may lead to extremely awkward formulations, blind alleys, and even inconsistencies. We will see below that this problem does arise within GR in the case of motion. The greatest safeguard against such a difficulty lies in the construction of an adequate model for the theory. One of the primary functions of such a model is to provide what I have called "conceptual coherence," to show how the concepts, formalisms, and measuring devices for the theory can add up to a unified and self-sufficient world-picture.

Einstein was clearly bothered by the same problem. Rods and clocks seem to have nothing to do with the four-dimensional tensor formalism of the theory, and their existence must be a quite separate postulate. Furthermore, their existence raises the objection mentioned above: it sets up a dichotomy between ordinary physical objects and the rods and clocks which have the special epistemological function of being standards, in opposition to the unificatory spirit of GR. Einstein considered the postulation of special rods and clocks to be an undesirable makeshift, but since he could find no better substitute to perform the vital function of measurement, he reluctantly included them in his ontology for GR. He writes in his Autobiography,[2]

One is struck by the fact that the theory introduces two kinds of physical things, i.e., (1) measuring rods and clocks, (2) all other things, e.g. the electromagnetic field, the material point, etc. This, in a certain sense, is inconsistent; strictly speaking measuring rods and clocks [qua physical objects] would have to be represented as solutions of the basic equations (objects consisting of moving atomic configurations), not as it were, as theoretically self-sufficient entities. However, the procedure justifies itself because it was clear from the very beginning that the postulates of the theory are not strong enough to deduce from them sufficiently complete equations for physical events entirely free from arbitrariness, in order to base upon such a foundation a theory of measuring rods and clocks,

and in replying to objections, he continues,[3]

For the construction of the present theory of relativity the following is essential:

(1) Physical things are described by continuous functions, field-variables of 4 coordinates. As long as the topological connection is preserved, these latter can be freely chosen.

(2) The field-variables are tensor components; among the tensors is a symmetrical tensor g_{ik} for the description of the gravitational field.

(3) There are physical objects, which (in the macroscopic field) measure the invariant ds.

If (1) and (2) are accepted, (3) is plausible, but not necessary. The construction of the mathematical theory rests exclusively upon (1) and (2).

A *complete* theory of physics as a totality, in accordance with (1) and (2) does not yet exist. If it did exist, there would be no room for the supposition (3). For the objects used as tools of measurement do not lead an independent existence alongside of the objects implicated by the field equations.

The discovery of a way of measuring the four-dimensional interval directly without using anything more than the entities included in the proper model of GR was first achieved by Marzke in 1959.[4] Since it was necessary to find some other way of making the measurements required by GR before that time, I will first consider what was involved in these other attempts. This will also have the value of showing just where and how ordinary physical rods and clocks night be introduced into the theory.

Since a 'rod' is entirely spatial and a 'clock' entirely temporal, we must first split up the space-time continuum into space and time parts before these can be used. It is a very important fact about our world that its four dimensions are divided into three spatial and one temporal. Mathematically, this appears as follows: for any given g_{ik} and coordinate system x^i we can introduce a new system X^i in which (at least locally) the metric tensor is diagonal, so that $ds^2 = g'_{ii}(dX^i)^2$. When this is done, it is always the case that one of these four g'_{ii} will have a sign different from that of the other three. We can call the component g'_{00} and specify that it is to be negative, while the others, g'_{11}, g'_{22}, and g'_{33} are all positive. This means that ds^2 will be positive if the interval is spacelike, and negative if it is timelike. The actual choice of index numbers and signs is a matter of convention, and some writers have made a different choice from that given above, which will be used henceforth; but the important thing is that the metric tensor must always break down in this way if it is to correspond to any possible *physical* space. If all the components had the same sign, or if they were divided two and two, we would have a mathematically possible construction, but not one that could be a physically possible space-time. For this reason it is improper to say that either SR or GR removes the distinction between space and time, despite their principles of special (Lorentz) and general covariance. The difference in sign always allows us to distinguish between spatial and temporal coordinates. On the other hand, this does not imply that the four-dimensional formalism is an unnecessary artifice. The distinction between space and time is only topological, in the following sense: although we must make a $3 + 1$ division of our coordinates in some way, the exact point of division is not determined and depends on the particular system, subject only to the limits of the light cone, which divides all points in the vicinity of a given point into those with a spacelike and those with a timelike separation. The difference from SR is that the division may be made in a different way at each point in space-time, governed by an indefinite number of parameters, rather than the single one of relative velocity that was the case in SR. Four-dimensional covariance has an additionally valuable unifying effect. In this representation electric and magnetic fields, mass, energy, momentum, and stress are at least no more different from each other than space and time are, and can and must be combined into a single (tensor) quantity when one uses this formalism. Without space-time it is unlikely that this 'latent unity' in our physical concepts would have been discovered, and for this reason Synge considers Minkowski (rather than the Einstein of 1908, who at first did not appreciate the significance of Minkowski's work) to be the true father of GR.[5]

With this background in mind, we may turn to the actual separation of space-time into space and time parts and see how rods and clocks might be used to measure them.[6] First of all, a clock at rest in a given system may be used to measure proper time τ, where $ds^2 = -c^2 d\tau^2$, if s is measured in units of length and τ in units of time. The condition of being at rest implies that $dx^\alpha = 0$ for all α, where α (and Greek indices generally) takes on the values 1,2,3. Then for two infinitesimally separated instants at the same spatial point, the interval $ds^2 = g_{00}(dx^0)^2$. Therefore $d\tau = c^{-1}(-g_{00})^{1/2}dx^0$, and we may integrate over x^0 for finite intervals of this proper time.

For spatial intervals the procedure is more complicated. We must go back to the fact that it is always possible to introduce in the immediate vicinity of a point a coordinate system X^i in which the metric will be Lorentzian. As discussed above, this possibility is bound up with the equivalence principle, and includes as its converse the fact that this mesh cannot generally be extended in any straightforward way for finite intervals. Let us choose X^0 parallel to the original x^0 so that $dX^\alpha/dx^0 = 0$ for all α (freedom of Lorentz covariance). In this system we can define a spatial interval dl by $dl^2 = dX^\alpha dX^\alpha$. If we use the laws for transformation of tensors to go back to the original coordinates x^i, we find that $dl^2 = \gamma_{\alpha\beta}dx^\alpha dx^\beta$, where $\gamma_{\alpha\beta} = (g_{\alpha\beta} - g_{0\alpha}g_{0\beta}/g_{00})$. $\gamma_{\alpha\beta}$ is thus a new three-dimensional metric tensor for the spatial part of the metric, and dl is the length measured by our infinitesimal rigid rod. It is important to note that unless all the $g_{0\alpha} = 0$, in which case the metric has a natural separation into spatial and temporal parts and may be called stationary, dl will *not* measure $g_{\alpha\beta}dx^\alpha dx^\beta$ (which in fact will not even be an invariant). Since the four-dimensional $g_{ik}dx^i dx^k = g'_{ik}dX^i dX^k = dl^2 - (dX^0)^2$, we can find $dX^0 = (-g_{00})^{1/2}dx^0 - (-g_{00})^{-1/2}g_{0\alpha}dx^\alpha$, which shows that X^0 is dependent on x^0 if not on x^α. We can get a similar result by interpreting spatial distance as half the round-trip travel time of a light signal between two infinitesimally separated spatial points, multiplied by c. In any case, the integral of dl will in general depend on the particular world line chosen, and integrals along the possible paths between two points at rest with respect to the metric will give the same result only in the completely static case where all the g^{ik} are independent of time. Insofar as it is necessary to use finite distances (in analyzing planetary orbits, etc.) we must assume that the metric is thus static, at least to the degree of approximation desired.

It is possible to define simultaneity for two points with an infinitesimal spatial separation as follows: we send a signal from B to A which is immediately reflected back to B. Since the distances from A to B in opposite directions are equal, we may say that the reflection at A, which occurs at x^0, is simultaneous with the midpoint of the measured interval from transmission to reception at B. But in general this value will not equal x^0; it will differ by the amount $\Delta x^0 = -g_{0\alpha}dx^\alpha/g_{00}$. Along any open curve, we may define simultaneity between neighboring points by this method. Now if we were to integrate around a closed curve, we would expect the integrated value of Δx^0 to be 0.

Otherwise our procedure would be inconsistent, since two events at the same (spatial) point with different values of x^0 would then be simultaneous. But in general we *will* get a nonzero value, unless the $g_{0\alpha}$ all vanish and the metric is stationary as above. Thus in GR we can not always carry out a consistent synchronization of clocks, even within a single frame.[7] In SR this could always be done, and the only problem was that of comparing two different frames. This difficulty reflects the greater complexity allowed in the geometry, but also the even closer connection of space with time and the necessity for four-dimensional space-time. Even if the spatial part $g_{\alpha\beta}$ of the metric were of the utmost Riemannian complexity, synchronization could always be carried out if space and time were distinct continua, since then the requirement that $g_{0\alpha} = 0$ would automatically be satisfied.

The use of actual rods to measure the infinitesimal spatial separation dl is extremely awkward. In any material used for the purpose rigidity depends on crystalline structure, and we can not proceed to indefinitely small rods without losing this structure. In the case of any actual rod, like the standard platinum-iridium meter bar, its length must be some multiple of the Bohr radius h^2/me^2, a fact which automatically brings in quantum considerations and raises the problem of whether any of these 'constants of nature' might change in time.[8] Thus a 'rigid rod' appears to be essentially a macro-object of considerable complexity. In the course of his excellent study of the operational foundations of GR, Basri has developed an ingenious, though somewhat artificial and complex, procedure for measuring spatial lengths. Instead of using finite bodies, he begins with particles, represented by world lines with no spatial extent, and systematically defines notions of coincidence, rigidity, collinearity, orthogonality, parallel displacements, and spatial geodesics, all of which are needed for a complete theory of measurement for multidimensional spacelike separations.[9]

Basri's approach may be perfectly satisfactory for its purpose, and it does avoid importing unwarranted classical notions into GR, so that it represents an improvement over Reichenbach and Grünbaum. However, it misses a crucial point. For if it is possible to measure all four-dimensional intervals directly and in the same way, with purely spatial and purely temporal intervals merely the special cases $dx^0 = 0$ and $dl = 0$, why should one be concerned to develop a different method for one of these special cases? It might be more convenient, but it can hardly be more basic within the theory. In particular, it would be advantageous if all measurements, even those of *spacelike* four-dimensional intervals, could be treated as time measurements. These could then be measured by some sort of standard clock, and 'clock' is a more general notion than 'rigid body'. Any cyclical process in nature might serve as a clock, and the limitations on such processes are not fully known.[10]

In fact we *can* carry out this program—provided that we are willing to pay the price. For what we must do is to accept as an axiom the SR principle of the constancy of the speed of light in all (locally) inertial systems. This enables us to set up a universal proportionality between distances and travel times of light. If we wish to measure the distance between two particles, whose world lines lie side by side, we may send a light signal from one to the other and have it reflected back to the original. The distance between the two particles (or world lines) may then be *defined* as half the total time between the transmission of the original signal and the reception of the reflected signal. If we are measuring time in seconds and distance in meters, we must then multiply by $c = 3 \times 10^8$ m/sec. (We will discuss the possibility of using the same unit for each below.) Synge has made the possibility of using only time measurements, and giving the dimension of every physical quantity as some power of time, the central feature of his "chronometric" approach to GR.[11] What is lost is the possibility of keeping the constancy of c as a testable hypothesis, rather than merely an assumed convention, and this is what bothers Basri. However, this assumption is essential for the whole four-dimensional approach. Unless there can be some universal way of comparing the magnitudes of space and time measurements, there is no sense in combining them in a single metric. The use of c as a conversion factor is the only method with any empirical support. While it may be challenged in external criticism of GR, it can hardly remain an open question within GR.

A further difficulty now arises. A certain interval of time elapses between the transmission and reception of the signal by the first particle. At exactly *what* time within this interval are the particles taken to have this separation $c\Delta t$? If the particles are at rest with respect to each other throughout this interval (i.e., if their world lines remain parallel) this ambiguity does not matter, but if they are in relative motion we can give no definite answer except by additional conventions. Furthermore, we could not even verify that the world lines were parallel. A series of light signals sent back and forth to test parallelism might give the same value for the distance every time; but during the periods when the signal was not being reflected or transmitted the world lines might gyrate wildly, only returning to their 'proper places' at these instants. Nor would the introduction of rigid rods help us here. We could not know that the points being compared at the opposite ends of the rod were really simultaneous. As we have seen from the discussion above, points with equal values of x^0 may not be simultaneous unless $g_{0\alpha} = 0$.

In addition, the spatial and temporal intervals which we have been considering have all been very small, so that they could function as reasonable approximations to the mathematical infinitesimals needed in Riemannian geometry. However, all the currently recognized experimental tests of GR are cosmological. The intervals which they consider cover the vast reaches of physical space and time, and even the macro-objects suitable for making measurements in our everyday world are inadequate here. We cannot integrate the distance element *dl* without specifying a particular path, and we can integrate only the proper time of an object which is at rest in a given coordinate system (all x^α constant). It would seem impossible to give a definite meaning to cosmological space and time intervals within the framework of GR. For this reason all discussion of possible experiments has been confined to the case of a stationary metric with $g_{0\alpha} = 0$. It is recognized that these are mere approximations, but it is hoped that they will be adequate to the limits of experimental error. Newman and Goldberg have made some progress toward defining large distances in the general case, using a method of 'geodesic deviation' which will be discussed below.[12]

The crux of the problem, as I see it, is again the use of an improper ontology. In classical physics space and time were absolute and distinct, and simultaneity was unambiguous. The basic entities were particles, which occupied only a single spatial point at any given instant, but continued in existence for long stretches (perhaps the whole) of time. Geometry was Euclidean, distances (whether large or small) were related to coordinate differentials by the Pythagorean theorem, and the distance between two particles at a given time was therefore unproblematical. The problems arising from the discovered need for a convention of simultaneity, and the impossibility of communicating with infinite speed had already led, in SR, to an ontology in which the four-dimensional space-time point, corresponding to the intersection of two world lines, was the basic element, and finite spatial extensions were on a par with temporal extensions. These intersections, which might be particle collisions, interference of photons, or interactions between light and particles, are called events. In such a world it is plausible that we should expect a unique result only for the separation of these events, and since this is a space-time separation, we should try to deal with the four-dimensional interval s (or ds) directly.

GR provides us with the following materials for interpreting its postulates: (1) any pair of points separated by a null or lightlike interval can be connected by a light ray; (2) any pair with a timelike separation can be connected by world lines corresponding to possible paths of material particles; one such world line is a geodesic or minimal path; and (3) points with a spacelike separation cannot be connected at all. This means that we can use light rays and test particles (in the sense discussed above) in making our measurements. However, anv assumptions about the crystal structure of extended bodies, the wavelength of spectral lines, the behavior of clock pendluums, or the elastic properties of watch springs, which at present are not explained within GR, are not permitted. Fletcher has pointed out that this avoids Reichenbach's problem about whether we can be sure that space-time will be Riemannian when measured with any of these classical or quantum devices.[13] Weyl had hoped to go a step further and use only light rays, but Lorentz discovered that there were distinct geometries in which light rays would follow the same null tracks but the timelike geodesics would be different.[14]

Marzke's approach is a purely geometrical construction of world lines and their intersections. If it is to have operational significance, one must grant that every world line needed is in fact realized by some particle or light ray, but there is no reason to expect that their supply is importantly limited, given the fact that any such measurement is only an approximation, albeit an arbitrarily good one. Marzke begins with the SR case of flat space-time and then shows how his procedure can be modified to take account of curvature. Given an arbitrary world line as the track of a particle which can transmit, reflect, and receive light signals at any time, we can always construct a nearby parallel. A timelike geodesic and such a parallel to it (which in curved space will not be a geodesic itself), with light bouncing back and forth from one to the other, can then serve as a "geodesic clock." The reference geodesic contains some device for counting the number of "ticks," or successive reflections of the signal, between any two events in its history. As long as the two particles of the geodesic clock remain parallel the ticks will occur at equal intervals.

We can now use our clock as follows: Suppose we wish to measure the interval S between some event A on our reference geodesic AZ and another event B off it. Assume for the time being that S is timelike, and B is future to A. We then just send out signals from our geodesic toward B at definite times, and one of these, leaving at C, will in fact reach B and be reflected back at some later time E. Now in the reference system in which the clock is at rest, B is simultaneous with some event D on AZ. Our clock will measure a time interval T from A to D. Assume that the spatial distance from B to AZ at D is L. Then the square of the four-dimensional separation is given by $S^2 = T^2 - L^2$ if we measure lengths and times by the same unit, either multiplying times in seconds by c or dividing lengths in meters by c. (We will use the former alternative.) The light signal then requires the same time L to go from C to B, and from B to E, according to our basic axiom. Therefore C must be located at time $T - L$ from A and E at $T + L$ from A. But $(T + L)(T - L) = T^2 - L^2 = S^2$! If C is found to be N_1 and E to be N_2 ticks after A, we may then measure S simply by $S = (N_1N_2)^{1/2}$, taking the interval between ticks as our unit. If the separation is null, the light signal can reach B only if C coincides with A; in this case $N_1 = T - L = 0$ and $S = 0$, the desired result. If the separation is spacelike, a signal from AZ sent out at some C *before* A could still reach B and come back at E; if so, $N_1 = T - L < 0$, $S^2 < 0$, and as in special relativity, we may interpret the 'imaginary time' S as a real distance in a system in which A and B are simultaneous. Indeed, B may occur before A in all reference systems; then both N_1 and N_2 must be negative and their product is positive, again giving a timelike interval. Since each world line defines a direction for its proper time, we can always tell whether the number of ticks should be taken as positive or negative. Consequently, as long as we can send light rays in the spatial direction of B indefinitely far into the past and future, and can identify which one actually was reflected at B, we can measure any kind of interval by this method.

Having defined the process of interval measurement in general, we need only to determine a standard unit. Here we have a great advantage. Instead of using objects like rods or naturally periodic processes which must be assumed to maintain their characteristics though time, we may choose *any* pair of well-defined events A and B. Since an event corresponds to a four-dimensional point, we need not worry about possible changes in time. In order to measure the interval between events P and Q, we simply choose a geodesic clock whose reference geodesic passes through both A and P. By the process above we get values N_1 and N_2 for the number of ticks to the transmission and reception points for the light signal to B, and similar values N_3 and N_4 (measured from P) for the signal to Q. It can be readily seen that the ratio of the intervals PQ to AB is $(N_3N_4/N_1N_2)^{1/2}$. Furthermore, we need not keep referring back to AB. Having calibrated PQ, we can use it to measure XY. The ratios cannot change, so the calibration operation must be transitive. This fact is especially useful for the case where AB and XY are themselves separated by a spacelike interval, so that there could not be a physically realizable geodesic connecting A and X. We simply find some PQ sufficiently far in the past or future so that it has a timelike separation from both AB and XY.

If the space is curved, we must use an approximation process. This is not a serious limitation; after all, any kind of space and time measurement involves some sort of approximation and can never demonstrate directly that a distance, say, is an irrational number. If we start with two initially parallel geodesics for our clock (e.g., two freely falling particles) the effect of curvature will be to make them move closer together or further apart, destroying the parallelism required in the theory of the clock. However, at arbitrarily small intervals we can replace the original parallel to our reference geodesic with a new geodesic which will start off parallel in its turn. If the interval from AB to PQ is T (measured along the geodesic from A to P), the average characteristic radius of curvature in the region is R, and the number of divisions (new clocks) is N, the overall proportional discrepancy $\Delta S/S$ will be T^2/NR^2. By making N sufficiently large, we can make this error as small as desired.[15]

The previous discussion indicates that it is not enough just to be able to measure distances, times, and/or four-dimensional intervals. If we are to know how accurate these measurements are, we must know the curvature of space-time in the region under consideration. Furthermore, this information is needed for another reason. We have already suggested that the goal of geometrodynamics is to give a complete interpretation of physical quantities in terms of geometrical ones. But the only possible geometrical properties of space-time, independent of the coordinate system used, are its various curvatures and algebraic combinations of them. From differential geometry we learn that the basic measure of curvature in Riemannian geometry, from which any others can be derived algebraically, is the Riemann curvature tensor R^i_{jkl}. In a general four-dimensional manifold this has 20 independent components when all four indices of the tensor are contravariant or covariant, and our problem will be to determine the value of each of these components at a given point.

One might try the following sort of experiment: suppose we have two test particles which are released from rest at the same height above the earth's surface, but separated from each other. They will fall freely, and their world lines will be geodesics in the earth's gravitational field, initially parallel to each other. They will each fall "straight down" with the same downward acceleration. However, if we assume that the earth's field is uniform and radial, the direction of "straight down" will differ for the two particles; in particular, they will begin to move closer together as they fall. Of course, this mutual acceleration of the particles (quite independent of the gravitational attraction that they may have for each other) will appear very small in comparison with the downward acceleration in a system in which the earth is at rest. (We will see that this smallness is related to the value of c.) However, it is extremely important in determining that we have a real gravitational field present. If the earth's surface were really flat, the separation of the two particles would remain constant. By choosing a coordinate system with uniform upward acceleration, we could transform away the field everywhere and the particles would remain at rest. We could gain no more information by using two particles than by using only one. In the actual situation, the mutual acceleration proves that the earth is round, or at least that its field is not uniform. We could find a system in which either one of the particles would remain at rest, but the other would still move.

This situation is novel in the history of physics, and brings out the operational significance of the truly relativistic feature of GR: the only genuine accelerations, which can serve to test the presence of fields, are relative accelerations. Unless we have at least two test particles, we can get no information about the relevant field variables. In both classical mechanics and electrodynamics (including the modifications from SR!) only one particle at a time was needed. In classical mechanics we might write as a general formula $kF^\alpha = m_i d^2x^\alpha/dt^2$, where m_i is the inertial mass of the given particle, k is the appropriate response parameter (such as passive gravitational mass) and F^α represent the components of the field in question. (Remember that Greek letters run from 1 to 3.) Likewise the tensor representation of the Lorentz force law for electrodynamics is

$$\frac{D}{D\tau}\left(\frac{dx^i}{d\tau}\right) = \frac{q}{m} F^i_j \left(\frac{dx^j}{d\tau}\right). \tag{1}$$

Here the Latin indices run from 0 to 3, q and m are the charge and inertial mass of the single test particle, F^i_j are the components of the electromagnetic field tensor, and the operator D/Du represents the "absolute derivative" according to the formula

$$\frac{DA^i}{Du} = \frac{dA^i}{du} + \Gamma^i_{jk} A^j \frac{dx^k}{du}, \tag{2}$$

where A is any vector, u any parameter, and Γ^i_{jk} are the Christoffel or "bending" coefficients, which do not form a tensor. In a suitably chosen coordinate system the Christoffel coefficients can all be made to vanish, so that the absolute derivative is here just the ordinary derivative.*

In GR we must introduce a new vector η^i, representing the separation of two nearby geodesics (which might in turn be characterized by parameters w and $w + \Delta w$) at the corresponding values of the time parameter x^0 (or t) for each. Thus η^0 is always 0, but the other components may change in time. The equation governing their evolution is

$$\frac{D^2\eta^i}{D\tau^2} + R^i_{jkl}\left(\frac{dx^j}{d\tau}\right)\eta^k\left(\frac{dx^l}{d\tau}\right) = 0. \tag{3}$$

While this equation does contain the separation of the two particles as well as their positions in the general coordinate framework, it does not contain their inertial mass or any other of their individual properties. This again is

* Here, and in general, I will not attempt to derive any of the formulas given. My task is not to repeat mathematics well known to people in the field, but rather to discuss the relative philosophical significance of the various equations. If the reader has never seen these equations before, or does not know how (for example) F^i_j is derived from the field vectors **E** and **H**, I can only refer him to any of the standard texts on relativity, differential geometry, and electromagnetism.

a reflection of the equivalence principle, that only geometric position governs the gravitational behavior of a body, and serves as an added stimulus to the geometrodynamical program of not requiring any nongeometrical properties. $\eta^i(x^0)$ is an observable, measurable by the procedure described above, and so are its time derivatives.[16]

To see how this procedure works out in practice, choose a coordinate system in which the reference particle characterized by w, *from* which the separation vector is drawn, always remains at rest. Then $dx^0/d\tau = 1$, and all the other components vanish. This in turn means that we can only get the components R^i_{0k0}, since the others will be multiplied by zero in our equation. The equation now simplifies to $d^2\eta^i/dt^2 + R^i_{0k0}\eta^k = 0$, since in this system the Christoffel coefficients vanish, and the reference particle feels no apparent force. In particular, choose Cartesian coordinates $x^1 = x$, $x^2 = y$, $x^3 = z$, with z the direction of the vertical. If we have the configuration originally described, with both particles dropped from the same height, we may take the separation to lie along either the x- or y-axis. If we make the assumption that η^3 will then begin and remain 0, or at least be far smaller than the horizontal separation η^1, we may use the approximation $R^1_{010} = -\eta^1_{,00}/\eta^1$. (Here and elsewhere the comma will denote ordinary differentiation with respect to the coordinates following it, here twice with respect to x^0.) If we start with two particles released from rest immediately above each other along the same vertical line, we can be rigorously sure that in such a field all components will remain 0 except η^3. However η^3 will increase since (speaking classically) the lower will be in a stronger apparent field and thus be given a greater apparent acceleration. Therefore, $R^3_{030} = \eta^3_{,00}/\eta^3$. By using a variety of other configurations and seeing how they evolve, we could measure other components for this coordinate system and thereby measure the whole gravitational field within that region. For this degree of approximation, we may use classical calculations rather than direct measurement to estimate these components. Wheeler has shown that for the case in question, $R^1_{010} = R^2_{020} = GM/c^2r^3$, $R^3_{030} = -2GM/c^2r^3$, where M is the mass of the earth and r the distance from its center.[17]

If the above analysis is correct, we see that the typical components of the Riemann curvature tensor, which will henceforth be considered the measure of any real nontransformable gravitational field, vary as the inverse cube of r for isolated central fields. This is the first r-derivative of what had classically been considered the field, and the second r-derivative of its potential. Since we know that the Christoffel coefficients are composed of first derivatives of the metric tensor and the Riemann curvature tensor includes its second derivatives, we may already suspect that the components of the metric tensor will play a role analogous to that of the classical potential, and the Christoffel coefficients to the apparent (coordinate-dependent) gravitational force. We may note that there is a rough analogy between the levels of differentiation. Classically, the potential might be measurable by some instrument whose zero-point was conveniently set, and it was certainly a useful concept. However, the magnitude of the potential had no real significance; at a given point it could always be transformed away to any other value by a mere change of scale (i.e., by the addition of an arbitrary constant of integration). Nevertheless, once the zero-point had been set, the variation of potential over the classical gravitational field had real physical significance. In GR the coordinate system functions somewhat like this scalar constant. It is not determined by the basic differential field equations, though it is needed before any numerical readings can be given. No particular one has any absolute significance, and its effects can be transformed away just as a potential could be set to zero at any point. The question naturally arises whether the Riemann curvature tensor might in turn be considered a mere "potential" for some still more basic entity. However, the Riemann curvature tensor plays such a fundamental role in giving a coordinate-free characterization of the geometrical properties of a Riemannian manifold that it leaves no freely eliminable factors. It does provide the desired invariants, so that further complication is unnecessary for GR as a theory of gravitation. One might be tempted to make this complication only if Riemannian geometry were not sufficiently general for this purpose, and to do so we might have to use the next more complex mathematical entity in the scalar-vector-tensor sequence.

However, it might be desirable to go still further in trying to use GR as the basis for a unified field theory, including electromagnetism. For we will see that information about the covariant first derivatives of the Riemann tensor can enable us to determine the electromagnetic field. If these derivatives could also be measured by means of light rays and the mutual separations of collections of *neutral* particles, we would have an operational basis for such a theory in which we would need no properties of the particles involved—neither their charges or their masses. If we still need to know the values of charges, an asymmetry is introduced which would make a complete unification, with a single "total field," much less plausible.

The method described above is not the only way in which the curvature can be measured. Wigner has shown how the one and only independent component in a simple two-dimensional universe could be calculated. A light signal is bounced back and forth between a clock-carrying particle and a mirror, and the successive times t_1, t_2, t_3 between reflections at the clock are recorded. The curvature is then $2(t_1t_3 - t_2^2)/t_2^4$.[18] Bertotti has given a generalization of this method, as well as an alternative based on the Doppler effect and frequency shifts for light in a gravitational field.[19] Using the notion of Fermi-Walker transport of an "orthonormal tetrad" of vectors along a curve, Synge gives a very general method involving a cloud of particles which need not be moving along geodesics.[20]

What is important here is not the details of any particular method of measurement, but rather the fact that the Riemann curvature tensor can be measured directly, and is not simply a postulated but unobservable entity. If its components, or algebraic combinations of them, are equated to more traditional physical quantities, such as momentum and energy density, the equations make predictions capable of empirical test. The only problem is that the measuring devices in the two cases are conceptually very different. The devices normally used for measuring energy are (at least) not just geodesics and light signals. Therefore, unless we try to make an ontology for a single level in which these kinds of devices can coexist side by side, we must expect to require conventions relating entities on different levels. I will be supporting the latter alternative.

In any case, the further details of measurement are not relevant at this stage in our argument. We have demonstrated that the conceptual resources of classical GR are sufficient to measure the interval, the metric tensor, and the Riemann curvature tensor. In addition, Basri has given an excellent analysis of how we may give operational significance to the particular coordinate systems which are customarily used in discussing important special cases of the field configurations.[21] We may therefore conclude that the requirement of Bohr and Rosenfeld has been satisfied; GR does indeed provide within its framework the means for measuring all quantities which appear essentially within the theory. We are left only with the task of giving a second interpretation of its devices in terms of those at some level which we take to be more obviously observational, such as the level of everyday objects and laboratory equipment.

Having completed our investigation of the general principles on which GR is based, and of the theory of measurement which it includes as a means of giving operational significance to its concepts, we must now examine the specific laws which appear within this general framework. I have suggested above that it would be reasonable to expect, on the basis of our knowledge of other field theories, that these should be of two kinds: (1) laws of motion, and (2) ponderomotive equations, or field equations proper. In all classical theories these laws had been logically independent of each other. Neither could be derived from the other, and without knowledge of both one could not apply the theory to any concrete physical situation. In GR they are no longer independent: it turns out that the field equations allow only one possible set of motions. However, this result is by no means obvious: it took years to derive it for a series of special cases, as we shall see below. Einstein did in fact treat the two problems as separate, and came up with distinct arguments and solutions for each one.[1] It is therefore a tribute to his genius that the law of motion laboriously derived from the field equations did indeed agree exactly with the law postulated earlier.

In discussing the problem of motion within the context of any theory, we must ask: (1) What sorts of things do the moving? (2) What is the arena or background against which they move? (3) What constitutes a possible motion against this background? Only when the answers to these questions are clear can we consider which of these possible motions are permitted. In GR we may say rather glibly that test particles and light signals are the things that move, but what does GR allow as a possible test particle? As with other theories, we might start with a negative definition—a test particle is any material body which generates a field negligibly small in comparison to the other fields in the region of interest. This definition is satisfactory for most applications, since these depend on approximations anyway. However, it is still far too vague to shed light on the peculiar character of GR. Before we can even consider deriving equations of motion from the field equations, we must further analyze "test particle" as a singular region of lower dimensionality than the manifold in which it is located, or as an extended body with fully specified internal hydrodynamical structure, or as a particular geometrical configuration of the manifold itself. We can allow "test particle" as an undefined primitive term only if we are willing to give up all hope of deriving the laws of motion from something else, as opposed to introducing them as an independent axiom. Furthermore, this deeper analysis must be in terms of the other concepts of the theory, rather than some extratheoretic notion like "absolute observable."

The arena is a particular four-dimensional Riemannian manifold. This is not the absolute three-dimensional Euclidean space of Newtonian physics,

11 The Law of Motion and the Field Law

nor is it Minkowski's four-dimensional flat space-time; but neither is it arbitrarily chosen. It is uniquely determined by the distribution of sources and the appropriate boundary conditions, according to the field equations. Its geometry is characterized by the Riemann curvature tensor, or more conveniently, once a particular coordinate system has been chosen, by the metric tensor. Possible motions are then represented by curves in this manifold, with the separation of points on the curve always null ($ds = 0$) for light rays and timelike for test particles. Even if spatial position does not change in the chosen coordinate system, we still have motion according to our definition as long as there is a lapse of time. From the standpoint of a rigorous empiricist or operationist, we might regard the world as the totality of these rays and particles, so that the only real, or physical, points occur at their intersections. On this interpretation a motion is a history or sequence of events (intersections) for a particular ray or particle, and the arena is the totality of such histories. It will be more convenient for the development of the theory to assume that the background manifold is continuous, with interphenomena (in Reichenbach's sense) similar to any observable phenomena.

In describing the Marzke method of measurement in the previous chapter, we required that we should have at our disposal an indefinitely large supply of timelike geodesics, so that any pair of points with a timelike separation could be connected by one of them. In addition, we assumed that these were given by the trajectories of free particles, i.e., particles not subject to any nongravitational forces. This in turn presupposed both an existential assumption, that free test particles do exist in large supply, and a solution to one case of the problem of motion. The assumed law is this: in the absence of electromagnetic or otherwise nongravitational forces, the trajectories of free test particles will be timelike geodesics in the given Riemannian manifold. From the class of possible timelike world lines, the geodesics are singled out as the class of physically realizable motions here.

What justifies us in making these identifications? In the first place, it is a mathematical fact, drawn from differential geometry, that in a general Riemannian manifold we can always determine a class of geodesic lines. This class has immediate geometric significance, quite aside from any possible physical interpretations, and it is the only class of lines which can be characterized in such a straightforward way. It is even possible to define geodesics in non-Riemannian manifolds for which there is no metric tensor, as long as we can use the so-called affine connection.[2] Since classical GR is committed to the use of Riemannian geometry, this generalization is relevant only to possible extensions of the theory.

On the other hand, we know from the Eötvös experiment that all particles respond in the same way to a given gravitational field, and move along the same trajectories regardless of their mass or composition. In other words, a class of paths in space-time is *physically* singled out as the trajectories of any particle in free fall. These paths are real and observable, quite independent of any geometry in which they may be represented. What GR does is to identify these physical paths with geodesics. But the class of geodesics, on the contrary, does depend entirely on the choice of geometry. Therefore, what the geodesic hypothesis does is to select out of the infinitely many possible geometric manifolds that particular one whose geodesics coincide exactly with the world lines of freely falling test particles, and to proclaim this as the real geometry of the space-time region.

Now the conventionalists are right in saying that we are in no sense *required* to make this choice of geometry. However, suppose we decide to choose some other Riemannian geometry (including Minkowski's "pseudo-Euclidean" geometry as a special case). In such a geometry a class of geodesics can still be defined, but we may still ask, what in the physical world corresponds to this class? If there are no readily identifiable motions that take place along these geodesics, it is not clear how we can determine them experimentally or use them in our physics. This situation should certainly be unsatisfactory for an empiricist. On the other hand, we must introduce considerable mathematical complications in discussing free fall in this geometry. By making the identification of geodesics with free particle trajectories we allow a simpler mathematics and a greater empirical testability for its concepts.

Nevertheless, the foregoing considerations would avail us but little unless we could determine when a particle was indeed free in the above sense. Fortunately, it is relatively simple to isolate a body from the influence of other forces, for the very reason that the latter are not universal and thus affect different bodies in different ways, unlike gravitation. Electromagnetic fields, for example, can be made to vanish in a given region by a suitable arrangement of conductors, insulators, etc., and in any case they have no effect on uncharged particles. All we need to do is to drop pairs of such neutral particles in a sufficiently good vacuum, and they will move along the desired paths. This is far easier than any attempt to identify "really inertial" systems, and no other known physical laws allow us to determine special paths in space-time on such a large scale. What we have done is to achieve a geometrization of the physics here by identifying the most prominent class of geometric lines with the most prominent class of physical trajectories, and as long as this identification can be used consistently with the rest of the theory, it is certainly the simplest route to our goal.

The role of the Eötvös result cannot be overstressed here. We can certainly imagine a world in which there were no universal forces, so that gravitational accelerations would depend on the properties of particular test particles. In such a world there would be no universal physical paths. We would have to single out a single particle, or perhaps a small collection of similar particles, and declare arbitrarily that its paths under appropriate conditions were to count as the geodesics of the geometry. There would be no physical justification for any particular choice, and under the circumstances it would probably be more reasonable to stick with Euclidean geometry, whether its straight lines were ever realized or not. Thus Reichenbach, Grünbaum, and their disciples are correct in arguing against the conventionalists that once a standard has been chosen, the geometry is determined.[3] However, they completely miss the point by speaking as if the standard were some *particular* kind of rods and clocks. The crucial fact is rather that *any* body will do. Furthermore, it is not any assumed metrical properties of the standard (such as invariance under transport) that are relevant, but simply its path in free fall. Once we can trace all these paths we have all the information we need to determine the geometry. We have not yet *measured* the intervals and curvatures, to be sure; but measurement is a separate operation. In measurement we *discover* the metrical features of the already determined geometry; whereas on Reichenbach's theory, we *invent* a geometry as an attempt to make the results of our various measurements consistent. The mistake of Reichenbach et al. is to make their method too general, so that it does not assume any particular physical laws and would be effective in a possible world far more chaotic than our own. But why should we suppress empirical information that seems very well established? The *method* of GR does indeed assume the null result of the Eötvös experiment, and develops from it. Since this also enables us to dispense with conventions about particular standards, we may conclude that the determination of the geometry, and the solution of the problem of motion in GR, depend less on conventions and more on facts than Reichenbach would lead us to believe.

Despite these considerations, one might still ask whether it is possible to construct a plausible physical theory in which these special paths were not geodesics. Dicke has shown that this can in fact be done if we introduce a second tensor field h_{ik} comparable to the metric tensor g_{ik}, but with no geometrical significance of its own.[4] However, if the effect of the h_{ik} is governed by a universal response parameter of its own (as required by the Eötvös result) we can always add it, perhaps with an additional multiplicative scalar factor, to the original g_{ik}. The result will be a new tensor g^{*}_{ik}, which may be interpreted as the metric tensor for a new space with a different geometry. In this new geometry the special paths will turn out to be geodesics, which they were not in the geometry described by g_{ik}. Much of Whitehead's theory of relativity might be interpreted in this way, with the

g_{ik} the fixed background Euclidean (or uniformly curved) space, which affects all bodies equally (i.e., not at all). The h_{ik} then function much like Whitehead's J-field, or "potential mass impetus."[5] However, this separation would seem to add very little to what we might learn by using the g^*_{ik} from the start.

Since a geodesic is defined as the shortest (or more generally, the extremal) path between two points in a manifold, we can use the calculus of variations to give its equations. Since the end points of the path are located at different times, it might seem that the variational procedure requires that a future event (the destination) helps to determine a present event (the current speed and direction of the particle). One might be tempted to ask, How does the particle know in advance where in space-time it is going? However, this is a misinterpretation. From the original variational principle we can derive a differential equation which will provide a local determination of the motion in terms of current data. However, this equation must still be supplemented by initial and boundary conditions. Suppose we have a particle at the four-dimensional point A. If we also know its present velocity, we can discover by calculation that its trajectory will pass through point B. On the other hand, if we knew that it would pass through B, we could work backward and calculate what its initial velocity must be. In neither case is there some special causal influence; rather, there is a symmetry between already known and yet-to-be-calculated information. For making predictions of future events, we will normally know the initial velocity and use the differential equation. However, if we believe that extremal principles like the geodesic hypothesis reflect a pervasive and important feature of Nature, we may use this to provide a deeper explanation of why a motion between A and B already observed was the only possible one.

The general variational principle for geodesics may be written in the form $\delta\int ds = 0$. This is a special case of the equation $\delta\int L(x^i,\dot{x}^i,q)dq = 0$, where q is some parameter to measure location along the curves from which the extremal is to be selected, and $\dot{x}^i = dx^i/dq$. The Euler-Lagrange relations that result from this are $\frac{d}{dq}\left(\frac{\partial L}{\partial \dot{x}^i}\right) - \frac{\partial L}{\partial x^i} = 0$. In the case at hand we have $L = (g_{ik}d\dot{x}^i d\dot{x}^k)^{1/2}$. It turns out that we can without loss of generality choose the path length s or proper time τ as the parameter q. Then $\dot{x}^i = dx^i/d\tau = u^i$, the components of the unit vector for the four-velocity, which is always tangent to the path. In this case $\delta\int(g_{ik}u^i u^k)^{1/2}d\tau = 0$, and since the integrand always has the value 1, the differential equation turns out to be $d(g_{ik}u^k)/d\tau - \frac{1}{2}g_{jk,i}u^j u^k = 0$, where the comma stands for ordinary (i.e., not covariant) differentiation with respect to x^i. If we introduce the Christoffel coefficients

$\Gamma^i_{jk} = g^{il}\frac{1}{2}(g_{jl,k} + g_{kl,j} - g_{jk,l})$, we may finally rewrite the equation of motion for free test particles in the form $\frac{d^2x^i}{d\tau^2} + \Gamma^i_{jk}\frac{dx^j}{d\tau}\frac{dx^k}{d\tau} = 0$. The first term is simply an ordinary acceleration with respect to the proper time of the particle. If we interpret the equivalence principle as asserting that every acceleration of a neutral test particle can be compensated by a gravitational force, we may interpret $\Gamma^i_{jk}u^ju^k$ as the appropriate "gravitational force." Since the Christoffel coefficients do not form a tensor, they can all be made to vanish at a point in some coordinate system. But our law of motion then tells us that in such a system the particle will have *no* acceleration. Thus we are able to give mathematical expression to the fact that any apparent acceleration for a single particle (as opposed to the mutual accelerations discussed above in connection with curvature) can be transformed away.

In fact, this differential equation for the law holds in more general non-Riemannian spaces. It is simply a special case of the general law of vector transplantation under parallel displacement in an affine space, applied to the tangent vector u^i or any multiple of it. However, under these circumstances Γ^i_{jk} is a more general connection than the Christoffel coefficients, and need not be symmetric in the lower indices. It cannot be derived from the metric tensor, but must be assumed as an irreducible entity. Although it is not a tensor, it cannot be made to vanish if it is antisymmetric, and if this is true, the law of motion demands nonzero accelerations.[7] Still more simply, we might argue—as Synge does—that this equation is the most natural generalization of Newton's law of inertial motion. In Newtonian mechanics the ordinary derivative of the velocity is zero in inertial motion; in GR the absolute derivative $Du^i/D\tau$ should therefore be zero. When worked out, this analogical approach gives the same law of motion.[8] However, Newton was thinking only in terms of a three-dimensional velocity, and the extension to the four-dimensional quantity u^i is risky at best. Furthermore, Synge's argument robs us of any deeper understanding of Newton's own law. It reduces the GR law to the level of a brute, inexplicable fact, rather than illuminating Newton's law as a special case of the more general geodesic principle. Nevertheless, this approach will be valuable when we do come to introduce nongravitational forces, since in this case the particles do not move along geodesics, and it would seem reasonable to set the product of the inertial mass and the absolute derivative of u^i equal to the tensor representation of the appropriate force.

At least in classical physics, the only nongravitational force which could be expected to exert a measurable effect on a single test particle was the electromagnetic force. We can argue by analogy from its covariant representation in SR that the law of motion for charged particles is

$$\frac{d^2x^i}{d\tau^2} + \Gamma^i_{jk}\frac{dx^j}{d\tau}\frac{dx^k}{d\tau} = \frac{e}{m}F^i_k\frac{dx^k}{d\tau},$$

where e is the charge of the test particle, m its inertial mass, and F^i_k is the tensor for the electromagnetic field, derived from the first derivatives of the vector four-potential A_i. One can prove directly that this expression is indeed covariant.[9] Alternatively, one can modify the variational principle described above by writing $I = \int ds - (e/m)\int A_i dx^i$ as the integral action whose variation is to be set equal to 0. This leads to the same equation.[10] Finally, Chase has shown that this law can be derived directly from the field equations, by a procedure similar to that for uncharged particles, which will be discussed below.[11] In any case the term on the right side of the displayed equation represents the deviation from the "natural" geodesic path caused by a force field which may be considered real since it affects bodies in different ways and cannot be transformed away, even locally, by purely geometric means.

We have already assumed that GR is to be evaluated in the first place as a reconstruction of *classical* physics, in the sense that it should not be required to account directly for quantum phenomena. However, quantum theory has in many cases given us a deeper understanding of accepted classical principles, which remain valid to the limits of the correspondence principle, as well as predicting new and unsuspected effects. If it can do the same for GR, there is no reason why this assistance should be refused. Wheeler has argued that one great accomplishment of quantum mechanics is that it explains why extremal principles should be expected as important physical laws, rather than simply taken as brute facts or "justified" by appeals to esthetic considerations or principles of sufficient reason. If this is true, any classical theory should be strengthened if its basic laws can be shown to be derivable from extremal principles, and any attempt to try to unify physics should give a high status to them. Following Feynman, Wheeler argues that we do not need to assume a priori that geodesics or any other paths should have a preferred status, but only that each possible path has some quantum-mechanical probability associated with it. In Feynman's theory, we may even take these probability amplitudes as equal (perhaps a more general use of sufficient reason!), as long as their phases are proportional to the action integral for the path. If we then take a sum of histories over all these possible paths, this phase difference will cancel by destructive interference the contributions from nonextremal paths, and we will get the same result as if we had assumed from the start that the path *must* be extremal.[12] This fact does not make the geodesic hypothesis less axiomatic or more derivative *within* GR; however, it does give it greater inherent plausibility within the larger context of physics.

Having considered the relativistic account of motion in a given field, we must turn to the second main problem of dynamics as mentioned at the beginning of the chapter: given a distribution of sources of a gravitational field, how can we determine the value of the field variables at all points in the space-time region under consideration? Just as before, we must break this general problem down into the following subquestions: (1) What sorts of things are we to take as sources, and what is their proper mathematical representation in the language of GR? (2) What mathematical entities are to be interpreted as describing the gravitational field, and how are they related to the source variables? (3) How are we to characterize, interpret, and evaluate the significance of the resulting field law of gravitation? Such a law must take the place of Newton's classical inverse square law of gravitation within GR.

Let us recall that Newton's theory assumed that the only source of gravitation was the scalar invariant known as mass and (in its basic mathematical form) that this was located in discrete particles whose contributions could be simply added. The form $F = Gm/r^2$ is not a differential equation, and it assumes action at a distance, rather than spatio-temporal contiguity of cause and effect. However, such an inverse-square law may be considered an integral of Poisson's equation $\nabla^2\phi = \rho$, where ρ is the density of this mass (or charge, in the case of electrostatics) and ϕ is the scalar potential whose gradient is the field $\mathbf{F}$. This is one of the simplest and most basic partial differential equations, especially in those regions where ρ vanishes and it reduces to Laplace's equation $\nabla^2\phi = 0$. Since GR is supposed to be essentially a field theory of gravitation, we must look to Poisson's equation rather than Newton's as our guide in searching for suitable analogies. Furthermore, we cannot even assume that an integral form comparable to Newton's exists.

By the principle of general covariance, GR is committed to using four-dimensional tensors, and any representation of sources must have tensor character. However, this still allows scalars and four-vectors as possible special cases. But we cannot use classical mass density as the only source variable. Even in SR, mass is not an invariant, since it is convertible with energy, and the total (rest plus kinetic) energy is only one component of a vector, with momentum providing the other three. Now if our only potential sources were isolated particles, it might be tempting to try to use this four-vector as our source variable. Nevertheless, even this is not enough. For in a continuum theory we explicitly wish to treat continuous distributions of matter, in which there may be a variety of internal pressures and shearing stresses. In the hydrodynamics of continuous media, these stresses are represented by a three-dimensional 2nd-order tensor. But here again we can achieve covariance only by combining this with the energy and momentum in a four-dimensional tensor. Its components may be interpreted as follows (once again, Greek indices run from 1 to 3, Latin indices from 0 to 3): T^{00} = density of total energy; $T^{0\alpha} = T^{\alpha 0}$ = momentum density: $T^{\alpha\alpha}$ = hydrostatic pressure in the spatial directions; $T^{\alpha\beta}$ = shearing stresses, where $\alpha \neq \beta$. Unless we choose a coordinate system in which the source is at rest, however, the density and pressure will appear in all the components, as in the equation for a fluid without shear,

$$T^{ik} = \rho_0 u^i u^k + \frac{p(u^i u^k - g^{ik})}{c^2},$$

where ρ_0 is the classical rest density, p the pressure, and $u^i = dx^i/ds$. From the electromagnetic field tensor it is also possible to write down a T^{ik} representing electromagnetic field stresses, energies, and the Poynting vector of field momentum.[13]

This determination of the T^{ik} also holds true for special relativity, where we describe only the inertial behavior of matter, without reference to the source of gravitation. However, we must bear the equivalence principle in mind. For *if* the equivalence principle is valid, and *if* we must introduce all these state variables in order to give a covariant representation of the inertial behavior of a material system, then we also need all of them to describe its *gravitational* effects. Specifically, the principle requires that the total T^{ik}, including contributions from incoherent collections of particles (dust), continuous fluids, and radiation, must be identical with the source of the gravitational field. It might indeed be possible to find some new scalar that satisfies covariance (such as T, the tensor formed by contracting T^i_k) and develop a scalar theory of gravitation using it as a source term, but this would not satisfy the equivalence principle. Dicke has argued that the weak form of equivalence, while it does require at least one tensor field, is still compatible with a combination of a tensor and a scalar field.[14] But rightly or wrongly, GR does accept the strong principle, and from the standpoint of simplicity it is desirable to see how far we can get with a single tensor theory of gravitation before considering such possible generalizations. The use of T^{ik} as the source term is also advantageous if our goal is to describe gravitation in wholly geometrical terms. For all the tensors used to characterize the geometry can be derived from the metric tensor, which is also a symmetric, 2nd-order, four-dimensional tensor. Just as Poisson's equation related one scalar (ϕ) to another (ρ) by means of the Laplacian operator, it would be desirable to find a similar operator relating the mathematically homogeneous g_{ik} and T_{ik}. Nevertheless, it should be mentioned that there is one considerable disadvantage for application to concrete situations. For if we deal with extended bodies, we must have a complete knowledge of the internal hydrodynamical state of the body, whereas in Newtonian theory it was enough to know its mass density at all points in order to calculate its gravitational effects.

At this point all that we can conclude about the field equation is that the matter tensor T_{ik} should appear as the right-hand side, while the left should be some combination of the g_{ik} and their various derivatives. To be sure, such a combination must also be a 2nd-order tensor; but without further restrictions there are still indefinitely many possibilities. In developing the integral form of his law, Newton at least could be guided by observations, but since our variables are not straightforward observables like classical mass and force, we do not have this recourse. Yet two guides remain: a principle of simplicity, and the very high accuracy of the Newtonian solution, which can then provide a comparison standard for the limiting case. Einstein's beliefs concerning the ontological simplicity (and thus ultimate intelligibility) of the world are well known. Furthermore, he regarded 'naturalness' or 'logical simplicity', as a means of limiting the number of possibilities allowed, and as an essential part of the 'inner perfection' of a theory.[15] With respect to GR, he wrote,[16]

Equations of such complexity as are the equations of the gravitational field can be found only through the discovery of a logically simple mathematical condition which determines the equations completely or (at least) almost completely. Once one has those sufficiently strong formal conditions, one requires only little knowledge of the facts for the setting up of a theory.

But what does this simplicity really amount to here? What are these "sufficiently strong formal conditions"?

I have argued above that simplicity is always a context-dependent notion, which gains concrete significance only against a background of definite assumptions. In this case the appeal is to analogies with the classical law that provides such a background. If we look at the geodesic law of motion, we note that the Christoffel symbols are dimensionally equivalent and formally analogous to the inertial and gravitational force fields in classical theory. Furthermore, they are simply additive combinations of the first derivatives of the metric tensor. Since the force fields were all first derivatives of the scalar potential ϕ, the analogy suggests that the ten g_{ik} should be regarded as the analogs of potential. The fact that the Christoffel symbols are not tensors and thus can be made to vanish in a suitable coordinate system is a mathematical representation of the fact that gravitational forces (as accelerating single particles, rather than pairs) can be similarly transformed away. But now we note that Poisson's equation involves only the second derivatives of ϕ, and is linear in them. This led Einstein to two of the formal conditions for the law: it must be a partial differential equation for the g_{ik} that (1) contains no derivatives higher than the second, and (2) is linear in these second derivatives.[17] One might like to add, again following Poisson, that it contain no first derivatives either and be linear throughout, but this might prove too restrictive: there is no assurance that a tensor could be found satisfying all three of these requirements, and unless the resulting expression had tensor character, it could not be equated to the T_{ik}. Again appealing to our general knowledge of physics, there are no known cases where third (or higher) order differential equations are required in basic laws, nor do we have the conceptual resources for interpreting them, but nonlinear equations and combinations of first and second derivatives are known to occur.

Remarkably enough, it turns out that these two requirements do limit the range of possibilities tremendously. In fact, the most general expression compatible with them is simply

$$c_1R_{ik} + c_2Rg_{ik} + c_3g_{ik} = T_{ik},$$

where the c's are constants, $R_{ik} = R^j_{ijk}$, the Ricci or contracted Riemann tensor, and R is the curvature invariant formed by contracting the Ricci tensor.[18] If there were no further requirements, and we were simply trying out possible laws in order of decreasing simplicity, it would be reasonable to try using only the first term on the left-hand side.

However, there is another important requirement, which perhaps deserves logical priority over the first two. For T_{ik} is not an arbitrary tensor. On the contrary, it is supposed to provide a complete representation of the properties of matter, and among these is the vitally important fact that it obeys conservation laws. Again following SR, the law of conservation of mass-energy and momentum may be expressed in differential form as $T^{0k}{}_{,k} = 0$, where the comma denotes ordinary differentiation, and the law of continuity in hydrodynamics as $T^{\alpha k}{}_{,k} = 0$.* Putting these together gives the single law $T^{ik}{}_{,k} = 0$. In generalizing to arbitrary Riemann spaces, as required by GR, we get $T^{ik}{}_{;k} = 0$, using the semicolon for covariant differentiation. We can only reproduce this property of matter if we choose a combination on the left side whose covariant divergence vanishes identically at all points. This requirement, recognized by Einstein, means that we must use $R_{ik} - \frac{1}{2}Rg_{ik}$, the Einstein tensor. The third term has vanishing covariant divergence anyway, and if we are concerned with simplicity, we can drop it until and unless we know that it is required.[19] For some time Einstein did believe that it was

* This of course is not the usual way of writing this law, but when currents and densities are interpreted as components of the T^{ik}, this expression is equivalent to the usual formulation.

needed, and that its constant had significance on the cosmological scale, but eventually found that he could drop it without loss of generality.[20] Eddington has shown that there are indeed other geometrical tensors with vanishing covariant divergence, as long as we are willing to drop requirement (1) and allow higher-order differential equations.[21] Furthermore, he suggests that only an unwarranted bias prevents us from doing so. However, the bias does seem warranted by both conceptual and mathematical simplicity, and the fact that accepting (1) leads to the *unique* solution $R_{ik} - \frac{1}{2}Rg_{ik} = \kappa T_{ik}$, seems all the more reason for using it, at least as a first attempt. The constant κ can also be evaluated by comparison with the Newtonian result, and found to equal $8\pi G/c^4$, or simply 8π, if a system of units is chosen in which G and c can be set equal to 1.

The requirement that both g_{ik} and T_{ik} should be symmetrical means that each can have at most ten distinct components, and our field law above may appear as ten independent partial differential equations, to be solved simultaneously for the g_{ik}. However, this is not correct. Since the tensor divergence of each side must vanish identically, we already have four constraint equations and thus only six independent equations left. This means that the system is underdetermined, in that we cannot get unique solutions for the g's. However, as Hilbert first recognized,[22] this is really an advantage from the standpoint of covariance. If our solution specified the ten g_{ik} uniquely, it would also have to specify the coordinate system, since the same g's and different coordinates would mean different underlying geometries. Yet covariance forbids us to single out any particular coordinate system in this way as having physical significance. In fact the remaining six equations are enough to specify the geometry, while giving us just the amount of freedom needed to choose in any arbitrary way that we like the four coordinates in which this geometric structure is to be described. In other words, the conservation laws for matter and the covariance principle are intimately related. It is worth noting that there is a somewhat similar situation in classical mechanics, where the conservation laws may be expressed as symmetry principles reflecting the assumed homogeneity and isotropy of space and time, with no individual points or directions being given physical significance.[23]

We can make a further important transformation of the field law. It turns out that $R = \kappa T$, i.e., the four-dimensional curvature invariant R is proportional to the invariant T formed by contracting the matter-energy tensor. Because of this we can write $R_{ik} = \kappa(T_{ik} + \frac{1}{2}Tg_{ik})$ in complete symmetry with the original field equation. But from this fact we see that if all components of T_{ik} vanish in a particular region, so that there is neither ponderable matter or energy-carrying fields there, we can write $R_{ik} = 0$, rather than $G_{ik} = R_{ik} - \frac{1}{2}Rg_{ik} = 0$. Since the Ricci tensor R_{ik} is easier to handle mathematically than the Einstein tensor G_{ik}, and may have a more straightforward conceptual interpretation, there are definite advantages in using this as our field law in all such empty regions. *

In Laplace's equation the vanishing of ρ did not mean that ϕ also vanished throughout a region. This was only one of many possible solutions, and the actual one depended on the boundary conditions. Likewise the vanishing of the Ricci tensor does not signify that space-time is flat. The Ricci tensor is indeed a kind of measure of curvature, with geometrical significance, but the proper overall measure of curvature—the proper means of characterizing underlying geometrical structure—is still the full Riemann tensor. Only when $R^i_{jkl} = 0$ is the space truly flat or Euclidean, and in general the Riemann tensor will not vanish when the Ricci does. In effect, this divides the 20 components of the Riemann tensor into two groups of 10 apiece. The 10 carried in the Ricci (or Einstein) tensors describe purely local effects, resulting from the existence or nonexistence of matter at the point in question; the other ten are global, reflecting effects from different regions and boundary conditions, even on a cosmological scale.

Thus the homogeneous or Laplace-like equation $R_{ik} = 0$ is indeed useful as well as relatively tractable. In particular, it enables us to avoid the problem of requiring a full description of the hydrodynamical state of an extended body. If the sources could be considered mass-*points*, then $R_{ik} = 0$ might be presumed valid at *all* points in space-time except for singularities corresponding to the world lines of these mass-points. Conceptually, this is precisely the same situation as was assumed in the original, integral form of Newton's law. Although this approach has proved adequate for analyzing the gravitational effects of the sun and planets, we will see below that it runs into difficulties and cannot be used as the basis of a general solution, although Newton's approach could be so used.

If we must return to the inhomogeneous equation in the presence of matter, we must come to grips with the problem of interpreting the nature of the relation between the terms on the left- and right-hand sides of the equal sign in the basic field equation. Only through understanding this can we appreciate both the physical and the conceptual content of GR. In general, there are three main kinds of interpretation.

*In the absence of matter, we can set the left-hand side of the equation equal to zero, rather than to T_{ik}. In a careful analysis, Levinson and Zeisler have shown that this gives us more freedom. Instead of the 2nd order tensor G_{ik}, we could also use a 0th, 4th, or 6th-order tensor, or some combination of them. However, the other restrictions mentioned still allow only *eight* possible 'laws of gravitation' for free space, with Einstein's own $G_{ik} = 0$ the only one using only 2nd-order tensors. For details, see Horace C. Levinson and Ernest B. Zeisler, *The Law of Gravitation in Relativity*, ch. 6, pp. 70–87, University of Chicago Press, Chicago, 1931.

1. One might argue that all the talk about "real curvatures" and "underlying geometrical structures" is simply a *façon de parler*. The only physically significant quantities are those incorporated in the matter tensor T_{ik}. Our only legitimate goal is to proceed from a given distribution of sources to accurate predictions of the motions of test particles. Mathematically, it is convenient to use such 'intervening variables' as the g_{ik} and the various quantities derived from them. But the field equation is meaningless except when conjoined with the equation of motion. By itself, it tells us only how to introduce the right mathematical tools. This is the line that would probably be taken by most phenomenalists and positivists. This approach had some justification in classical physics, where one could indeed go from masses to masses or from charges to charges without ever referring to fields and potentials. However, this depended on the existence of an exact and convenient integral form of the field equations. Thus, any advocate of this position should attach highest priority for finding an integral form of the GR equation. As yet this has not been done, and if there is really no way to eliminate the geometrical terms, only blind prejudice, philosophic dogma, or hope can prevent one from giving them further significance. This interpretation is also that of Mach. Although he remained skeptical (largely on methodological grounds) of the whole approach of GR, he saw that such an elimination would be far more significant than in classical physics. Since the classical fields and potentials had nothing to do with space-time structure, the latter had to be presupposed as an independent and absolute element. Now since the g_{ik} and their derivatives do fulfill all the roles formerly ascribed to classical space and time, a successful elimination of them would exorcise these absolutes forever. Space-time could indeed be regarded as nothing but the totality of a system of causal relationships among material objects, both nearby and long-range. We will return later to the question of how Mach's program might be accomplished, and whether there are any indications of success.

2. On a second interpretation, both sides of the equation are equally significant, with neither having ontological priority over the other. Space-time curvature is real, and so is matter; but they are distinct and forever irreducible. Once again, this was precisely the assumption made in classical physics. However, there they were wholly independent of each other, with only an actor-arena relationship. Neither had any causal influence on the other. Finally, Euclidean space and time were so lacking in conceptual richness and flexibility that it was inconceivable for them alone to play any of the complex roles of matter.

However, GR does have a law apparently equating a physical (phenomenological) and a geometrical quantity. This is quite different from

Poisson's equation, where mass density and potential were equally physical concepts, with no geometrical or other external interpretation. The law was just one of the internal relations among the variables of the theory. But here we must assert that to every physical structure or variation there corresponds a geometrical structure or variation—and vice versa. This view is reminiscent of traditional philosophic parallelisms (such as psychophysical) and might be called materio-geometric parallelism. But it still doesn't explain why this parallelism should occur. It can only postulate it as a wonderful cosmic coincidence, or the reflection of some pre-established harmony. An appeal to causality doesn't really help. Because of the symmetry of the equation, we could equally well say that matter causes curvature or that curvature causes matter. Furthermore, it is a strange sort of causality that relates two such radically different elements, like that assumed between the mind and the body. All it seems to do is to reduplicate the ontology.

Despite these objections, interpretation (2) is probably the one used in practice by most people in the field; at least it was the most common before the development of geometrodynamics. Because it is so all-inclusive, its ontology is the most flexible, and it does not depend on the successful completion of any reductionistic programs which are still only hopes. It does accept the basic idea of a geometrical description of gravitation, and at the same time the apparently distinct properties of matter. Furthermore, these objections overlook an important ambiguity in the term 'curvature'. In GR it does appear that T_{ik} is the only reasonable candidate for the name 'matter'. However, there are numerous geometrical tensors derived from the g_{ik} which might be called 'curvature' in one sense or another. If the Einstein tensor G_{ik} were chosen, it would appear as a mere reduplication of concepts, but if the R^i_{jkl} were chosen, it would not. Since matter confined to a small local region could still affect the values of the Riemann tensor in remote regions of space-time, there might indeed be some justification for saying that the matter causes the curvature.

3. The third interpretation is close to the heart of geometrodynamics. Here only the left side of the field equation has ontological significance. Matter is not an independent entity, standing on a par with space-time; it is rather the *name* that we give to certain kinds or aspects of curvature (most clearly represented in the G_{ik}) which manifest themselves to our senses in striking ways and thus provide the possibilities for observation and measurement embodied in the T_{ik}. The geometrical model of the world is complete in itself, and does not need to be supplemented with such 'foreign entities' as 'matter' or 'extended material objects,' which would destroy the conceptual coherence of the model. This position was perhaps best expressed by Schrödinger, who wrote,[24]

I would rather you did not regard these equations as field equations, but as a definition of T_{ik}, the matter tensor . . . Matter does not *cause* the geometrical quantity which forms the first member of the above equation to be different from zero, it *is* this non-vanishing tensor, it is described *by it.* (Schrödinger's italics)

And elsewhere he asserted,[25]

The great achievement [of Newton's laws] was to concentrate attention on the *second* derivatives—to suggest that *they*—not the first or third or fourth, not any other property of the motion—ought to be accounted for by the environment. It is very often more essential just to point out the kind of conception that matters and to emphasize it by claiming for it a name of its own, than to find out in detail the laws which control it. . . . Matter is no longer the content of the receptacle space-time, is no longer the actor who performs in time on the stage of space. Matter and space-time have really merged into one. It is sometimes said that matter determines the curvature of space-time. But the most advisable is, I think, to reserve the expression of *matter* to indicate the object of our direct observation and to regard *curved space-time* as the picture or *model* we form of this object in our minds. . . . Our observations on matter determine *us* to equip the model we have chosen with such features (curvature) as make it fit in with observation.

This was also the position of Eddington, Borel, and Meyerson.[26] To express the point in the conceptual framework of this treatise: GR chooses the various curvatures possible in a four-dimensional manifold as its model, with space-time as its content and the various tensor curvatures as its form. Since only one such manifold is used, and no nongeometric elements are introduced, the coherence of the model is guaranteed. It can then be put forward as a possible ontology for describing completely the behavior of matter. The field law thus is *not* a statement within the *theory proper*, describing relationships assumed to hold among the elements of the model; instead its character is that of a *coordination relation* or correspondence rule, enabling us to *translate* back and forth between descriptions at one level (the macroscopic and phenomenological properties of matter) and another level (the curvatures of space-time). If all statements about matter can in fact be translated into statements about curvatures, we may say that the level of observable matter is *grounded* in that of space-time; and if all the observationally significant features and distinctions of matter can be reproduced in a natural way by distinctive kinds of curvature, we may even be able to say that matter has been reduced to space-time curvature.

Would this automatically satisfy the program of Plato and Descartes, or allow us to speak of the 'elimination' of matter? Here we must be very careful about jumping to conclusions. Čapek, for example, accepts the mathematical possibility of making such a complete translation, but insists that the conception of space required, particularly with its incorporation of time and becoming, as well as its indefinitely complex curvatures, is so remote from the classical one that comparisons are meaningless.[27] Certainly Descartes' three-dimensional Euclidean space, with time excluded as a radically different sort of entity, is far simpler and capable of supporting a geometrostatics at best. However, let us recall that Descartes' primary motive in reducing matter to space was to achieve intelligibility, which for him meant treatability by the methods of analytic geometry. Given the development of mathematical techniques and the awareness of conceptual possibilities in his era, any appeal to curvatures or manifolds of higher dimensionality would have been unintelligible. Since mathematicians have gradually learned how to handle and interpret these manifolds, while still recognizing their basically spatial character, there is no reason why a follower of Descartes should not expand his notion of intelligibility accordingly. The limited properties of Cartesian space were more the result of historical accident than some 'true' or absolute meaning of 'space'. Čapek is quite right in recognizing that the ordinary conception of space still retains its Cartesian connotations (and may even reflect the structure of our perceptual apparatus, in Kantian fashion). But this only means that popular expositions of the principle are likely to be misleading, not that a physicist or philosopher can never achieve intelligibility by interpreting matter in terms of space-time curvature.

There may even be a conceptual advantage. Mast makes the interesting (and I think correct) point that the use of G_{ik} for matter is much closer to the classical pre-Newtonian point of view, where matter was understood as an underlying (normally continuous) substratum *out of which* individual objects were constructed by the imposition of suitable *forms*.[28] Here we have a similar situation, where the underlying Riemannian curvature, while remaining continuous and nonzero throughout, can occasionally manifest itself in specific regions in distinctive ways, corresponding to a nonvanishing Einstein tensor. This is a distinction of forms, not the introduction of a new element of content. On the other hand, matter as such was not a basic concept in the Newtonian theory. The elements were instead individual material objects endowed with mass, and 'matter' (if it had any use) was nothing more than a collective name for these objects. The concern of GR is not the elimination or geometrization of *material objects*. Indeed, they have no special status in the theory, for a T_{ik} can be generated equally well by such things as electromagnetic fields, and the use of this tensor does not require us to separate out the sources of its terms, or say what kinds of observable things they refer to (nay, it forbids us to do so!). GR, therefore, seeks to geometrize matter in general and not just specific forms of it, however they may present themselves to our senses and measuring apparatus.

Let us return to the notion of intelligibility, for this is normally the motive for those who adopt interpretation (3).[29] Anyone taking this approach must assume that the geometrical quantities, particularly the G_{ik}, have an intrinsic intelligibility, or that the T_{ik} do not, or both. We will return later to the meaning of the various geometrical expressions. For the latter question, one might say that (1) terms like mass, pressure, etc., are irreducible in describing the behavior of matter unless we replace them wholesale by something like geometrical terms. They cannot be reduced to other physical quantities like charges and electromagnetic fields. However, they have no explanatory force. They are postulated to describe the facts of observation, but give no deeper penetration into the real situation. Alternatively, one might say that (2) the quantities appearing in T_{ik} are just phenomenological macro-observables. There is a correct (and presumably sufficiently intelligible) physical description of the situation, but it involves such things as elementary particles, which must be described in wholly different conceptual terms. Although the macro-variables in the T_{ik} are in principle reducible to these, we have no idea how to carry out such a reduction in detail, nor are we even sure what the proper concepts at this level are. Lacking such an analysis, we may as well try to reduce them to geometrical quantities, perhaps in the hope that these may even prove appropriate for the supposed micro-level.

Nevertheless, intelligibility is not wholly on the side of geometry. As Schrödinger himself pointed out, the choice of which geometry to use and which structures to emphasize in it is determined by our knowledge of matter. We cannot, in the way that Descartes might have hoped, develop our geometry in a wholly a priori fashion and only later correlate features of it with observables. Had we taken this approach, we might well have attached more importance to the Ricci, rather than the Einstein, tensor. And indeed the conservation law which prevents us from doing so is often considered the most intelligible feature of matter. This is certainly true in such particle reactions where we look for conservation and symmetries in before-and-after states, while treating the intervening processes as an inscrutable black box. One might then say that space-time and matter combine to form one single, real, all-inclusive entity, but that it is neither especially spatial nor especially material.

Einstein's attitude is a curious amalgam of interpretations (1) and (3). It is clear that he did not accept position (2) as anything more than a temporary working hypothesis. He intensely disliked "the illegal marriage between the artificial energy-momentum tensor of matter and the curvature tensor," and thought that the use of the general field equation was "in bad taste, since we do not know what T_{ik} really is, and we mix a geometrical tensor on the left side with a physical tensor on the right side."[30] He wrote,[31]

The right side is a formal condensation of all things whose comprehension in the sense of a field-theory is still problematic. Not for a moment, of course, did I doubt that this formulation was merely a makeshift in order to give the general principle of relativity a preliminary closed expression. For it was essentially not anything *more* than a theory of the gravitational field, which was somewhat artificially isolated from a total field of as yet unknown structure.

One might think this would have led him to adopt (3) wholeheartedly. But Einstein was not trying to put forward such a radically different ontology. He had been influenced very strongly by the methodology and program of Mach. Furthermore, his choice of the term general *relativity* suggested that he hoped to carry further the program of eliminating absolute space and time, which special relativity had started. He saw GR as the theoretical embodiment of Mach's principle, and was sensitive to the charges of back-sliding from SR leveled at him by Bridgman and others. Furthermore, he kept his belief in the electromagnetic theory as advocated by Lorentz and others, holding out only a slight hope that geometry might lead to a better understanding of the structure of elementary particles.[32] Einstein's own resolution was to concentrate on solutions of $R_{ik} = 0$, treating matter as singularities (in analogy to the Newtonian treatment)—though, without a general classification of allowable singularities, he never really felt happy with this approach. For a singularity is just that, a region where things behave in strange, peculiar, "singular" ways. Since the field equations are no longer valid here, we cannot even have a complete field theory, in the sense used above. I suspect that Einstein's vacillation between (1) and (3) reflects the very eclectic epistemology which he proudly and rather self-consciously proclaimed.[33] His rationalist impulses incline him to (3), while his empiricist or positivst impulses drive him back to (1), and he has too much respect for the merits of both positions to resolve the conflict completely.

In any case, there is still one more crucial problem which interpretation (3) must face before it can be fully satisfactory. For (1) the crucial difficulty was to find an integral equation that related sources to test particles directly; for (2) it was to find some interpretation of the field equation as a real causal connection between two commensurate but mutually irreducible terms. Now in (3) we have interpreted the field equation as a coordination between elements at two different levels. But once the field equation is removed, what remains in the theory proper, except perhaps for the geodesic law of motion? We apparently are left with saying that these curvatures simply *exist*, as a brute and inexplicable fact. The question, why does physical space-time have the structure (and thus cause the motions) that it does?, seems unanswerable in these terms. In contrast, the classical question, why does a certain distribution of force fields exist? could always be answered by locating its sources, through manipulation of the appropriate field equation. Schrödinger's claim that it is enough just to single out certain geometric structures merely skirts this issue. In fact, until the formulation of the initial value problem in contemporary geometrodynamics, there was no adequate answer. Only by getting away from the somewhat static four-dimensional version of the field equation and seeing ways in which one geometrical structure can generate another in time was it possible to recognize the real dynamical content of the theory.

There is a sense in which any kind of field that is represented as a function $f(x^i)$ of the coordinates used to describe a spatial manifold might be regarded as a *property* of that manifold, or, more precisely, a property of the points comprising it. However, this terminology is highly misleading, since the fields are not usually assumed to describe any inherent characteristics of the point or higher-dimensional region in question. Instead they serve only as the basis of a mathematical framework enabling one to express relationships between different parts of the field. If we extend the notion of 'property' this far, then everything with any spatio-temporal character might be regarded as a property of space-time, though for all other substances there is only a limited class of things that can meaningfully be counted as properties. Points and regions may indeed have properties and relations; but since they are geometric objects, they can have only geometric properties and relations. I have suggested above that recognition of this fact was an important motivation for the nineteenth century search for ethers—the kinds of substances which could take fields as their properties. If we do not wish to include fields (qua substances) or ethers in our ontology as separate entities, quite independent of but coextensive with space-time, we must show that the quantities replacing them are truly geometric in character. Only then can the third interpretation discussed above be even conceptually justified.

It is clear that GR requires the tensor character as a *necessary* condition for reality or ontological significance, since all coordinate systems are equally legitimate, and anything that can be transformed away by a mere change of coordinates can hardly be regarded as a basic element of the world. However, indefinitely many tensors can be constructed, and we need further conditions if our ontology is to be more definite. The third, or geometrodynamical interpretation, seeks to restrict this class by allowing only tensors that have a clear-cut geometrical significance. This stipulation presumably rules out T_{ik} and any others which are constructed by appealing only to the observable behavior of matter, but it is still not enough. Only if a tensor points to some striking feature of the structure of the manifold, which differentiates that manifold from all other possibilities, can it be an essential part of a geometrical ontology. In Riemannian geometry the metric tensor is fundamental, in the sense that all other structural tensors can be derived from algebraic combinations of it and its derivatives. However, it does not point to all these features in a way that reveals their structural importance. Thus while it may have formal or epistemological priority, it does not provide a complete ontology. In effect, it tells one how to find the geometrical treasures, but doesn't say what they are or even what they look like. The Riemann tensor, which is the basic *structural* description, might be thought sufficient. However, by itself it contains too much information in that it treats every detail on a par, when in fact we may be more interested in combinations, sums, or averages of its components. Compare the situation in thermodynamics, where it might be very nice to know the behavior of each particle of a gas (and indeed we *could* calculate everything else from this) but where we are really more interested in the distribution of temperatures and pressures that result from large-scale sums of energies and momenta. Analogously, we wish to include some of the possible contractions of the Riemann tensor, which are in effect sums or averages of components; but just as not every such sum over the micro-variables is important in thermodynamics, not every combination is a significant geometrical feature.

It is essential to realize that we are referring to structural features inherent in the manifold itself, rather than figures which might be described against it as a background or framework. Thus in the case of two-dimensional Euclidean geometry, we would consider the plane itself, not the triangles, circles, etc. which might be drawn in it. Without this we are forced back to the problem facing Plato and Descartes. For each was able to introduce structure only by cutting up the manifold into separate regions using two-dimensional boundary elements. In Plato's case they were free-floating perfect geometrical figures; in Descartes' case they were interconnected and algebraically describable. However, we have seen above that each ran into difficulties in the physical interpretation of these boundaries, which ultimately had to appear as some sort of singularity in the three-dimensional manifold. Descartes really had no choice: since his space was perfectly homogeneous and flat, there was no other possibility for structural distinctions. Plato's chōra, to be sure, was very inhomogeneous. However, it was truly "metrically amorphous." It had no structure at all or at least no intelligible structure, and for Plato this amounted to the same thing. But if the seamless four-dimensional Riemannian manifold has enough *inherent* structural distinctions to enable us to ground all the diverse facts of observation (at least in the area of gravitation) without requiring foreign entities to subdivide it, we may expect that this crucial stumbling-block for the previous theories has been overcome.

Just what structures can be considered inherent? Aside from the overall global facts about the topology of the manifold, the only obvious candidates are the various curvatures, which in general can vary from point to point. Up to now we have rather cavalierly used 'curvatures' as a general class name for whatever the various tensors derived from the Riemann tensor are supposed to describe. But why are we entitled to use this name, with its many connotations? We had originally introduced the Riemann tensor as a measure of the relative change in the coordinate positions of two nearby particles moving through a given space, a notion that seems to belong more to kinematics than to geometry proper. A more common definition in affine geometry relates it to the change in the components of a vector which is carried by parallel transport around a closed loop surrounding a small element of area in the vicinity of a given point. Whether or not the vector returns congruent to its original position depends on whether or not R^i_{jkl} vanishes at that point, and this in turn is supposed to measure whether or not the manifold is flat or curved there. Nevertheless this still seems quite far from the ordinary definition of a curvature as the inverse of the radius of a sphere or circle in a given plane that is tangent (to the second order) to the manifold at that point. Such curvatures are of course scalars, not higher-order tensors. Fortunately, the two are very closely connected. It turns out that we can choose two orthogonal unit vectors in any pair of directions, and consider the surface formed by the set of geodesics tangent to all possible linear combinations of these vectors. This surface is characterized by a scalar known as the Gaussian curvature, and the only tensors which appear in the mathematical expression for this curvature are R_{ijkl} and g_{ik}. Furthermore, the Riemann invariant $R = R^i_i$ is proportional to the sum of these curvatures for all possible surface directions in the overall four-dimensional manifold, agreeing with the principle that the contractions of the Riemann tensor should represent sums or averages of curvature. Furthermore, the R_{ik} appear alone in an expression for the sum of these sectional curvatures over the surface between each of the original set of orthogonal directions and a new one; and finally, the G_{ik} appear alone in the expression for the mean curvatures (sum of the sectional curvatures) of the three-dimensional manifold orthogonal to one of the original directions.[34] If this counts as a full listing of the quantities that might reasonably be called curvatures, it may be considered to give an equally full inventory of the tensors needed to characterize the structure of physical space-time. The fact that the Einstein tensor G_{ik} also appears here is important. Since its components do have direct geometric significance, they no longer appear as an awkward or unnatural combination, introduced solely because they were the only quantities that could be equated to the T_{ik}.[35]

If we are simply concerned to find as many invariants—i.e., magnitudes independent of any coordinate system—as possible, we need not stop here. In fact, Géhéniau and Debever have claimed that there are fourteen invariants which can be derived from the metric and Riemann tensors.[36] However, their analysis is purely formal, and without some reasonable conceptual interpretation the role of these invariants in geometrodynamics would seem to be subsidiary.

In spite of the foregoing, one might still argue that we have only related the various geometric tensors to measures of the curvature of lower-dimensional manifolds; we still have not given an adequate account of what it means for a four-dimensional (or perhaps even a three-dimensional) manifold as a whole to be curved—curved with respect to what? Even though the manifold in question is clearly non-Euclidean or non-Minkowskian, one might claim this is not sufficient to establish real curvature, unless the term is stretched far beyond its usual meaning. Thus Whittaker wrote,[37]

> It is an unfortunate custom, to apply the word "curved" to any space whose geometry is not Euclidean, because curvature, in the sense of bending, is a meaningless term except when the space is immersed in another space, whereas the property of being non-Euclidean is an intrinsic property which has nothing to do with immersion.

In a similar vein Eddington wrote, as if it were a matter of course,[38]

> When we use the phrase "curvature" in connection with space-time, we always think of it as embedded in this way in a Euclidean space of higher dimensions.

Even though he admitted that these higher dimensions have no real existence, he considered them valuable for visualizing the manifold and thereby understanding its structure more fully. For Whittaker the question was primarily semantic, since he never supposed that we can in any way envisage this higher space. For Eddington the term 'curvature' was also meaningful only in terms of immersion or embedding; nevertheless his language ("We always think . . .") indicates that it is quite possible to do this envisaging, and therefore the term is appropriate.

Robertson has argued strongly that from any operational standpoint no distinction can be made between curvature and departure from Euclidicity, and insists that we should give up the appeal to visualization, despite its obvious usefulness in analyzing the curvatures of ordinary curves and surfaces.[39] It is clear that this is also the approach of Einstein and of Wheeler, who likens the notion of curvature to the problem of fitting together a table of airline distances into a shorthand notation correct for any pair of cities.[40] From any standpoint that attaches ontological significance to the four dimensions of space-time, it is certainly important that curvature can be fully characterized without appeal to higher dimensions, since these would have to appear formally on a par with the others, yet be radically different conceptually.

Nevertheless, if we look at Eddington's subsequent practice, it appears that the 'thinking' he had in mind was not a matter of formal mental pictures, but of finding a more intelligible mathematical representation. Mathematically it is probably simpler to deal with flat spaces of more than four dimensions than with non-Euclidean four-spaces. And if the notion of higher dimensions with regard to physical space seems incomprehensible, one might say the same about curved spaces that cannot be embedded. There is a sense in which we can envisage the whole of Newtonian space and time; this is presumably what led Kant to regard them as objects (as well as forms) of pure intuition. Yet it is doubtful that we could do this for a general Riemannian space, even though we could recognize by local measurements and observations that we were living in one.[41] Finally, there is no obvious explanation within GR of why space-time should have four dimensions, no more and no less—the formalism itself makes no reference to dimensionality. Indeed, Good, going further than Eddington, even speculates that physical space-time might really have more than four dimensions, and only our physical limitations prevent us from perceiving this (like Spinoza's infinitely many other attributes).[42]

According to Schlaefli's theorem, any n-dimensional Riemannian manifold can be embedded in a flat space of m dimensions, where $m = \frac{1}{2}n(n+1)$,[43] and where

$$ds^2 = \Sigma_{i=1}^{m} \epsilon_i (dz^i)^2, \ \epsilon_i = \pm 1.$$

For $n = 4$, $m = 10$. In saying that the higher space is flat, I do not mean that it is strictly Euclidean; rather it is a generalization of Minkowski's space, where some of the dimensions will be timelike (i.e., opposite in signature). In any case, there is no transformation that can convert space into time or vice versa. But it does not follow that all ten dimensions are always needed. If the manifold is already flat, we need only the original four. For this reason, the appeal to embedding provides a natural ordering for manifolds in terms of the complexity of their curvature, with four dimensions required for the simplest and ten for the most complex. I have proved elsewhere that in the general case of spherical symmetry, only six dimensions are needed; and indeed the especially simple de Sitter solution can get along with only five.[44] We know that not all mathematically possible g_{ik} are possible candidates for physical space, but only those for which the covariant divergence of the corresponding T_{ik} vanishes. It might turn out that all the metrics still permitted by this requirement would require very few extra dimensions. If this could be proved, it would provide an additional and quite interesting way of stating this restriction in geometric terms. Two other advantages are possible: in cases where we have complete spherical or cylindrical symmetry, we can suppress two of the spatial dimensions and exhibit the relation between the remaining two dimensions of space-time as a surface in a three-dimensional space. This concentration on two dimensions allows us to talk about recognizable *shapes* of the manifold, and suggests analogies to these shapes at higher dimensions. In addition, the possibility of embedding allows us to define new concepts which might help characterize our geometric structures in an especially revealing way. Just as we can compare the distance between two points along a geodesic on a two-dimensional surface with the displacement between them in a corresponding Euclidean three-space, we might define a higher-dimensional displacement, at least in terms of the spatial or the temporal dimensions taken separately. For these reasons I think it is worth while to take Eddington's proposal seriously.

In SR the constancy of the speed of light in all legitimate (i.e., inertial) coordinate systems is a central axiom, and its value c thus acquires the status of a universal constant. However, we have made little or no reference to it throughout the foregoing. What is its significance in GR? In the first place, the speed of light is *not* constant if there are gravitational fields. Even though the law $ds = 0$ holds for all light rays, we can no longer expect to represent ds by a simple $ds^2 = dl^2 - dt^2$, except in the immediate vicinity of a given point. The constant coefficients must be replaced by functions, which need not have a constant ratio. Secondly, if we allow very general coordinate systems "with no immediate metrical significance," as GR requires, it is not clear that we can have a meaningful notion of "the speed of light" in all of them. Finally, c appears to be an essentially electromagnetic, rather than gravitational concept. It plays no role in Newton's theory, and in Maxwell's first theoretical derivation it is related to the electromagnetic permittivity and permeability of free empty space—free, that is, of electromagnetic fields or media, regardless of what gravitational fields might be around. It is therefore tempting to follow Wheeler's move of considering c a mere conversion factor between two sets of units (e.g., meters and seconds) which can both be used equally well for measuring all kinds of intervals. It thus needs no more explanation than 5280, the number of feet in a mile, and is really dimensionless. Since we are using four-dimensional language, it is legitimate and far simpler to set $c = 1$.[1]

Nevertheless, we should not be too eager to ignore it. For in SR c is not only an actual velocity, that of light in vacuo; it is also a *limit* which cannot be exceeded (though it may be equalled) by the velocity of any other signal. Lindsay and Margenau have shown that we can begin by holding on to c as an undetermined constant, rather than suppressing it by setting it equal to unity, without any assumptions that it be related to light or anything else. If we do this, we find that it can be calculated from other parameters in the system being described, and that once again it sets a maximum limit on the speed of transmission of any signals, presumably including any possible gravitational waves.[2] GR thus does preserve the SR postulate that there is an upper limit on all physically significant velocities, and therefore there can be no instantaneous action at a distance.

If any coordinate system must still break down into three spacelike and one timelike dimensions, it follows that the four coordinates are not completely equivalent, and there might be some advantages in keeping distinct units for them. More importantly, the actual value of this limit speed does play a central role, at least in the application of the theory. We have already mentioned as a requirement on GR that it should reduce to Newton's laws in the limit, or as a first approximation. Since we know that observationally

Newton's laws are a very good approximation, this stipulation seems reasonable. But conceptually it is very puzzling. For we do not have two commensurable theories, phrased in the same general terms, whose predictions agree in certain areas and disagree in others. Returning to our notion of conceptual geography, we know that the theories are in some way comparable, since they can both be applied to the same physical situations and give observable predictions. Given that notions of mass, potential, and force seem to describe the most significant features from the standpoint of gravitational mapping, we expect that something analogous to these should be found in GR. Yet our analogies were strictly formal, and based primarily on the order of the derivatives that appeared. The metric tensor does play a role more like that of potential than any other Newtonian variable; and the Christoffel symbols do bear the closest resemblance to Newton's gravitational force. However, Newton's potential is a scalar, Einstein's a tensor with 10 distinct components; Newton's force has 3 components and is independent of velocity, while the Christoffel symbols include no less than 40 and depend on the velocities; Newton's inertial mass is conceptually different from gravitational mass, and in a rigorous presentation of his law of motion both should appear with different symbols; Einstein makes them conceptually identical, and neither need appear in the law at all. How then can we possibly compare them? It seems highly misleading to think of Newton's laws as a special case within the conceptual framework of GR. Kuhn has made the same point in comparing SR with Newtonian mechanics.[3]

It is therefore important to see just how this "approximation" is carried out and what assumptions are made. Let us break down the law of motion as

$$\frac{d^2x^i}{ds^2} = \Gamma^i_{jk}\frac{dx^i}{ds}\frac{dx^k}{ds} = \Gamma^i_{00}\frac{dx^0}{ds}\frac{dx^0}{ds} + 2\sum_{\alpha=1}^{3}\Gamma^i_{0\alpha}\frac{dx^0}{ds}\frac{dx^\alpha}{ds} + 1\sum_{\alpha,\gamma=1}^{3}\Gamma^i_{\alpha\gamma}\frac{dx^\alpha}{ds}\frac{dx^\gamma}{ds}.$$

We now interpret ds as proper time (measured in seconds). Let us try to simplify our analysis by taking an illustrative special case, where the calculations can be carried through more readily, in order to show the extent of the approximations usually made. Specifically, we will assume that the metric tensor has separable spatial and temporal parts in the coordinate system chosen, and that it is isotropic in all three spatial directions. We can therefore write $ds^2 = \phi dt^2 - \psi d(l/c)^2$, where ϕ and ψ are arbitrary functions of the coordinates and $dl^2 = dx^2 + dy^2 + dz^2$. Also let $v = dl/dt$, $\beta = v/c = d(l/c)/dt$. On these assumptions we find that $dt/ds = (\phi - \psi\beta^2)^{-1/2}$, $d(l/c)/ds$

$= \beta(\phi - \psi\beta^2)^{-1/2}$. Substituting back into the original law of motion, we get

$$\frac{d^2x^i}{ds^2} = \frac{1}{\phi - \psi\beta^2}\Gamma^i_{00} \text{ (1 term)} + \frac{2}{3^{1/2}}\frac{\beta}{\phi - \psi\beta^2}\sum_{\alpha=1}^{3}\Gamma^i_{0\alpha} \text{ (3 terms)}$$

$$+ \frac{1}{3}\frac{\beta^2}{\phi - \psi\beta^2}\sum_{\alpha,\gamma=1}^{3}\Gamma^i_{\alpha\gamma} \text{ (9 terms).}$$

This calculation is not intended to be strictly rigorous. If we sought rigor it would be more accurate to use the vector components β^α, and include them in the summation. However, the point is to show that the coefficients of the Christoffel symbols can be grouped by powers of β, a dimensionless measure of the relative velocities of the particle in question and of light. The magnitude of each component β^α cannot be larger than that of β itself, so that the use of β directly gives us an upper limit for the amount of correction or approximation that will be governed by each term involving some power of β. The numerical coefficients, which are all of the order of magnitude of one, have no particular significance. Now it is an important empirical fact about the world that for all bodies large enough to exhibit measurable gravitational behavior which is not swamped by electromagnetic or nuclear effects, β is a very small number. In particular, this is true of the sun and planets. Thus unless some of the other Christoffel symbols are greater than Γ^i_{00} by a factor of $1/\beta$, we can ignore the third and even the second group of terms in practical applications. In the limit as $\beta \rightarrow 0$, the right-hand side of the equation reduces to $(1/\phi)\Gamma^i_{00}$. If we can make the further assumption that the fields are very weak, so that ϕ is close to 1, we then get $d^2x^i/ds^2 \cong \Gamma^i_{00}$. Comparing this with Newton's $d^2x^\alpha/dt^2 = g^\alpha$, the vector for the gravitational field, we see the justification for interpreting the Christoffel symbols as forces. A further comparison allows us to write $g_{00} = -1 + 2\Phi/c^2 = -1 + 2Gm/c^2r$, where $\Phi = Gm/r$ is the Newtonian gravitational scalar potential, for the simplest spherically symmetric field.

Thus we are able to compare a mathematically complex entity with a mathematically simpler one, but only because of a contingent empirical fact not inherent in the formalism. *Sub specie aeternitatis*, we could have chosen our ordinary units of space and time so that the ratio between them would be close to one. Had we been light rays or elementary particles, we might well have done so. But it is no accident that the standard units (meters and seconds) are such that the "conversion factor" between them is a very large number. In terms of these units, the velocities of all ordinary experience differ from 1 by no more than a few powers of 10. Now let us consider a 'possible world' in which all the velocities of material bodies (the v's) remained the same in our units, but light traveled much slower, so that c would be smaller and the β's greater, even though we could still keep the requirement that nothing travel faster than light. In such a world we could no longer make our approximations; the terms with β's in the numerator could exert substantial effects, and our mathematical calculations for any kind of motion would be far more complicated. On the other hand, what if light traveled much faster than it actually does? Consider the limit as $c \to \infty$. In this limit, all the β-terms would vanish and only the first term would remain. Indeed, the metric tensor itself would reduce to $ds^2 = \phi dt^2$, ϕ could always be set equal to 1, and there would be no necessary linkage between space and time. Even if we started with the tensor formalism and assumptions of GR, we could prove that they were not required and that Newton's theory would be perfectly accurate, rather than merely a good first approximation. A similar situation occurs more obviously in SR where Newton's laws are again perfect in the limit. Thus we may conclude the following: in GR as well as SR, it is the existence of a maximum velocity that sets limits on the accuracy of Newton's laws, while the fact that they work as well as they do results from the fact that light travels as fast as it does. Even if one attempts to conceal the magnitude of this velocity by a manipulation of units, its effect crops up in some other form.[4] Thus, c has gravitational as well as electromagnetic significance.

If we are willing to consider the metric tensor as a measure of curvature, we might (rather crudely) interpret ϕ as curvature of time and ψ as curvature of space, as d'Abro suggests.[5] If we then attend to the expression $\phi - \psi\beta^2$, we note that β enters asymmetrically, by diminishing the effect of ψ but not of ϕ. Since $\phi - \psi\beta^2$ does appear in the law of motion, it is tempting to follow d'Abro in saying that most ordinary gravitational effects reflect the curvature of time, and only at high speeds is the curvature of space important, even if the magnitudes of ϕ and ψ are comparable. However, this is an oversimplification if taken too seriously. We still must deal with the Christoffel symbols, which are constructed out of the first derivatives of ϕ and ψ.

Since the calculation is straightforward and the results interesting, I shall present them here. For the given metric, I have found the Christoffel symbols to be

$$\Gamma^0_{00} = \frac{1}{2\phi}\phi_{,0};\ \Gamma^\alpha_{00} = \frac{1}{2\psi}\phi_{,\alpha};\ \Gamma^0_{0\alpha} = \frac{1}{2\phi}\phi_{,\alpha};\ \Gamma^\alpha_{0\alpha} = \frac{1}{2\psi}\psi_{,0};\ \Gamma^0_{\alpha\alpha} = \frac{1}{2\phi}\psi_{,0};$$

$$\Gamma^\alpha_{\gamma\gamma} = \frac{-1}{2\psi}\psi_{,\alpha};\ \Gamma^\gamma_{\alpha\gamma} = \Gamma^\gamma_{\gamma\alpha} = \Gamma^\alpha_{\alpha\alpha} = \frac{+1}{2\psi}\psi_{,\alpha}.$$

All other components vanish, with different Greek letters referring to different indices 1, 2, 3. If we substitute these back into our original law of motion and let $A = (\phi - \psi\beta^2)^{-1}$, we can write the equation explicitly as

$$\frac{d^2x^0}{ds^2} = \frac{A}{2\phi}\left[\phi_{,0} + \frac{2\beta}{\sqrt{3}}\sum_{\alpha=1}^{3}\phi_{,\alpha} + \beta^2\psi_{,0}\right]$$

and

$$\frac{d^2x^\alpha}{ds^2} = \frac{A}{2\psi}\left[\phi_{,\alpha} + \frac{2\beta}{\sqrt{3}}\psi_{,0} + \beta^2\psi_{,\alpha}\right].$$

Since ψ does appear by itself in the denominator for the accelerations in the space components x^α, it cannot be ignored even at low speeds if ψ is noticeably different from one. We should also note two other curious results: (1) the space derivatives of ψ never appear in the expression for $x^0 = t$, and the time derivative of ϕ never appears in the expression for x^α; (2) if the solution is static in these coordinates, so that both ϕ and ψ are independent of x^0, the first and third terms drop out of the x^0-equation, and the 'acceleration' is directly proportional to β; the second term drops out of the x^α-equation, and the only effects attributable to *changes* in ψ appear multiplied by β^2. Since, except in very strong fields, the functions ϕ, ψ, and the A/ψ factor will remain close to one, while the derivatives describe the larger observable effects that we attribute to gravitation, we can attribute the bulk of the 'gravitational force' to $\phi_{,\alpha}$ rather than $\psi_{,\alpha}$.

Having discussed the significance of the size of units and the role of the speed of light, we must turn to the idea of dimension in GR, and the question of whether all dimensions can be reduced to those of length. We have already seen how the use of c as a conversion factor enables us to measure times in terms of distances. Conversely, we could refer to distances by the corresponding times. If our goal is to effect a complete unification of physics and geometry, it is essential that we be able to avoid reference to apparently foreign entities like masses and charges if they cannot be expressed in geometrical terms. However, there are two quite distinct senses of 'dimension', and much confusion has resulted from vacillating between their different connotations.

The first sense of 'dimension' is a purely geometrical one, though generalizable to quite abstract 'geometries'. This is the sense used when we say that space has three dimensions, that space-time has four dimensions, or that a surface has two. Here all the dimensions are assumed to be similar in kind, and perfectly commensurate with each other. They can be fitted together into a single manifold in which each appears in the same way, and simple arithmetical combinations of magnitudes in different dimensions have a straightforward meaning. Thus if we add a distance travelled along the *x*-axis to a distance along the *y*-axis, the result is a distance, that of the total path covered. In contrast, it is at least not obvious what the sum of a distance and a mass could possibly refer to. Even if the manifold is not homogeneous or isotropic, the differences in different dimensions can all be described in the same terms. Given a set of points and the mutual distances between pairs of them, Wheeler has shown that there is a straightforward procedure for determining the number of dimensions required to make all these distances consistent with each other. Each time we find it impossible to fit all the distances together with a given number of dimensions, we must add another.[6]

Of course, this is hardly the procedure followed in practice, and the 'distances' referred to might just be arbitrary numbers without operational significance. However, the method above brings out the point that it is a truly remarkable empirical fact that *only* four dimensions are required for physical space-time. Given the whole continuum of points, it is certainly possible that far more might have been needed. In such a world we might still be able to represent the line element as a quadratic form, but it would require the summation of many more terms. Virtually all theories which involve space and time do start with their dimensionality as a basic assumption about their global or topological structure, and so far this has never been in disagreement with the results of measurements, even if it is not inductively inferred from them.

Attempts have been made to account for the dimensionality of space on various grounds. Kant recognized that the three-dimensionality of space was intimately connected with the possibility of inverse-square laws of force, but at the price of regarding force as a mysterious influence spreading out over concentric spheres of increasing size, like the energy associated with a wave.[7] In any case Kant's purpose was more to elevate inverse-square laws to the a priori status he assumed for space. Other similar analyses argue that only in such a world can there be the stable orbits necessary to support planetary systems and thus human life, but these arguments assume the validity of Newtonian mechanics.[8] From another direction, Weyl has shown that only in a four-dimensional space-time world is a theory like Maxwell's, with its general assumption of gauge invariance, possible, and only then can the facts of wave propagation be accounted for.[9] Furthermore, given that space-time as a whole is four-dimensional, only a separation into three spatial and one temporal dimensions can account for the facts. Only then can an observer separate a causal past and future by introducing a light cone. If all four were spatial, there would be no propagation; if the dimensions were divided two by two, there would be perfect symmetry of space and time, and "past and future would be melted into one world domain."[10] Lenzen claims that given the same laws of gravitation and electromagnetism (which certainly make no explicit reference to dimensionality), there would be no causality in the two-by-two world; motion should suddenly occur in a region without any cause being present in that region or entering it, while in a wholly spatial world there would be more than causality—the solutions would be overdetermined.[11] Nevertheless, since all these arguments are empirical, they prove only the importance of 3 + 1 space-time through its intimate interconnection with our most general physical laws. They do not and cannot show that there is anything intrinsically necessary about this structure.

Let us now turn to the second sense of dimension. This is the one intended when we speak of dimensional analysis, ask for the dimension of a physical magnitude, or claim that length, time, and mass are the three irreducible dimensions for physics. We see that these are really *kinds* or *classes* of dimensions in the first sense, since several of the latter may be included under, e.g., length. In distinguishing these dimensions_2, we assume explicitly that they are qualitatively different and incommensurable. We cannot add quantities with different dimensions and expect a meaningful result. Finally, dimension_2 has no geometrical significance at all; its origin lies rather in theoretical mechanics. It gives rise to an important rule, which Törnebohm calls the "rule of homogeneity": if a set of expressions for physical quantities appears as a sum or in an equation, with each preceded only by a $\pm$ sign, they must all have the same dimension_2.[12]

But if this is true, it is important to know just how many dimensions are required. If the dimension_2 of each quantity is to be represented as a product of exponential powers of the basic dimensions, it is clear that the more basic dimensions, the more possible combinations, and thus the less likelihood that any two quantities will have the same dimension. If it is considered desirable to be able to make as many additive combinations as possible, and in a well-developed mathematical theory this is clearly an advantage, it will be desirable to have as few dimensions_2 as possible. However, this problem cannot be solved unless we have some criterion for determining when two dimensions really are incommensurable, or in more epistemological terms, when we should distinguish them within the framework of a physical theory.

It has been the standard practice of physicists to take mass, length, and time as the three basic dimensions, and to write the dimension of any other quantity as $M^{\alpha}L^{\beta}T^{\gamma}$, where α, β, and γ may be either positive or negative integers or, at most, simple fractions. This analysis applies both to dynamical variables like momentum (written $M^1L^1T^{-1}$) and constants of nature like the gravitational G (written $M^{-1}L^3T^{-2}$). According to Newton's second law, every force must have the dimension $M^1L^1T^{-2}$. Thus if we have a force law in which only one quantity has an unknown dimension, we can use this fact to calculate its dimension. Thus from Coulomb's law in the form $F = q_1q_2/r^2$, we can infer that charge has dimension $M^{1/2}L^{3/2}T^{-1}$. Despite this, it was often considered more convenient to treat charge as a separate dimension of its own, since the dimension of so many other electromagnetic quantities bore a simple relation to that of charge, and in classical physics electromagnetism and mechanics were largely independent of each other.

Campbell has extensively criticized the idea that length, time, and mass are somehow more basic than other magnitudes and should serve as the basis for a system of dimensions, or even that one should try to find some minimum set and interpret the others in terms of them. His arguments are primarily conceptual and operational, based on the assumption that if anything should count as basic, it is those quantities which are measured directly, rather than calculated from others. On these grounds weight might be given priority over mass; and since except in the (unusual) case of rectangular solids, we do not normally determine volume by multiplying the result of three length measurements, we might either give it a dimension of its own, or represent it by (mass)/(density), for example, if the latter two were taken as primary.[13] And he rightly points out that two quantities which may have the same dimension in the standard system may be very different in their meaning and role within the theory, as well as in their operational origin.

Nevertheless, Campbell reveals the way in which he misses the point when he says that the special role of M, L, and T is just an "historical accident" stemming from their central role in Newtonian mechanics.[14] For contra Campbell, I would claim that 'dimension$_2$' and 'difference in dimension$_2$' are not observational notions at all, but theoretical ones. One does observe differences in function or role, to be sure; and one does carry out both direct and indirect measurements by very different methods. However, the decision as to whether such differences should be interpreted as reflecting differences in dimension$_2$ depends entirely on the theory. What dimensions$_2$ refer to are rather the most general and irreducible distinctions in the descriptive framework provided by the *model.* As we have seen before, the use of models may require an abstraction, idealization, or oversimplification of the full empirical reality. Nevertheless, it is the model that provides the basis for representing the coherence and mathematical interconnections of the observable phenomena. If a model has many separate dimensions$_2$, it can describe a great number of different things and make a host of qualitative distinctions, but it will have few ties binding them together; if a model has few dimensions$_2$, it can make few qualitative distinctions and naturally tends to replace them by quantitative ones, but its laws will be far more comprehensive and its mathematics more powerful. If two theories for the same body of phenomena have models with different numbers of dimensions_2, any comparison by appeal to observation or measurement must judge their adequacy *as a whole*. There is no immediate or intuitive notion of what *must* count as a dimensional distinction. As for being a historical accident: (1) It is a historical fact that Newton's mechanics was indeed the first successful physical theory whose model had a completely mathematized form. Space, time, and matter were in fact the primary conceptual distinctions in the theory, and as interpreted by Newton they were fully tractable mathematically. Descartes had had space, possibly time, and mass not at all as his dimensions_2, and his theory was not adequate for the phenomena. Historically, it is reasonable to say that the concept of dimension_2 did not *exist* before the development of mathematical physics. (2) There is nothing accidental about the role of space, time, and matter in Newtonian mechanics—they are central concepts. If one accepts the possibility that an adequate theory of mechanics might have been formulated with a wholly different conceptual basis, the only accident is that Newton's was the theory actually constructed, rather than one of these others.

It follows therefore that the L, T, M system of dimensions has no absolute significance, but reflects the tremendous success of Newtonian mechanics when applied to a vast range of phenomena. But are all three necessary, even within this conceptual framework? Törnebohm has given a careful analysis of the formal aspects of the problem.[15] Given a quantity with the dimensions $M^{\alpha}L^{\beta}T^{\gamma}$, we can replace it by a quantity with dimensions $M^{\alpha}L^{\beta+\gamma}$, provided only that we multiply it by a constant K whose dimensions are $(L/T)^{+\gamma}$. In the three-dimensional system, L/T has the dimensions of a velocity. If we are to be consistent in converting all quantities to the two-dimensional system so that units are not changed and numerical values can be compared, we must use the same velocity in all conversions. All velocities will then appear as dimensionless numbers, but the choice of unit remains arbitrary. If there is any velocity which appears in physics as a universal constant, this will be an obvious choice, and in the new system its numerical value will be one. This can be said to define the most 'natural' system of units. Of course, the speed of light c satisfies these requirements, so we simply multiply everything by c^{γ}. Now the whole argument above has made no reference to the specific qualitative features of L and T. It is therefore possible to do precisely the same thing with M and L. Starting with $M^{\alpha}L^{\beta}$ in the two-dimensional system (which is not necessarily $M^{\alpha}L^{\beta}T^{0}$ in the three-dimensional system!) we can replace it by $L^{\alpha+\beta}$ simply by multiplying by a constant K' with dimensions $(L/M)^\alpha$. It may be less obvious what might count as a suitable candidate for our L/M constant. However, we note that the gravitational constant G, with dimensions $M^{-1}L^3T^{-2}$, gets dimensions $M^{-1}L^1$ and is multiplied by c^{-2} in the two-dimensional system. As the other universal constant, the only other that appears in prequantum physics, it is ideally suited for the purpose. We multiply everything by G^α and thus apparently have a system in which (1) every quantity has the dimension of some power of L; (2) both the speed of light and the gravitational constant are now dimensionless numbers with the value 1, so we need not try to account for numbers like 3×10^{10}; and (3) quantities for which the sum of the exponents $\alpha + \beta + \gamma$ was the same can now be added together, no matter how different the individual exponents were in the original three-dimensional system. Since the argument is perfectly formal and symmetrical, we need not have chosen L, with its connotations of length, as our one remaining dimension; M or T would have done just as well.

All this seems just too good to be true. And it is too facile and magical: one suspects that a rabbit has been pulled out of a hat. Yet the fact remains that for any given mass m, we can always find a length proportional to it by multiplying by G/c^2. This new $m^* = Gm/c^2$ will of course have a different and much smaller numerical value, if m is measured in grams and m^* in cm, but m^* is still a length. Likewise we can "convert" all charges to lengths by multiplying by $G^{1/2}/c^2$. But what, if anything, does this accomplish? Törnebohm points out that it provides no operational definitions, and no advantages for measurement. He concludes,[16]

> This does not mean that all measurements have been reduced to length measurements, so that the whole of physics is experimental geometry. Dimensions do not determine how magnitudes are operationally defined. In fact they are *purely syntactical* concepts and have *nothing* to do with the *semantics* of the physical symbols. The significance of the reduction to the one-unit basis is *merely* that it effects a *formal* simplification of the relativistic equations. (Italics mine.)

In contrast Fletcher, one of Wheeler's associates, arrives at a wholly different conclusion after examining these conversions. He states,[17]

> The consequence of all this is that there is *nothing but lengths* in geometrodynamics. All physical quantities can be interpreted as lengths, areas, volumes, etc. or their reciprocals. (Fletcher's italics)

Just as Campbell goes too far in tying the notion of dimension$_2$ too closely to observation and measurement, Törnebohm goes too far in the opposite direction of considering it in overly formal and syntactical terms. If there is no *conceptual* reason, in terms of the properties postulated for the model, to reduce all the dimensions$_2$ to one, his construction is not only a pointless game, but a potentially dangerous one. For by cutting down on the number of possible distinctions we can make, it may *force* us to overlook some very important features. Furthermore, dimensional homogeneity seems to be at best a necessary, rather than a sufficient condition for allowing us to combine different terms additively. We must still give a conceptual account and justification of the meaning of any proposed combination, treating each case individually. And we also run the danger of extending the meaning of 'length,' in using it to represent everything else, so far that many of its important connotations are no longer warranted. Törnebohm's reduction is more important than he believes, but much more can and must be done with its 'semantics' in order to support Fletcher's conclusions. We must in fact *construct a model* in which no entities other than combinations of lengths appear, so that there is no material (as opposed to formal) need for distinctions of $dimension_2$. If (and only if) this can be done we can use Törnebohm's results to guarantee formal consistency.

Such a conceptual justification does exist for the elimination of T as a separate $dimension_2$, and the conversion from 3 to 2 such dimensions. But this is accomplished only by converting time into a fourth $dimension_1$ of length. This is possible and desirable in GR and even in SR in the Minkowski formalism. For if we are to use any four-dimensional representation, we would like all these dimensions to appear in the same way, without the radical distinctions implicit in a separation into $dimensions_2$. The four-$dimensional_1$ world-picture satisfies these requirements. Velocities, for example, appear here as the *slopes* of world lines, and 'slope' is clearly a dimensionless notion. But the possibility of putting space and time together in a coherent way reflects the important similarities that hold between them when they are considered separately. Both are assumed to be continua which together provide a framework for the *location* of objects. Space and time are not *constituted* by objects, but *occupied* by them. For both, we can choose a zero-point arbitrarily and then use all real numbers, both positive and negative, to represent other points. Indeed, in Newtonian physics intervals could extend infinitely far in any direction. In occupying them, a body, event, or process simply cut off a segment of each of these continua, corresponding to its spatial extent and its temporal duration. Finally, both were equally appropriate measures of both the size of an extended body and the distance separating it from others, even if the intervening region were assumed to be completely empty. It was in fact recognition of this symmetry between space and time that led Whitehead to take the four-dimensional event, rather than the three-dimensional body, as the basic 'substance' of the world, despite his rejection of the rest of the conceptual framework of GR. Furthermore, we can even satisfy an operationist, if we stick to the

Marzke method of measuring all intervals. For we have seen that this works equally well whether the interval is spacelike or timelike.

What can be said of masses and charges along these lines? One can learn to speak without feeling self-conscious (I speak from personal experience) of masses and charges as lengths, and give them numerical values in centimeters. But one must still ask, *where* are these lengths located, and how are they related to the other lengths that describe the size, shape, and duration of the body in question? For mass (and charge) appear very differently from space and time. At least in classical physics, mass is constituted by bodies; there is no underlying 'mass-continuum' that they may be said to occupy. When two bodies are compared in the dimension of mass, the difference between them is not another real mass, as the spatial separation is another real distance; it is just a numerical difference. Although the numerical values of all masses are commensurable, once a suitable operational standard has been introduced, they do not fit together to form a single continuum. Only positive (real) numbers are allowed (charge does allow negative numbers), and the zero-point is not arbitrary. A zero value means nonexistence of the mass, and in Newtonian physics this requires nonexistence of the body itself. Each body in effect defines its own zero-point.

Now if we are to conclude from the comparison of space and time above that the ultimate ground for the elimination of time as a distinct dimension$_2$ is the possibility of treating it as just another dimension$_1$ of space, it seems that we must choose between two alternatives for the elimination of mass and charge.

I. We may try to regard mass as a fifth dimension$_1$ of space, and charge as (perhaps) a sixth. In describing a body, we can then say that its mass and charge are simply its extent in these additional dimensions, quite analogous to its height, width, depth, and duration and measured in the same units. The full 'size' of the body for geometrical *and* dynamical purposes would then be given by a set of six numbers instead of four. However, this interpretation leads to immediate difficulties. (1) All the expressions in the formalism refer only to a four-dimensional manifold. World lines, for example, are only four-dimensional, and none of the tensors used has components corresponding to these additional dimensions. If we were to add such components, we would change the formal structure of the theory as well as making it far more complicated mathematically. These putative higher dimensions thus seem to be shut off from the original four. (2) While it is quite possible to conceive of a four-dimensional world if one remembers the temporal peculiarities of the fourth dimension, we have no idea just what these other dimensions could represent, especially if they are assumed to have some of the connotations of space or time. Since they certainly do not make the space flat or Euclidean, even the advantages mentioned by Eddington in connection with embedding do not apply here. (3) By the very character of mass and charge, they are very strange sorts of dimensions, not conceptually comparable with the other four, even if it were possible to make some formal (mathematical) combination of them. If we try to use mass density rather than mass, we could at least get a possibly continuous distribution throughout space-time. However, mass density has the dimension M/L^3 which converts in the one-dimensional system to L^{-2}, the reciprocal of an area, so we cannot use this as a basic dimension for our purposes. All these would seem good reasons for regarding the difference between mass (or charge) and the other four as sufficient to require a new dimension$_2$.

II. The other alternative is to try to incorporate mass and charge into the original four dimensions of space-time, so that they appear as characteristic (four-dimensional$_1$) lengths or intervals, rather than distances in a new dimension. But bodies already have characteristic *geometrical* lengths, describing their size and shape. What then is the *orientation* (or direction) of these new dynamical lengths with respect to the old ones of width, depth, etc.? Between which points in the body should they be measured or at least represented in the corresponding model? If the body is spherical, we might interpret mass and charge as radii. But how are they then related to the geometrical radius of the body? There is certainly no obvious way of representing them as length vectors in space-time.

Neither in classical physics nor in GR do we have a law applying to all bodies which relates their perceptual, observable geometric structure to their dynamical parameters governing their generation of and response to causal influences. It is possible to give a complete description of the sizes, shapes, and configurations of a collection of bodies without ever referring to their masses or charges, and (subject to certain limitations) we can describe their dynamical behavior without reference to sizes and shapes. In trying to interpret dynamics geometrically, we appear to be superimposing two individually complete but quite distinct geometric structures on top of each other, and trying to fit them into one space when there are really two. And even if our main interest is in dynamics, we cannot simply ignore the observable shapes and sizes; all our geometrical notions are ultimately derived from them, and we need the distances and times used in describing them.

It appears then that we are thrown back to precisely the same predicament in which Descartes landed. For Descartes certainly realized that bodies did not all have the same density, and that mass (or rather weight) was not proportional to volume. Since his philosophical assumptions required him to interpret mass (or any other physical magnitude) in geometrical terms, he arrived at the unsatisfactory notion of 'atomic' shapes filling containers to different degrees. At this stage of the argument the interpretation in terms of characteristic lengths appears no better conceptually. Let us recall Burtt's interesting observation that Descartes himself suggested that mass and other quantities might be treated as dimensions.[18] However, given his notion of geometry as necessarily three-dimensional and Euclidean, there was no way for him to regard them as additional dimensions$_1$. And if we accept the reasonable principle that all dimensions$_2$ incorporated into a model must have the same (or at least some satisfactory) degree of intelligibility, he could not really introduce them as dimensions$_2$. His monistic conception of intelligibility, effectively ruled out any other dimension$_2$, and only when Newton extended the notion could they appear. If then we are to justify Fletcher's grandiose claim at all, we must demonstrate that the space and geometry involved is so rich and complex that these characteristic dynamical lengths as well as the observable sizes and shapes have a direct and 'natural' geometric significance, as proper measures of structural features of space-time. Classical GR never really addressed itself to this vital question, and thus Törnebohm had reason to suspect that his results had only syntactical significance. However, classical GR was still willing to accept 'foreign' dimensions, albeit reluctantly. Geometrodynamics *does* attempt to answer the problem of how to interpret these lengths, by appeal to the topological properties of the more complex space-time assumed within its model, as I will try to show below. However, I believe that even here its answer is not wholly satisfactory.

If we are to evaluate the empirical adequacy of GR, we must turn from these very general considerations to particular solutions, i.e., the values of g_{ik} corresponding to particular configurations of the possible source terms. Now it is a mathematical truism that a differential equation does not determine a unique solution by itself. We must also specify precise initial and/or boundary conditions. In all of classical physics it was assumed that the laws and conditions were logically independent of each other; they could be specified quite separately, but in combination they somehow generated the solution desired. In GR we have the additional task of specifying an appropriate coordinate system, with no universal requirements on the operational significance of the particular coordinates chosen. However, in classical physics it was usually possible to write down a general solution of the differential equation; the introduction of boundary conditions then amounted to little more than the specification of what appeared in the general as arbitrary parameters.

Because of the great generality and mathematical difficulty of Einstein's field equations, as discussed above, GR has normally followed precisely the opposite procedure. What one does instead is to *start* with boundary conditions. Since the metric tensor does have direct geometrical significance, any geometrical boundary conditions (such as symmetries) can be incorporated as restrictions on the ultimate form of the metric tensor. From this one calculates the various combinations of derivatives leading to the Ricci tensor. Only then are they compared with the specifications for T_{ik} according to the field equation. If this still does not specify the g_{ik} completely (no arbitrary functions, etc.) we finally appeal to comparisons with the corresponding Newtonian solution. This last step also has the advantage of enabling us to give a conceptual interpretation of any remaining parameters. Let us see how this older 'plan' works by considering its application to the best known and most important special case, that of spherical symmetry.

There are many advantages in considering spherical symmetry. In the first place, we know that, to a good approximation, the sun and planets may be represented or modeled as geometrical spheres when taken as sources of gravitational fields. Since this is the area where we most readily expect to find observable gravitational effects, it will be most useful in empirical tests. In the second place, it is probably the mathematically simplest form still capable of representing a physical situation, except for the Minkowski metric for empty space. If we choose as our coordinates some measure of time and radial distance and two angle variables, spherical symmetry means that the actual values of the last two never appear in the equations, the g_{ik} must be functions only of the t and r, and the angle variables must appear in the expression for ds^2 only in the combination $d\theta^2 + \sin^2\theta d\psi^2 = d\Omega^2$. Thirdly, we have a simple way of representing the 'shape' of such solutions. Consider, for example, a three-space with cylindrical symmetry and coordinates r, z, and θ. Then any function $F(r,z,\theta) = 0$ may be represented as a surface of revolution gained by rotating the corresponding curve $F(r,z) = 0$ around the z-axis. In the GR case we have two suppressed dimensions, and can rotate any $f(r,t) = 0$ to get a hypersurface of revolution in four-space.

There are a variety of ways of writing the most general form of g_{ik} for spherical symmetry, depending on which coordinate system is chosen. Synge gives six, each with certain advantages which depend on the problem at hand.[19] All can be derived from each other by appropriate coordinate transformations, and Basri has shown how the coordinates used can be determined operationally.[20] The isotropic coordinates discussed above may be used, giving $\phi = (1 - m/2\rho)^2(1 + m/2\rho)^{-2}$, $\psi = (1 + m/2\rho)^4$, $\rho^2 = x^2 + y^2 + z^2$ as the "Schwarzschild exterior solution," the metric outside of a spherically symmetric distribution characterized by the parameter m. The calculation has been based on the assumption that $R_{ik} = 0$ throughout this region, and by the Newtonian comparison m can be identified with the 'geometrized mass' $m^* = Gm/c^2$, where the latter m is the ordinary mass measured in grams, for example. It is usually most convenient to use 'curvature coordinates,' in which we have $ds^2 = -d\tau^2 = -\phi dt^2 + \phi^{-1}dr^2 + r^2d\Omega^2$, $\phi = 1 - 2m/r$. The latter form is actually more general than the simple Schwarzschild solution, if we replace the given ϕ by other functions of r. In particular, the substitution $\phi = 1 - r^2/R^2$, where R is a constant, corresponds to the "de Sitter universe" of cosmology, and $\phi = 1 - 2m/r + q^2/r^2$ corresponds to a source with charge as well as mass, the Reissner-Nordström solution. The q is again 'geometrized' charge $q^* = G^{1/2}q/c^2$. In the latter case T_{ik} does not vanish, since the electrostatic field which the charge sets up carries energy and thus must be treated as a source term in the field equation. However, the T_{ik} can be calculated in this case, and we get the result by merely substituting into the somewhat more complicated inhomogeneous field equation. Other solutions for cases of spherical symmetry have been found, even if they cannot be written in this simple form, such as the interior solution inside a star of uniform density. We should also note that the r and ρ which appear in the equations above are not operationally direct measures of length, such as would be obtained by laying a ruler along a radial line. We can indeed write down a metric corresponding to such a direct interpretation of the radial coordinate, but the resulting tensor will be less simple and useful. In fact r does satisfy the condition that for fixed r and t, the area of the corresponding sphere will be $4\pi r^2$ (The Euclidean value), and ρ will likewise give us the Euclidean volume of the sphere.[21] As Robertson has pointed out, if we use the directly measured length in a non-Euclidean space, we must add correction terms to the Euclidean values for area and volume.[22] By the principle of covariance, any measure of radial length is equally legitimate.

The Schwarzschild solution is thus a rigorous and exact solution of the field equations for the (possibly idealized) case where the source is a perfect sphere of uniform density endowed with total mass m (in the limit, a mass-point). No approximations have been made or required. Since we have solved the first part of the dynamical problem by calculating the metric, it would seem that we should simply substitute these values of g_{ik} into the law of motion, and then solve the resulting differential equations of motion, to determine particle trajectories or orbits in this metric. If we use isotropic coordinates x, y, z, t these equations take the form discussed above upon substitution for ϕ and ψ. If we use curvature coordinates, we find that we can orient them so that all motion is confined to a plane. Then the differential equations for t and the remaining angle give immediate first integrals, which introduce constants of the motion k and h. The constant $h = r^2\ (d\psi/ds)$ (here ψ is the angle variable) is in fact proportional to the classcial angular momentum, and its conservation law also holds for the Schwarzschild solution. But once this is done, we no longer have to solve the most interesting 2nd-order equation, that for r—which we certainly did need in the Newtonian case. We simply divide the expression for the line element by ds^2 and substitute $d\theta/ds = 0$, $d\psi/ds = h/r^2$, $dt/ds = k/\phi$, giving a *first*-order equation for r, specifically $(dr/ds)^2 + h^2\phi/r^2 = k^2 - \phi$.

But now a curious fact emerges. If we substitute for ϕ the Schwarzschild value $1 - 2m/r$, and multiply the whole expression by $\frac{1}{2}\mu$, we get

$$\frac{1}{2}\mu(k^2 - 1) = \frac{1}{2}\mu\left[\left(\frac{dr}{ds}\right)^2 + r^2\left(\frac{d\psi}{ds}\right)^2\right] - \frac{\mu m}{r}\left(1 + \frac{h^2}{r^2}\right).$$

If we identify μ with the mass of the test particle whose motion in the field is being investigated, we note that the left-hand side of the equation is a constant, and the bracketed term on the right is *formally* analogous to the classical definition of kinetic energy. This suggests that we might consider the remaining term a 'potential energy', enabling us to write $E = T + V$, i.e., the constant total energy equals the sum of kinetic and potential energies. It was also a standard move in classical mechanics to remove $d\psi/ds$, the angular velocity, by introducing the constant angular momentum. This enabled us to replace the other kinetic term by an apparent potential term, the 'centrifugal potential', whose derivative gave the 'centrifugal force'. If we follow this same procedure, we can write the total apparent potential for the Schwarzschild case as $V(r) = -\mu m/r + \mu h^2/2r^2 - \mu m h^2/r^3$. If we then define the 'total apparent radial force' by $F(r) = -dV/dr$, this force will be $F(r) = -\mu m/r^2 + \mu h^2/r^3 - 3\mu m h^2/r^4$. In these expressions the first term represents the classical Newtonian attraction, the second the classical centrifugal repulsion, and the third is the sole relativistic correction, an attraction proportional to r^{-4}.[23] It is important to remember that radial velocity is measured with respect to proper time rather than the coordinate time t, and that r is not precisely the same as Newton's, as mentioned above. However, subject to these limitations the formal similarity remains striking. In particular, since the relativistic correction also depends on h, we see that for purely radial motion ($h = 0$), the apparent force will be simply the Newtonian inverse-square attraction. Since μ, m, and h all have units of length in our system, and will normally be very small in comparison with r, we can see why the Newtonian approximation is so accurate. Now suppose that we are interested in the overall shape of the orbits in such a field, as given by $dr/d\psi$, rather than the actual motions in time. It turns out that this too differs from the classical expression only by the addition of a small correction term.[24]

These considerations are important because of the significance of the

Schwarzschild solution in all attempts at testing GR empirically. Text after text, commentary after commentary repeats in chorus the claim that there are precisely three empirical tests of GR, and that it stands up reasonably well to all three. The equivalence principle does not count here, since it represents a fundamental assumption that leads to the use of the whole formalism, rather than a specific prediction derived within the framework of the theory. However, we still have the precession of the perihelion of Mercury, the bending of light rays by the sun's gravitational field (observable during an eclipse), and the gravitational red shift. It is not my business to determine just how accurately our best observations agree with predictions here, but rather to point out how much (or little) they really test.

Schild has made a careful study of the gravitational red shift in relation to the equivalence principle.[25] He concludes that it can be predicted directly from the equivalence principle, without any appeal to the full covariant formalism of GR or any particular solutions. Schild concludes that the principle rules out any gravitational theories based on a flat space-time, unless one is willing to accept additional arbitrary and ad hoc assumptions within such a framework. However, there are indefinitely many theories using a curved space-time that are compatible with the principle and might even incorporate it in a more natural way. No appeal to the red shift would enable us to single out any one (like GR) among such theories. At best our observations of the red shift would then provide additional confirmation of the equivalence principle, which can also be tested directly, and thereby provide further justification for giving up flat space-time. Once we do operate within GR, we can predict it directly using the Schwarzschild solution; still this just shows that GR is *one* of the possible curved-space theories, not that it is the only one. In addition, the red shift simply describes the behavior of natural clocks in a gravitational field, and makes no reference to any motions, whether of light or of particles. Thus it tests only the g_{00} component of the metric tensor, treating it as if it were a scalar, and relates coordinate time to proper time.

The bending of light and the perihelion shift are in one sense more general, since they involve the other components of the metric tensor and thus bring in the whole structure of space-time, rather than time alone. But in another sense they are more specific. They do not refer to arbitrary fields, in which g_{00} might have any value, but only those with spherical symmetry. They cannot be derived from principles as general as equivalence within the framework of GR, but only from the particular Schwarzschild solution. Insofar as other curved-space theories might well have a different field law, metric, or law of motion, their predictions for spherical symmetry might well be different, and the observed results could rule some of them out. Nevertheless, once we have actually made the calculations, we could arrive at mathematically identical results from any number of quite different conceptual standpoints, including flat-space theories. We could even be so crude as to take these results as mere constraints, and adjust any available parameters in these other 'theories' to make sure that they give exact agreement. This in fact was the point of our previous discussion of 'potentials' and 'forces'. For conceivably we might even keep classical physics, with its Euclidean space and central forces, and simply postulate the existence of a small additional inverse-fourth attraction. No new parameters are needed; the term involves only mass and angular momentum. It is not clear how we might interpret such a new force, or justify it conceptually beyond its ability to save the phenomena. Nevertheless, this might seem a much smaller price to pay, rather than taking the revolutionary step of replacing one whole conceptual framework with another.

If these results had really been considered important and well-known anomalies within classical gravitational theory, it is likely that theories like that outlined above would have been tried at least. In fact they were not. The gravitational red shift and bending of light are too small to be noticed unless one is actually looking for them, and since classical physics gave no reason to expect them, no one did bother to look. Likewise the perihelion of Mercury, which must be observed over a long time before a significant shift can be noticed, was so small that it was more convenient just to live uneasily with it, rather than postulating a new force. Only recently has Dicke suggested a possible use for my "pseudo-classical" expression for the potential, interpreting some ten percent of the correction term as the effect of a mass quadrupole moment resulting from the flattening of the solar disc. In fact, Dicke's approach is not quite identical to the hypothetical one I have considered, and while it may be adjusted to give the perihelion shift for Mercury, it does not agree with the Schwarzschild predictions for the other planets, and the still scanty evidence favors Einstein here.[26]

I conclude that "the three classical tests" are tremendously overrated if they are taken to provide an empirical foundation for GR. At best, they test special cases, and indeed cases where the field and motion laws of GR, its supposedly fundamental laws, were used only indirectly. The effects are all extremely small and hard to measure precisely, especially in comparison with the much larger phenomena ordinarily attributed to gravitation. They could all be explained (in the formal sense of mathematically derived) equally well within many other systems, though not Newton's own. And they are certainly not the basis for nor the main impetus to GR. Whether or not Einstein knew of their existence in 1915, he certainly did not develop GR as a theory to explain *them*, either individually or together. If they are central to anything, it is to the ad hoc "theories" which came along after the fact and tried to reproduce them. Newton's theory of gravitation gave such good results observationally that Einstein could hardly suspect where (if anywhere) GR might predict discrepancies large enough to be tested. The existence of these effects, which did include the one remaining observational difficulty for Newton's theory, is thus only a happy accident.

Should we then conclude that GR rests on a shaky empirical foundation? Not really. Given the actual magnitude of the speed of light, we know that GR should agree with most classical predictions of gravitational phenomena within foreseeable margins of experimental error. But if this is so, then the vast body of observations which confirm classical physics may also be taken as confirming evidence for GR. What is important is that there is no area where GR is known to disagree with observations, and in most others we have good reason to assume it innocent until proved guilty. To me the more striking fact is that two theories with such radically different conceptual and mathematical approaches can agree over such broad areas. One might very well have expected the contrary.

In the absence of other test cases not involving the structure of the universe as a whole, it follows that the main arguments for accepting GR rather than Newton's theory must remain conceptual rather than observational. Each appeals to different principles of intelligibility and ideals of natural order. Each is defensible, and each has certain defects from the standpoint of the other. But an empiricist will still ask whether these are the only three tests, or why there should not be more, in the hope that the decision could ultimately be made on this basis. Part of the problem of finding new experimental tests lay in the fact that gravitation, in its Newtonian version, was far less interesting than electromagnetism conceptually. Every gravitational problem was solved within Newton's theory in the same way, adding or integrating over individual sources using the inverse-square law in its integral form. Boundary conditions played a very small role, understandably enough, since gravitation as a universal force could not be shielded out by introducing boundaries. If they were material, they had to function as additional source terms. Electromagnetism had a vast variety of special cases, with mathematically different solutions of Maxwell's equations reflecting different boundary conditions. This in turn led to a large number of new concepts and distinctions lacking for gravitation.

With this in mind, I suggest (following Wheeler) that the difficulties in appreciating the real content of GR and knowing where to look for further empirical tests stem from an inadequate *plan*, as I have introduced the technical term in Chapter 4. We had the formalism, but no idea how to *use* it effectively to generate particular solutions or understand its inner dynamic.

In trying to exploit the full richness possible in a tensor language, we used only the paucity of distinctions taken from the scalar language of Newton's theory, and thus could only scratch the surface. The derivation of the Schwarzschild solution began by taking the situation most simply described in *classical* terms. The field law used, $R_{ik} = 0$, actually assumes the absence of matter throughout the region in question. The mass m which plays such an important role in Newton's law here enters only surreptitiously in the boundary conditions as a constant of integration. What is most natural and basic for Newton appears rather specialized and arbitrary within GR. It therefore seems desirable to look at the laws and possible values of g_{ik} and R_{ik} in more directly *geometrical*, rather than classical terms, and introduce a new set of distinctions on this basis. To put it more crudely: only by forgetting Newton and what he told us gravitation was all about can we develop GR fully, and then from this broader perspective find new and perhaps wholly different areas for possible empirical test.

We have seen above that Newton's inverse-square force law and corresponding Gm/r potential had a universal character, applicable to all physical situations. However, the possibility of such a universal integral form depended on two crucial features: (1) The gravitational field generated by any body depended only on a single parameter characteristic of that body, its mass; other properties like its charge, and any velocities or other characteristics of its state of motion were quite irrelevant. (2) The basic partial differential equation for the field (Laplace's equation) was linear, so that the sum of any two solutions to the equation was also a solution. If each of the 'partial solutions' was taken to represent the gravitational field from one particle, their sum was then the total gravitational field from the whole configuration of particles. For better or for worse, neither of these features obtains in GR. The use of T_{ik} rather than simply m as a source term already incorporates all those properties of matter or nongravitational fields which can carry energy, in order to satisfy the equivalence principle; ideally, it should provide a complete, rather than just a partial description of 'matter' in the broadest sense.

The second difference, involving a change from linear to nonlinear equations, is perhaps even more important. For even the homogeneous equation $R_{ik} = 0$, which gave us the Schwarzschild solution, is nonlinear. The Christoffel symbols are indeed linear combinations of first derivatives of the potentials, but the Riemann tensor uses multiplicative combinations of the Christoffel symbols as well as derivatives of them, and the Ricci tensor is formed by linear combinations of components of the Riemann. If this were not so, we might hope to represent the metric tensor corresponding to a collection of point masses as a sum of Schwarzschild solutions in the form $\phi = 1 - 2\Sigma_i(m_i/r_i)$, where the m_i are the masses of the particles and the r_i are the distances from each of them to the point where the field is being measured. Formally, this would be the closest analog of the total Newtonian potential. However, it is a *mathematical* fact that when the equations are nonlinear, such a sum will normally not be a solution. And as yet no one knows any other natural algorithm for combining separate solutions.

The nonlinearity of the field equations is a consequence of the requirement of general covariance. Bergmann writes,[1] "It is impossible to envisage a set of differential equations which is covariant with respect to general coordinate transformations, yet linear." In classical mechanics and electromagnetism, this problem did not arise, since both theories used only the Galileo or Lorentz group. "The group of the general relativity is the first one which demands that the simplest invariant law be no longer linear or homogeneous in the field-variables and in their differential quotients."[2] But if this is true, it means that covariance is much more than simply a decision to use a certain kind of mathematical language, i.e., general second-order tensors, to describe phenomena such as gravitation; it places an important *physical* restriction on the kinds of theories that can be acceptable. For linear and nonlinear theories are completely different conceptually. With a linear theory both sources and effects are all quite independent of each other: they may happen to coexist, and in this case their magnitudes can be added according to the superposition principle, but none of them is modified at all in the combination. Each retains its own identity, and would make precisely the same contribution if the others were not also present. Not only can we add them, but given a total field, we can always break it down into parts in a unique way, and ascribe each of these parts solely to the influence of a particular source. On the other hand, in a nonlinear field the whole does not equal the sum of its parts. These are intimately interconnected rather than being independent, for each is modified by the presence of the others. Given a total field at a point, we must treat it as a whole, and not expect to break it up into many independent parts.

These differences are physical not only in the sense that they require models with different conceptual distinctions. They may be expected to lead to different observational results. We recall that in isotropic coordinates the spacelike part of the Schwarzschild metric could be written in the form ψdl^2, where $\psi = (1 + m/2\rho)^4$. Now it turns out that if a distribution of point masses can be assumed initially at rest at some instant specified by $t = t_0$, we can in fact write $\psi = (1 + \frac{1}{2}\Sigma\alpha_i/\rho_i)^4$, which looks analogous to the linear combination that we supposedly ruled out earlier. The problem is that the α's are *not* identical with the m's, as they would be in a linear theory. We have instead $m_i = \alpha_i\Sigma_{j\neq i}(1 + \alpha_j/\rho_{ij})$ where ρ_{ij} represents the distance between the ith and jth particle. We could also solve for the inverse, giving the α's in terms of the m's. However, the net effect is that any attempt to break down the total field must be in terms of the α's, rather than the m's. Since both the m's and α's must be positive in sign, the m's will be larger than the α's, and the total ψ will be smaller, perhaps to an observable degree, than if we used the m's. In summary, the appropriate source parameter for particle i is m_i if it is acting by itself, with no other sources around; but if there are others, its source parameter is α_i, and its individual contribution is altered by those of the others.[3]

If this change in 'apparent mass' could be experimentally tested, it would provide a much stronger test for GR versus any classical linear theory (such as the inverse-fourth force idea) based on the Schwarzschild solution for a single particle, and it would bring out the radically new character of GR. The problem, of course, is that we cannot exercise laboratory control over concentrations of mass large enough to show these effects in a measurable way. The masses that we encounter either come by themselves or, like the sun, dominate the region in question; or else they come in a group. We cannot pull them apart and put them together to measure the difference. Furthermore, Wheeler has remarked that the nonlinearity of GR is a rather peculiar sort. We have already noted that the equivalence principle requires that with any field, we can always find a coordinate system in which the metric is Lorentzian locally (i.e., in the immediate vicinity of a given point). This postulate was in fact the basis of our using the two-dimensional g_{ik}, rather than some more complicated g_{ijkl}, to describe the metrical structure of space-time.[4] But in such a Lorentz metric the equations will appear linear. Our field then will appear locally linear; nonlinearity is a global property that shows up only when we investigate larger regions of space-time. If we confine ourselves to local measurements, we cannot detect it.[5]

It is obvious that nonlinearity has great disadvantages for anyone who is interested in particular solutions. Each case must be solved separately, with the source configuration taken as a whole, and the mathematical difficulties are so formidable as to require falling back on Newton except in very special cases. Along with the inadequacies of its original plan, these difficulties are primarily responsible for the lack of progress in theoretical research between 1925 and 1955. On the other hand, GR has one remarkable advantage. For in GR the laws of motion (e.g., the geodesic law for the motion of free test particles) do not have to be postulated as separate principles, independent of the field equation; in fact, they follow necessarily from it. This situation was unparalleled in classical physics, and it can be shown that in any linear theory one must postulate separate laws of motion; by themselves the field equations allow indefinitely many possible motions. Thus nonlinearity is a *necessary* condition for being able to derive equations of motion, but whether it is also sufficient seems open to question. Bergmann concluded confidently in 1949, "All theories which are covariant with respect to general coordinate transformations share the property that the existence and form of the equations of motion is a direct consequence of the covariant character of the equations,"[6] and tried to sketch how the derivation might be carried out in the general case. This is a far stronger claim than the conclusion that it requires nonlinearity. If true, it would give the covariance principle that much more physical content, since it would place immediate restrictions on possible motions. Nevertheless, the actual GR derivation in all special cases carried out so far has appealed to a more specific feature of Einstein's actual theory, and in 1956 Infeld, one of the leading figures in this research, insisted more cautiously that there was no general criterion for determining when the laws of motion could be derived from a nonlinear field theory.[7]

Historically, of course the geodesic law was first postulated as a separate principle. It was Einstein's replacement for Newton's first law of motion (which of course widened the range of 'inertial motions') and he no more suspected that it was derivable in GR than it had been in classical mechanics. The connection is even more remarkable because in an important sense the law of motion is more basic than the field equation. It is an attempt to embody the content of the equivalence principle in geometrical terms, and does not use the many other assumptions required to get the field law. It is therefore probable that the geodesic law is compatible with other field equations based on somewhat different assumptions, but not clear whether it could also be derived from them. If not, the two laws would function as preconditions of each other. The actual connection was first proved for a quite special case by Einstein and Grommer in 1927, and since then most other cases of interest have also been solved.[8] Not only the geodesic law, but the Lorentz law of motion for a charged test particle in the combined gravitational-electromagnetic field, can be derived in this way.[9]

The crucial feature appealed to in these derivations is the vanishing of the covariant divergence of the Einstein tensor, $G^{ik}{}_{;k} = 0$. We have seen earlier that it was precisely this property that led to the choice of G_{ik} as the geometrical representation of matter, since it was desirable that $T^{ik}{}_{;k} = 0$. In our earlier discussion it was enough to recognize that these four additional constraints gave us only six equations for ten unknowns (the g_{ik}) which in turn reflected the covariance requirement that any coordinate system and its corresponding g_{ik} is equally good for describing the geometry. But now let us look at the special character of these constraint equations more closely. With the T's, they appear as conservation laws, which already place constraints on the possible motions, whether or not they determine them fully. Now any test particle, even if it is considered small enough not to affect the total field significantly, is not a foreign entity. Its description must satisfy the same law as any other T^{ik}, whether it is represented as a continuous distribution in space or as a singularity, using Dirac's delta function. In contrast, Newton's field law had only mass, with no restrictions whatsoever on it, as a source term. If one is not fussy about rigor, it is relatively simple to show how this requirement leads to the geodesic law.[10] If one imposed four constraints in a covariant but otherwise arbitrary way, one would not expect a unique law of motion, or at least not the same one as Einstein's. If one used the vanishing of the divergence but not the general covariance group, one would not get the geodesic law; for example, the Lorentz group would give only Newton's first law and purely rectilinear motion. It is the combination of the two conditions that gives the desired result. Nonlinearity appears indirectly in that the vanishing of the divergence of G^{ik} depends on its mathematical character which involves multiplicative combinations of the Christoffel symbols.

In an effort to carry out the derivation more rigorously, three main lines of approach have been taken, depending on whether the test particles are interpreted as (1) point singularities in the field, (2) extended objects with internal hydrodynamical structure, or (3) geometrical configurations within the continuous structure of space-time. Einstein, Infeld, and their co-workers chose the first. In discussing the meaning of the field equations and the significance of T_{ik} above, we considered Einstein's objections to using such a "foreign tensor" as a source. Since the same objections applied to its use for test particles, and Einstein did not have the fully developed third interpretation, the one actually taken up by geometrodynamics, he was forced to fall back on singularities. Singularities are undesirable at best from the standpoint of a field theory that hopes to be complete, but they can still be useful approximations. It is an important fact about GR, closely related to its nonlinearity, that we cannot introduce a collection of singularities in an arbitrary way. We have seen above that they are not independent, and their effects can modify each other to the point where the equations involving the approximation of independence are no longer satisfied. It turns out that between two points with a timelike separation, there is only one line of singularities in space-time connecting them that will still satisfy the field law, and this turns out to be precisely the geodesic line connecting them. The equations of motion thus appear as conditions for the existence of solutions.[11] In practice, a difficulty arises from the fact that any test particle with nonzero mass sets up a field of its own which modifies the original field in its vicinity in a nonlinear way. This in turn requires a complicated sequence of successive approximations.[12] In an important sense, this first analysis is closest to that of Newtonian mechanics, where point particles could also be represented as singular points where the mass density goes to infinity.

The second approach, as developed by Fock and Papapetrou, avoids the problem of singularities and is able to use continuous distributions throughout. However, it requires an interpretation for each of the components of T^{ik}, and uses special coordinate conditions which restrict the generality of the covariance.[13] The third approach rejects "the problem of motion" insofar as it breaks down the total structure of the field into a moving object and a background against which it moves. Certain structures within this total field may be identified (in cross-level coordinations) with ordinary test particles, but in reality we have only a single field which evolves in time according to the field law.[14] We will return to this line of analysis later.

Still other approaches to the derivation are possible, all leading to the same general result, the geodesic law. In keeping with the attempt to combine Schwarzschild solutions, Lindquist and Wheeler have considered a world composed of particles in a uniform lattice. Each has its own sphere of influence, in which the metric is approximately Schwarzschildean. But the boundary conditions between these spheres function as a dynamical constraint, determining a uniform expansion of the whole system akin to that predicted in some cosmological theories of the expanding universe.[15] It has also been possible to treat the two-body problem in the case where both masses are finite but one is much larger than the other, by an analysis based on perturbation theory, the analog of that used in planetary theory in comparing the effects of the sun and those of other planets.[16]

Up to now, we have been considering the various benefits that accrue to a theory of gravitation based on the equivalence and general covariance principles. However, there is a darker side to the picture. If we accept the unlimited freedom to choose coordinate systems that general covariance requires, we must be prepared to reject as ambiguous or meaningless many of the most important concepts and distinctions that were used throughout classical physics, both in interpreting the dynamics of classical theories and in connecting them with observations, insofar as these concepts appealed to particular preferred coordinate systems. In being too unrestricted, the notion of 'coordinate system' becomes so vague and general that very few physically interesting quantities are invariant in all of them, at least without changing the definitions given in classical physics. This was undoubtedly among the reasons that led Bridgman and Fock to reject or disparage the role of general covariance and replace it with specific coordinate conditions, as we have discussed earlier.

Throughout classical physics such notions as mass, energy, and momentum (both linear and angular) played a very central role. Whatever difficulties might arise in determining their magnitudes operationally, they were at least clear and well defined within the model and its corresponding formalism. But their greatest value, which distinguished them from all other concepts that might have been defined within the theory proper, lay in the fact that they obeyed conservation laws. They were macroscopic quantities that represented integrals of the local variables describing the detailed behavior of physical systems. There might be a great deal of reshuffling among the energies and momenta of the elements composing such a system, but when these were all added or integrated together, the total remained constant through time. This was a tremendous advantage, for if we were primarily interested in before-and-after phenomena we could neglect all the intervening steps and treat the system as a 'black box'. This was especially useful in such areas as collision problems, where the actual mechanism was quite unknown and presumably very complicated, for we didn't need to worry about it. Without conservation laws we might have to know every step in detail.

Wigner has pointed out that most of the classical conservation laws could be expressed in geometric terms, for they reflected the symmetry properties of Euclidean space and time.[17] Conservation of energy was thus bound up with equivalence of all instants in time, momentum with spatial homogeneity, angular momentum with spatial isotropy or equivalence of direction, etc. If a closed system, not subject to any external forces, could allow these quantities to vary, we would have to assume that space and time were themselves causal agents, responsible for the forces producing these changes, since by assumption nothing else could. But it was essential to the idea of space and time in classical physics that they should be completely passive and neutral, a fixed framework in which all motions could be compared.

When we introduce Riemannian space-time and generalized coordinates, we lose all these features. The manifold may be very different at different points and directions locally, whether or not these effects average out on a cosmological scale. Furthermore, we *do* expect space-time to be a causal agent. The gravitational effects (or what were described as such in Newtonian physics) are still perfectly real; all we have done is to ascribe them to the structure of space-time instead of to an independent force of gravity. Finally, if the resulting space-time were to have any symmetries that might show up in the Riemann tensor as its fundamental geometrical description, we would still be required by covariance to allow coordinate systems in which these symmetries would not show up at all. If our classical concepts and laws are to remain meaningful, we must define them in a more complex and less intuitive way. Still, they are too valuable to give up without a fight.

Synge has pointed out that conservation laws can apply to many things besides physical quantities, and that their existence in general depends on a mathematical fact: the integral of the *ordinary* divergence of some quantity (vector or tensor) over an n-dimensional 'volume' is equal to the integral of the quantity before taking its divergence over the $(n - 1)$-dimensional hypersurface enclosing that volume. This is the generalization of Stokes' theorem.[18] From this it follows that if the divergence vanishes throughout a region of space-time, the integral of the quantity over a three-dimensional hypersurface enclosing that region must also vanish. If, for example, we specify the region as a 'cylinder' in space-time with its ends corresponding to the values t_0 and t_1 of the time coordinate at each point, and the 'sides' sufficiently far away from the system that no influence comes in or out, and the quantity is zero along those sides, then the integral over the space at t_0 must equal the integral over the space at t_1. We may then conclude that the quantity, as integrated over all the space in a given region (which may extend to infinity if necessary), is a constant independent of time. The value of that constant, which is an integral of motion for that system, will be the value of the conserved quantity.

We have already made two illegitimate references to conservation laws so far. In setting up the field equation, it was decided that T_{ik} should serve as the source term and represent matter in the most general sense, but this did not determine what geometrical quantity should be equated to it. The Einstein tensor was in fact chosen primarily because its *covariant* divergence vanished identically, and we wanted the covariant divergence of the matter tensor to vanish as expressing conservation of energy and momentum. Strictly speaking, this justification for the requirement was both unwarranted and false. In SR, $T^{ik}{}_{,k} = 0$ does express this conservation law, but for SR ordinary and covariant divergence are identical. Conservation laws are based on ordinary, not covariant, divergences. The decision to use covariant, rather than ordinary, divergence in GR stemmed only from the fact that it was a tensor, while ordinary divergence may or may not vanish, depending on the coordinate system in Riemannian geometry. We will also see that T^{ik} *by itself* does not obey a conservation law in GR. Secondly, in discussing the laws of motion in the previous section, we claimed that their derivability was related to the four constraints $T^{ik}{}_{;k} = 0$, since these were supposed to be conservation laws, and such laws provide one way of representing laws of motion.

It is really not that surprising that the T^{ik} should not be conserved. In discussing nonlinearity in the previous section, we observed that a total effect cannot be represented as a simple sum of separate contributions. The remaining contribution, beyond what one might expect from such addition, must then come from the complex interaction of the fields themselves. By the equivalence principle, if such gravitational fields do affect motions, they must also have inertial properties, and it is to *inertial* properties—the relations among the energies, momenta, etc., of the parts of a system—that the conservation laws refer. Since they can do mechanical work, such fields must carry transferable energy, and these energies must also be included in any overall conservation law.[19] In a given coordinate system, such a field shows up in the form of nonvanishing Christoffel symbols, and unless the field is uniform, it cannot be transformed away by a coordinate mesh covering extended regions of space-time. The only difference between these energies and those deriving from other fields or 'brute matter' is that the former are hidden in the structure of the geometric framework, while the latter are explicit parts of the T^{ik}. What we must find is some representation in which they appear in the same way.

Formally, the situation develops as follows: Since the volume element dV over which we hope to integrate a density in order to get a conserved quantity is given by $g^{1/2}$ times the product of the coordinate differentials, it is convenient to use tensor densities (tensors multiplied by $g^{1/2}$) rather than straight tensors. Furthermore, it proves desirable to have tensors in mixed form, i.e., T_i^k. What we would like to find is a set of quantities Θ_i^k such that the ordinary divergence $\Theta_{i,k}^k = 0$. We can then regard the Θ_i^0 components as a measure of the energy and momentum densities. If we then integrate them over the whole spatial region at some time $x^0 \equiv t = t_0$, we expect that the result will be a vector for the total energy and momentum, and thus represent the total inertial properties of the system being studied. If we start from our supposed 'differential conservation law' $T_{i;k}^k = 0$, we find $T_{i;k}^k = g^{-1/2}(g^{1/2}T_i^k)_{,k} - \frac{1}{2}g_{jk,i}T^{jk} = 0$. The last term will not vanish unless the metric tensor is constant. Our problem is then to find some set of quantities t_i^k such that $g^{-1/2}(g^{1/2}t_i^k)_{,k} = -\frac{1}{2}g_{jk,i}T^{jk}$. We can then write $\Theta_i^k = T_i^k + t_i^k$, since $[g^{1/2}(T_i^k + t_i^k)]_{,k} = 0$. We could then say that just as the T's represent the energy-momentum density of all nongravitational fields, the t's are the energy-momentum density of the gravitational field.[20]

Now it is possible to do this; the problem is that in fact it is too easy. It turns out that there are infinitely many expressions for t_i^k, all of which satisfy this requirement. We might begin by using the field law to write out T_i^k, and then calculate both its ordinary and covariant divergences, in terms of the g_{ik} and their derivatives. We might then try to find the t_i^k in these terms from the difference between the two divergences. However, this approach gives us only an equation for the divergence of t_i^k. Since it is a mathematical fact that the divergence of any curl is identically zero, we can add an arbitrary curl term to any proposed t_i^k without affecting the result. It thus appears that unless we impose further restrictions, the notion of gravitational field energy is wholly ambiguous and indeterminate, and if it must be included in any conservation law, that law must share the same indeterminacy. In an attempt to overcome this ambiguity partially and to find a more general form, it is customary to introduce 'superpotentials' h_i^{kj} whose ordinary divergence may be set equal to either the t_i^k or the Θ_i^k. These h_i^{kj} have a mathematically simpler form than the t_i^k, and the additional advantage of allowing us to apply Stokes' theorem a second time. We may replace our integral of the Θ_i^k over a closed three-space by a surface integral of the corresponding superpotential over its two-dimensional boundary.[21] However, as Trautman has pointed out in a careful analysis, there are also indefinitely many of these superpotentials. Each of them, and the field tensors that they lead to, has advantages for certain kinds of problems, and each reproduces *some* of the *conceptual* features of the notion of energy density that seem worth preserving when it is generalized to gravitational fields. Yet none of them can claim universal validity; there are always aspects that are handled better by one of the other definitons.[22]

It may in fact be well and proper to conclude that energy density should not be considered a single concept at all within GR. From this viewpoint there are really a host of distinct and useful notions, each of which can be embodied in its own conservation law. It was only an accident that in the oversimplification resulting from the restriction to flat space-time each such concept always gave the same numerical value, and thus there was no need (or means) of distinguishing them. Such a situation is hardly unparalleled in the history of science. A set of concepts is developed that appears particularly important for an area, or for the best theory available at the time to describe that area. Then an extension is made, either by extrapolating to new ranges of the variables of the theory, analyzing new kinds of conceivable situations, or introducing a more general formalism, as we have done here. The original concepts are no longer perfectly applicable or even meaningful in this extension. We may try to hold onto the original definitions, or their closest formal analogs, but the resulting concepts may be uninteresting, with few lawlike connections to the others in the theory. It is far better to look at the totality of roles which the concept played in the old theory, and see how many of these can be filled by a new notion. It may happen that one candidate stands out ahead of the others by this test. If so, it is the obvious heir to the original name. If not, we must do one of two things: (1) we can *choose* one of them as being the 'real' energy (or whatever concept we are considering), and thereby reject any of the older laws or connotations which it now fails to satisfy; or (2) we can recognize a family of legitimate concepts where before there was only one, using either subscripts (like energy$_1$) or wholly new terms. This may make our theory more complicated, but at least it enables us to say more things. I have discussed elsewhere a similar case, where consideration of the possibility of time travel could lead to distinguishing several possible causality principles for SR, which otherwise appeared identical.[23]

Now the t_i^k are indeterminate in another important sense. We know that the T's do form a tensor, so that if they vanish in one coordinate system they must vanish in any other. This is not true of the t's. We can prove directly that they are not a tensor under any of the proposed definitions, nor should we expect them to be. By the equivalence principle, any gravitational field (and thus the energy and momentum that it might carry) can be transformed away in the vicinity of a point by a suitable choice of coordinates. But in another system the Christoffel symbols and the t's defined in terms of them will not be zero. Thus any notion of field energy appears to depend not only on the formal definition chosen, but on the particular coordinate system. Nevertheless the t's *do* behave like a tensor if we restrict ourselves to linear transformations, as in the Lorentz group of SR, instead of allowing the more general class. This suggests that if we can make some restrictions on coordinates and transformations, it might still be a useful concept.

Møller has devoted a great deal of his research to this problem. In 1959 he felt able to conclude that a uniquely correct expression for the gravitational field energy density followed from the requirement that this density should have as many properties as possible in common with that from other physical fields. Specifically, he set three requirements for the total energy density Θ_i^k: (1) It should behave like a tensor under linear transformations. As mentioned above, this was satisfied by Einstein's choice, the so-called pseudotensor, and many others. (2) Θ_i^0 should behave like a vector density under arbitrary transformations of the three spatial coordinates, as long as the one time coordinate is held fixed. This would enable us to change from Cartesian to polar coordinates, for example. (3) If we have a closed system in which the metric tensor approaches that of Minkowski as we go to spatial infinity at least as fast as does the Schwarzschild solution, the integrals $\int\int\int \Theta_i^0 dx^1 dx^2 dx^3$ should form the components of a four-vector corresponding to the classical total energy and momentum. In particular, the integral of Θ_0^0 should give the same value as the mass m in the Schwarzschild solution.[24] Here (2) refers to densities, and is in differential form. Requirement (3) is in integral form, and functions as the conservation law proper. What it relates are the macro-observables for the whole system.

However reasonable these requirements may be individually, it is a curious fact that they seem to be quite independent of each other, and while most proposed Θ_i^k satisfy two of them, it is very hard to satisfy all three simultaneously. Einstein's choice fails on (2), and Møller finally decided that his own choice failed on (3), leading him to turn toward other approaches to the problem.[25] But it is interesting to see what happens if (3) is satisfied, regardless of any other properties it may have. In the first place, (3) is meaningful only under quite special conditions: the space must not be closed, it must be asymptotically flat at infinity, and it must approach this flatness sufficiently quickly. In other cases the integral is in general not well defined. However, if these conditions do obtain, the notions of *total* energy and momentum do make sense, and behave like their SR counterparts. And this is true despite the fact that the corresponding densities have no clear meaning or geometrical character. The coordinates near the center of the system, and the metric expressed in terms of them, are still arbitrary, and we can always transform these densities away locally. However, in eliminating them at one point in space, we cause them to pop up in another in such a way as to keep the total constant and independent of the coordinate system.[26] We must therefore conclude that energy and momentum in the classical sense are essentially *global*, rather than local, concepts. Put in other words, we can say just how much energy there is in a physical system as a whole; but we cannot say exactly where in this system, or where in space, the gravitational field energy is located. There is no unique distribution function, and whatever one we assign will depend on our choice of coordinates.

The situation is quite unparalleled in classical field theories, where there was a complete symmetry between local properties (densities) and global properties (integrals). Given an arbitrary density distribution, we could integrate its total effect; given such a total quantity, we could always find a corresponding distribution. As we have seen above, this was closely bound up with the linearity of such theories. In GR we may have a density without an integral, or an integral without a density. Local and global properties do not imply each other. Their existence depends on quite different circumstances, and a different set of conceptual distinctions may be appropriate in each realm. In geometrodynamical terms, they are both still properties of space-time, and our problem becomes one of finding just what restrictions each set of properties places on the other. Aside from the asymptotically Schwarzschild case considered here, it may be possible to define total energy, energy density, or both in other special cases, as Thorne has tried to do for cylindrical symmetry.[27] But in the absence of any more general existence theorems, we must remain skeptical of the general significance of integral conservation laws and the corresponding conserved quantities within GR.

It is worthwhile to recall at this point that the equivalence principle was first introduced with the relatively modest goal of explaining the equality of inertial and passive gravitational mass. In view of the discussion above, we may well ask if it still does this job. In SR, the mass that is convertible with energy is clearly inertial mass, since the theory makes no reference to gravitational forces and is concerned only with the response to forces, whatever their origin. This equivalence is carried over into GR, where T_{ik}, rather than the single component T_{00}, is taken to describe the complete local *inertial* behavior of matter and nongravitational fields. Since T_{ik} also appears as the source term in the field equations, we may say that inertial and active gravitational T_{ik} are equivalent, but not mass as such—there is no reference to masses in these expressions. Likewise the principle really asserts the equivalence of inertial and gravitational *fields*, with no reference to any masses by which they might be multiplied to get total forces. As for the equivalence of active and passive gravitational masses, this followed classically from Newton's third law. However, this derivation required instantaneous action between separated bodies. In a field theory where no signal can be sent faster than light, we have an asymmetry of causal influences.

Within the GR that ultimately stemmed from this equivalence principle, none of these three masses has a clear-cut meaning. From the discussion above, we see that if we take the integral of Θ_0^0 as a definition of inertial mass or energy, that integral will be well defined only for special isolated systems and as a property of the whole, whereas every part of a Newtonian mechanical system had an inertial mass of its own. If we take the integral of T_0^0, our answer is not a proper scalar invariant, since its relation to the other components of T_i^k depends on the coordinate system. If we take the integral of any other combination of the T_i^k alone, we cannot satisfy the conservation law, even if in the last two cases we can always perform the integration. For active gravitational mass, the natural candidate is the m that appears in the Schwarzschild solution, since this serves as a measure of the deviation from flatness caused by the presence of that mass. But we have already seen that in the presence of other masses, it is α, rather than m, that determines this deviation and thus α is the 'effective active gravitational mass'. Furthermore, we have mentioned briefly (and will discuss more below) that if the body carries charge, we must add a q^2/r^2 term to $\phi = 1 - 2m/r$. If *total* deviation from flatness is to be our measure, we must conclude that charge contributes negatively to active gravitational mass, which is quite at variance with classical notions. Finally, passive gravitational mass does not really exist as a concept within GR. It has already been encompassed in the geodesic postulate. Besides, it could appear only as a limit concept. Any particle with finite mass does not just respond passively to existing gravitational fields, so that we could measure a pure response parameter. It sets up a field of its own, which interacts with other fields and causes the test particle to move on a geodesic of a different space from that which would have obtained without its presence. Rather than saying that two conceptually distinct quantities are in fact equal, it is better to say that *any* mass is actively gravitational.

Møller has required that when the appropriate calculations can be made, the values of all three masses should come out equal. This may be useful in fixing the value of constants of integration, and Jammer has shown that it does hold true in certain simple cases.[28] Nevertheless, it should be recognized that the classical notions of mass, energy, and momentum may have only a very limited application in GR. Since most applications of the theory have been to situations which were *previously* described in such classical terms, it is natural to expect analogs for them. However, there is no reason why they should be important in the inherent geometrodynamical model of GR. They are intimately related to the classical model of separate and distinct particles, each of which had an identity described by characteristic parameters that were quite independent of its environment. By now it should be increasingly clear that the model for GR must be based on interdependence, rather than independence.[29] Since these concepts provided the basic description of the dynamics of classical physics, it is clear that GR must introduce something to take their place. But this something can be understood only by analyzing the canonical variables appropriate for the dynamics of GR, as we shall see below in connection with its new plan.

Finally, before turning to the new interpretation of GR as it has developed since 1955, we must consider one more problem in the classical form of the theory, namely, the role of singularities. In analyzing the problem of motion, Einstein had appealed to the use of singularities, rather than phenomenological representations of extended matter, as the lesser of two evils. Singularities are not only esthetically unappealing, but they prevent us from having a 'complete' theory, in the sense defined in Chapter 8. We may avoid the appeal to foreign, nongeometric tensors, but only at the price of introducing regions where the continuous structure of the metric field breaks down. The field thus does not cover the whole of space-time. Einstein's own justification for their use was really more like a hope; he suggested that the appeal to singularities was a temporary makeshift comparable to the artificial isolation of gravitation from other fields. If it were possible to develop an adequate theory of the total field, including electromagnetism and everything else, such a theory might also be complete and thus fully unified.[30] What appeared as a singularity from the crude perspective of pure gravitational theory could be given a finer analysis in terms of the concepts of unified field theory.

Whatever the ultimate prospects of such a theory may be, their use in GR is not satisfactory—unless it is impossible to avoid them, a difficulty which might serve as a major criticism of GR. For the term 'singularity' covers a great variety of distinctions, each of which could have different physical significance. Singular regions of space-time can have 0, 1, 2, 3, or 4 dimensions; they may appear as the limit of regions of increasingly sharp curvatures, like horns or cusps, or as sudden 'holes' in an otherwise relatively flat manifold; they may affect all geometrical quantities, or only some of them; and they may have devastating effects on some physical processes, but none at all on others. If within the conceptual structure of GR there were some natural means of distinguishing which singularities are allowed and which are forbidden, or perhaps even if we could find some accurate and effective general criterion, however arbitrary and nonintuitive it might seem, the problem would not be as important. However, there seems to be no such criterion.[31] And even if one existed, we would still have to know just what forms of matter corresponded to each kind of allowable singularity. Without such a criterion, if we allow any one we have no real grounds for ruling out any others. Thus it is desirable to remove any apparent singularities if at all possible.[32]

One suggested distinction stems from the fact that coordinate systems are arbitrary, but truly geometrical invariants like the scalar curvature R are not. Thus if R or some other invariant goes to infinity at a point, we may conclude that this is a genuine geometrical singularity, independent of any particular system, which will affect any motion or physical process approaching that point. On the other hand, if R remains finite, we may suspect that the apparent singularity in the metric tensor can be removed by a coordinate transformation, and its physical effects will at least be less spectacular. The Schwarzschild solution provides a good example of both cases. Since $g_{00} = -\phi = -(1 - 2m/r)$ and $g_{11} = 1/\phi$ in the usual coordinates, they will go to infinity or zero (in which case the corresponding g^{ik} will go to infinity) at *both* $r = 0$ and $r = 2m$. The scalar curvature R also goes to infinity at $r = 0$, but remains finite at $r = 2m$.

Kruskal has found a system of isothermal coordinates such that $ds^2 = f^2(u,v)(-dv^2 + du^2) + r^2(u,v)d\Omega^2$, where f^2 remains regular throughout the whole region from $r = 0$ to $r = \infty$. Since it never changes sign, u and v retain their roles as radial and time coordinates, respectively, though in the original coordinates r would appear temporal and t spatial in the other side of $r = 2m$. Nothing very unusual happens at this point, which corresponds to $u = v$; but at $r = 0$, corresponding to $v^2 - u^2 = 1$, we encounter the unavoidable singularity.[33] I have shown that it is possible to remove singularities for a much more general class of ϕ's, rather than just the Schwarzschild value, though here it may prove necessary to use two or more separate coordinate patches, smoothly joined together.[34]

Nevertheless it does not follow that the $2m$ singularity should be forgotten as wholly unimportant, or that there was something very wrong about using the (r,t) coordinates in the first place. (As mentioned above, they *do* have some operational significance in terms of the area of a sphere.) For this is a curious point of symmetry. Consider the isotropic coordinates (ρ,t). Since $\phi = (1 + m/2\rho)^4$, the *spatial* part of the metric at least is regular for all of ρ, even though g_{00} vanishes at $\rho = \frac{1}{2}m$. Now r and ρ are related by $r = \rho + m + m^2/4\rho$. As ρ goes to infinity, the values of the two radial measures converge. But as ρ goes to 0, a curious thing happens: r will decrease only up to $\rho = \frac{1}{2}m$, where $r = 2m$. For smaller values of ρ, r *increases* again, going back out to infinity as $\rho \rightarrow 0$. Furthermore, if we replace ρ by $\rho' = m^2/4\rho$, the relationship of r to ρ' will be formally identical. The same thing appears in other coordinates. Einstein and Rosen replaced r by u, where $r = u^2 + 2m$. As u goes from negative to positive values, r goes in to $2m$ and back out again.[35] Finally, if we return to the notion of embedding in a higher-dimensional flat space, we find that our three-dimensional hypersurface needs a four-dimensional Euclidean space with the extra 'embedding dimension' w. We find then that $r = 2m + (w^2/8m)$. This curve, a parabola, can then be rotated around the two axes of symmetry to give the 'shape' of the spacelike hypersurface, a hyperparaboloid of revolution.[36] For $r < 2m$, there is no defined value of w. If we follow along w, we must go back out again along r.

These considerations suggested to Einstein and Rosen a means of removing the Schwarzschild singularity. Instead of having one space that curves up sharply and comes to a cusp at the point $r = 0$, they represented the solution by two perfectly symmetrical spaces, both of which asymptotically approached Euclidean space at great distances, joined together by what they called a 'bridge', centered at $r = 2m$. This value was the radius of the largest sphere that could fit into the narrowest part of the bridge at its center. In trying to go beyond this value, one simply moved onto the other sheet of the total space, and $r = 0$ corresponded to the point at infinity on this other sheet.[37] At no point did the manifold have a singularity; to reach $r = 0$ we would have to go infinitely far. Wheeler has taken over this idea of a multiply-connected topology and put it to more general use. By allowing the two Einstein-Rosen sheets to be parts of a single space, but very far removed from each other, he interprets the 'bridge' as a handle on the space, or a 'wormhole'. Thus it appears that if we give up the requirement that space-time should have a Euclidean *topology* and allow multiple connections, we can get rid of this class of singularities.

However, there is an important catch in this scheme. We have been concentrating entirely on the spatial part of the metric at a single initial time. Thus our talk of moving in and out along other radial coordinates and seeing what happens to r is entirely metaphorical; all these paths are entirely spacelike, and thus cannot represent the path of a light ray or test particle. If we allow the system to evolve dynamically in time, it no longer appears static, as it did in the (r,t) coordinates. Specifically, we find that the throat of the bridge or wormhole shrinks in diameter, and after a finite lapse of proper time $\tau = \pi m$ it pinches off entirely. Thus a geometry that initially appears nonsingular does develop a singularity in time.[38] And what happens *after* the singularity is reached is still quite unknown. For these reasons Wheeler concludes that the attempt to derive equations of motion by treating Schwarzschild masses as singularities is strictly improper.

Since an initially promising attempt has failed, one may well ask whether every geometry that is not already trivially flat must also become singular at some point and in some way. At present no general answer is known, though many particular cases have been studied. Komar has argued that such singularities must appear if one uses a coordinate system corresponding to 'geodesically parallel' spacelike hypersurfaces, but it is not clear what might happen in another representation.[39] Brill has shown that at least one case remains regular, that of a 'time-symmetric' gravitational wave which implodes from spatial infinity, reaches a moment of symmetry, and then explodes back out to infinity in a perfectly symmetric manner. Furthermore, such a wave carries energy which is positive definite, like that of a real mass.[40] However, this has the disadvantage of requiring an open space, extending to infinity and approaching Minkowski values there, just as was required for the definition of total energy for closed systems. Both because of its better agreement with astronomical observations and its better incorporation of Mach's principle, a closed-universe solution is preferable. Most nonsingular cases are ruled out here. Taub has shown that singularities arise for spherically symmetric gravitational waves, and Papapetrou has ruled out periodically varying nonsingular fields.[41] A similar situation occurs in cosmological solutions for the expanding (and then contracting) universe, whether it is taken to be filled with dust, radiation, or a combination of both.[42] In the face of this Wheeler offers as the most reasonable conjecture, that every solution describing a space which is closed in a proper sort of way eventually develops a singularity.[43]

This prospect is somewhat depressing. However, we can still turn to Einstein's hope. We have not yet tried to incorporate electromagnetic fields, and it may turn out that these can avoid singularities. In any case, we must still inquire what happens when we actually approach, reach, and pass a singularity. It may well be that the introduction of a new kind of law of motion, making reference to quantum effects, will solve the problem, just as quantum considerations overcame several apparent inconsistencies in the classical laws of motion. Wheeler suggests that the situation may be like that of a wave which evolves continuously until it crests and breaks up into foam, where we need more than the laws of wave motion for a complete explanation of the phenomenon.[44] In fact, the problem of singularities may *force* the development of a quantum geometrodynamics, despite the fact that GR was developed in the early years of the Bohr quantum theory, and was designed to encompass only the classical fields.

To sum up: the main burden of the arguments in this chapter has been to show that GR, despite its inherent strengths and successes in its limited areas of application, has many peculiar features which may appear as significant weaknesses. To overcome them it is essential to rethink the theory in its own terms, rather than borrowing concepts, modes of analysis, and principles of interpretation from very different areas where the analogy is not really appropriate; and to try expanding its scope from being a mere alternative to Newton as a theory of classical gravitational fields. Without these steps, it can appear to be a self-contained and dead theory—interesting, perhaps, and accurate as far as it goes, but not a significant area of present research and progress. For a long period between 1925 and 1955 it lay relatively dormant, and many physicists today, even if they do not reject GR, feel that they can safely ignore it. In terms of the scanty observational returns, the effort put into it seems too great. It is my personal belief that much of the effort in contemporary physics is unpromising in a different direction, with experiment far outstripping theory, and the use of incoherent models which can never give a fully satisfactory explanation of their own phenomena. Contemporary geometrodynamics has recognized its tasks, and has made remarkable progress toward solving them in recent years. Insofar as it can succeed, it is the most promising source of real intelligibility. In the remaining chapters, I will examine this new conception of GR in its important details, before putting together the whole ontology that finally emerges from its model.

PHYSICAL THEORY: POST-EINSTEIN GEOMETRODYNAMICS 4

We have completed by now our investigation of classical general relativity, or Einstein's theory of gravitation as it was understood up through the early 1950s. Had this book appeared at the time of Einstein's death in 1955, it would have ended here, on a somewhat melancholy note. For we have seen above that many difficulties and unanswered questions still remained in the theory, without any very promising procedure for resolving them. In addition, it seems likely that Einstein was never fully satisfied with his efforts to modify the original GR and develop a unified field theory, whether or not he recognized that his final attempt here led to physically incorrect predictions. For all its mathematical and conceptual grandeur, GR seemed to be dying a slow death, condemned not as false but as impractical and unnecessary, except perhaps for a very limited class of physical problems. Because of his skeptical attitude that orthodox quantum mechanics was no more than a useful algorithm, Einstein himself had begun to fall out of favor among the new generation of physicists, as more the grand old man of an earlier age than a significant contemporary researcher. Such far-reaching ideas arising from GR as the principle of general four-dimensional covariance, once thought essential to all physics, were given an increasingly smaller role.[1]

However, in the very next decade there began a great deal of new research in relativity which has effectively reopened the field. If this has not yet completely vindicated Einstein's dreams and hopes, it has at least shown that Einstein's theory contained far more riches than were originally suspected. Although different authors may have used the word in slightly different ways, I will use "geometrodynamics" (GMD) as a collective term for these new developments which have so extended our understanding of the possibilities of the theory. They may be summarized under three general headings, which we will consider in turn.

(*A*) GMD provides a new *plan* for GR, one which replaces the old static picture with one in which the truly dynamical and causal features stand out more clearly. In doing so it provides new ways for using the formalism of the theory in solving problems and suggests (through its analogies with other theories) new conceptual distinctions.

(*B*) GMD presents classical GR as an *already unified* field theory, by showing that there is a one-to-one correspondence between certain kinds of geometrical features and values of the electromagnetic field tensor, which incorporates the electric and magnetic fields into a single mathematical entity. When Maxwell's equations (in the appropriate generally covariant form) are then combined with Einstein's equations, we then have a theory representing gravitation and electromagnetism as parts or aspects of a single tensor field. Insofar as this program can be carried out fully, and we do not

The New Dynamics and the Role of Time 14

recognize any other fields (as classical physics did not), we have a total field theory. If we can then show that no singularities or unanalyzable concentrations of charge and mass are also needed, we will indeed have a unified field theory, which would then be the culmination of classical (prequantum) physics.

(*C*) This in turn requires the construction of a new *model*, which will provide material unification as the counterpart of the formal unification in (B). This requires showing that the Riemannian space-time manifolds allowed by GR are sufficiently rich that we can construct every physically interesting quantity out of space-time alone, without ever introducing any nongeometric elements. The program here is one of successive elimination of all other factors until space-time, the mere "arena" of classical mechanics, becomes the all-inclusive "everything." These results in turn may open up new areas in experimental gravitation and cosmology, and finally provide better answers to the absolute or relative status of space, time, and causality, and whether Mach's principle, that original stimulus to Einstein's thinking, is really incorporated into the theory.

Turning first to the plan of GR, we see that the old plan had an essentially static character which proved misleading to many interpreters. In keeping with the covariance requirement, both the geodesic law of motion and the field equations appear in four-dimensional form, with no apparent distinction between the space and time coordinates. We may recall that the usual starting point in getting particular solutions was not a distribution of sources, as in Newtonian gravitational theory. Indeed, in such important particular solutions as the Schwarzschild, the source term was taken to vanish, except perhaps at some singular point. One started instead with boundary conditions, usually of a global or cosmological character, chose a coordinate system which incorporated these conditions in a simple or mathematical way, and then wrote down a general four-dimensional metric tensor, using the field equations only to specify any otherwise arbitrary functions in this metric tensor. What is important to realize is that such boundary conditions had to be four-dimensional *already*, e.g., it was not enough to say that the solution was spherically or cylindrically symmetric, since these are only three-dimensional. One normally added, sometimes as a parenthetical afterthought, that the solution was supposed to be static or unchanging in time as well, or at least that any change was uniform and well specified, as in the expanding universe cosmological solutions.

The net effect of such a plan is to remove from the start the possibility of using the field equations to *learn* anything about dynamics, or the way in which a configuration might evolve in time. One was expected to know this important information about the whole history in advance, or else one could not work out the solution. One might discover, for example, that in the Schwarzschild solution the relationship of proper to coordinate time varied with *spatial* position (producing different red shifts, etc.), but that this ratio was constant in time at all points of space. This is in striking contrast not only with classical mechanics but with quantum mechanics, where the Schrödinger equation provides an answer to the typical problem: given the state of a system at one time, what is its state at some other time?

In order to use the geodesic law, it was necessary to know the complete four-dimensional metric tensor. Only then could one calculate the Christoffel symbols for the apparent gravitational force and thereby select that world line which corresponded to the motion of a test particle. Of course, in classical mechanics one also knew (or rather assumed) the geometry of space and time as always Euclidean. Nevertheless, one would need to know the forces only at time t to determine the positions at $t + dt$; one certainly did not need the distribution of forces *for all time* to make such calculations, and indeed the future might be completely indeterminate beyond the time whose state was to be calculated, without affecting any results. It is indeed reasonably simple to arrive at the Schwarzschild metric tensor, and most of the efforts for applying it to planetary motions have been concerned solely with tracing out the paths which seemed to be fixed for all time in this apparently fixed, static 'block universe'.

Given this assumption of the full four-dimensional metric, certain philosophers like Costa de Beauregard,[2] Meyerson,[3] and (in a different way) Smart[4] and Quine[5] have seen in GR further evidence for the idea that the world is really a fixed four-dimensional continuum, however internally variegated it may be, and that 'becoming' in time, with its connotations of mysterious change, should be replaced by extension in the temporal dimension as a mode of 'being' in this static world, not fundamentally different from extension in any of the spatial dimensions. Čapek, on the other hand, follows Bergson in claiming that the interconnection of space and time requires instead a dynamization of space as an evolving entity, whose future may remain to some extent unpredictable.[6] Space takes on some of the character of time and appears as a necessary breadth of duration, rather than time taking on the characteristics of space. In terms of the new plan of GR, I will argue that neither side is wholly correct, and that the world-picture from GMD is neither more nor less essentially static and four-dimensional than that of classical mechanics.

The mistake of those who defend the 'static' interpretation is a natural one. If we say that the geometry determines the paths of free particles, and we are only allowed to use 'geometry' in the four-dimensional sense, such a geometry must of course appear static—with respect to what could it change?—and the particle trajectories as permanent grooves in it. Furthermore, the basic tensor equations are so general and abstract that it is difficult to discover any particular features by looking at them directly. Therefore interpreters have turned to the few particular solutions, and often made unwarranted inductive generalizations from them to the characteristics of GR as a whole, especially concerning their lack of any interesting dynamical features. In the pure Schwarzschild case, where we postulate that there is only one gravitating source mass, it is perfectly reasonable to consider this forever at rest, so that the metric will appear independent of the time coordinate. But of course this assumption of changelessness is physically absurd. We know that there are other masses in the universe, and that they are constantly accelerating with respect to each other, if only because of their gravitational attractions. The Schwarzschild solution may be a perfectly legitimate approximation for experimental test, but it is hardly a source of philosophical morals.

How else can we learn more about the implications of GR for the static-dynamic controversy? One approach is to construct as many particular solutions of the field equations as possible. This is perfectly consistent with the old plan described above. By showing the variety of possibilities within GR, the construction of more solutions can falsify overhasty generalizations and provide greater possibilities for empirical test, since of course any experiment must refer to some particular solution. This can be done in two ways. (1) We may begin with simple physical configurations, such as a small group of source particles arranged in some definite pattern or a certain kind of gravitational radiation, and give precise global boundary conditions (symmetries, approach to flatness at infinity, etc.), and try to find the corresponding metric compatible with the field equations. (2) We may investigate classes of Riemannian geometries compatible with the field equations without any direct concern for possible physical interpretations. Thus the homogeneous equation leads to the set of Riemannian geometries for which the Ricci tensor vanishes everywhere. This class may be mathematically interesting in its own right, regardless of whether they are interpreted as possible spaces in which the matter tensor vanishes. But we then pass into pure mathematics, rather than mathematical physics, though it is always possible that a strange-looking solution might be given a physical interpretation. Work following the first approach is severely hampered by mathematical difficulties, and most of the simple solutions have already been worked out.[7] The second approach still stimulates considerable research, as in the work of Harrison[8] and Ehlers and Kundt.[9] However, this line may also lead to diminishing returns, and one may hope for a more general understanding than anything derived from this piecemeal basis.

In attempting to understand the inner dynamics of a new physical theory, it is natural and proper to look for analogies with theories which have previously been well analyzed and developed. This approach is essential in the construction of models, for even if some disanalogies are present, a knowledge of where, when, and why the resemblances break down can be extremely enlightening. Given the fact that GR was developed as a theory of gravitation, physicists have put much effort into finding analogs and points of comparison with the only other successful theory of gravitation—Newton's. However, as mentioned above, this has ultimately proved sterile and even misleading. It therefore seems worthwhile to turn to the form, rather than the subject matter of the theory as a source of analogies. We know that GR is a classical field theory, and that it deals with matter by considering it as a variegated continuum. Under the circumstances, we may turn to Maxwell's electrodynamics, which has been developed in immense conceptual and mathematical detail as a classical field theory which satisfies a Lorentz covariance principle; and to hydrodynamics, which deals with the mechanical behavior of fluids or continua, and for which the notion of a tensor was originally developed. Power and Wheeler have made an interesting table comparing the typical concepts and research problems of GR and hydrodynamics, and have shown that in 1957 most of the GR analogs were not only unsolved but even unstudied or unrecognized.[10] A similar situation existed until recently for the comparison with electrodynamics. Given the fact that the basic set of field quantities, the Riemann tensor, has twenty independent components, one may suspect that eventually GR may yield far greater richness and more interesting distinctions than either of the others, but the surface has barely been scratched so far, and we hardly know where to look for the appropriate problems. Let us then see what analogs we can find in electromagnetic theory.

In electrodynamics it is possible to write down Maxwell's equations in four-dimensional form if we introduce the antisymmetric tensor F^{ik}. Since its diagonal elements vanish, there are only six independent components, and it turns out the remaining $F^{0\alpha}$ correspond to the electric field **E**, the $F^{\alpha\beta}$ to the magnetic field **B**. But the equations were originally expressed as relationships between the three-dimensional magnetic and electric field vectors, and it is normally much more convenient to use them in this form. Now to speak of the dynamics of such a vector field normally means to speak of its time evolution, the way in which these three-dimensional vectors change in time, rather than with respect to any of the space coordinates. In Maxwell's theory the time and space derivatives do appear separately, and this gives an interesting result: since $\partial \mathbf{B}/\partial t$ is related to the curl of **E**, and $\partial \mathbf{E}/\partial t$ to the curl of **B**, then from knowledge of **E** and **B** (and source charges, if any) at any one time t we can find their value at $t + dt$, and by integration, throughout the whole of time. This is a very striking fact. It means, first of all, that the theory is deterministic. If there are no charges to act as sources or sinks for the field, and there are no boundaries to confine it, so that both fields extend freely through space, then from information about their value at one instant of time we can know the whole past and future. This result then is the analog for Laplace's theorem about classical mechanics, the source of most claims of scientific determinism: if an intelligence knew the positions and momenta of all the particles in a closed system (the universe as a whole if necessary) at a single instant, then using only Newton's laws he could discover the entire history of that system. Is there a comparable statement in geometrodynamics, and how might it be formulated?

This problem of dynamics and determinism is important to GMD for another reason. In discussing the 'third interpretation' of the field equations in Chapter 11, I claimed that this involved giving up T_{ik} as a separate element of the theory, and treating it simply as a name for the phenomenological representation of certain geometrical quantities. The equations thus appeared more like coordination relations between entities at different levels. But surely GR is more than a dictionary for translating from one language into another! And if nothing but geometry is left, what equations hold among the properly geometrical elements, if they all appear on the left-hand side in Einstein's equation? We must somehow break up the four-dimensional G_{ik} into smaller parts, more nearly analogous *in their dynamical roles* to Newton's **x** and **p** or Maxwell's **E** and **B**.

There is still another reason for the desirability of such a separation into separate geometrical aspects. Given any geometry, as embodied in the four-dimensional Riemann tensor, there is one and only one T_{ik} corresponding to it in a given coordinate system. For a given T_{ik}, there is likewise one and only one G_{ik}. However, since the G_{ik} and R_{ik} are only sums or 'statistical averages' of the more basic R^i_{jkl}, there are indefinitely many geometries for a given matter tensor, and to get this or the metric we must add further nonlocal boundary conditions. In particular the case $G_{ik} = R_{ik} = 0$ still includes so many interesting possibilities that it is essential to have a tool to examine their structure and evolution directly.

We know that we can never convert space into time or time into space by a mere coordinate transformation; in any system that we use there will always be three spacelike and one timelike coordinate at every nonsingular point. Furthermore, in the vicinity of such a point we can always introduce a locally Lorentzian system. Therefore, it is reasonable to expect that we should be able to break up space-time into a family of wholly spacelike three-dimensional hypersurfaces, each characterized by a single timelike parameter. To be sure, none of these can be called an instant of absolute simultaneity, and there are infinitely many ways of dividing the space-time up, depending on the choice of parameter, which may be a scalar function of the three space-coordinates, rather than a constant, as in the Lorentz transformations of SR where the uniform velocity was the parameter. What is important is that this can be done at all. It is certainly possible that we might know the complete three-geometry on one or two of these hypersurfaces, without already knowing the full four-geometry. But if Einstein's equations, when rewritten in the appropriate form, are sufficiently powerful that we can *calculate* the full four-geometry from this more limited information, we will have shown (1) just what the dynamics of geometry itself are, according to GMD; and (2) that Einstein's theory is deterministic in just the same way as Newton's, where by knowing position and momentum at one time we could determine it for all other times. In GMD our information must be primarily geometrical. Finally, we can show (3) that GR is no more essentially four-dimensional than classical mechanics, since we do not need to *begin* with the four-geometry.

Wheeler, following the approach first developed by Cartan and Lichnerowicz and extended by the canonical formalism of Arnowitt, Deser, and Misner, and using the analogies with electrodynamics, develops these equations from a variational principle.[11] It is well known that in source-free GR, where the matter tensor vanishes, the Lagrangian for the gravitational field is simply the curvature scalar R, so that we can write $\int R(-g)^{1/2}d^4x$ as the quantity to be extremized. Variation of the metric coefficients leads to $\int (R^{\alpha\beta} - \frac{1}{2}g^{\alpha\beta}R)\delta g_{\alpha\beta}(-g)^{1/2}d^4x$. This is maximized by setting the quantity within the first pair of parentheses equal to zero, which gives the field equations directly. Likewise for electromagnetism the Lagrangian $L = -(8\pi)^{-1}F^{\alpha\beta}F_{\alpha\beta} + j^{\alpha}A_{\alpha}$, where j^{α} is the current-charge density and A_{α} the vector four-potential, such that $F_{\alpha\beta} = \partial A_{\beta}/\partial x^{\alpha} - \partial A_{\alpha}/\partial x^{\beta}$. It will be typographically convenient here and in this chapter to change the index convention previously used: henceforth Greek letters will be used for summations over the four dimensions (0,1,2,3) and Latin letters for summations over the three spacelike dimensions (1,2,3). The definition of $F_{\alpha\beta}$ automatically satisfies half of Maxwell's equations, allowing us to use such a potential in the first place; and varying the potentials gives us $F^{\alpha\beta}{}_{;\beta} - 4\pi j^{\alpha}$ as the coefficient which must vanish for an extremum. But these variations provide the remaining Maxwell equations!

A variational principle is useful for other purposes besides deriving field equations. However, if we are to use it effectively, we must also specify which of the variables appearing in the Lagrangian is to be fixed at the limits of integration. This specification is equivalent to providing initial conditions (and often boundary conditions as well) along with the differential equations for the field. In this approach we have a considerable amount of freedom. In a given problem we may choose any combination of the variables to be fixed at the limits and the others varied to find an extremum as long as our choice is (1) self-consistent and (2) sufficiently complete to solve the problem at hand. The fact that we have two limits of integration, rather than one, gives us the greater flexibility in carrying out variations. However, it is possible to take the two limits so close to each other that their separation becomes vanishingly small. If we require that our variables be continuous, this restriction implies that they must have approximately the same value at both limits. If so, then instead of specifying data on both we can give the data on one and the ratio of the discrepancy on the other over the separation—in the limit the derivative on the first. This latter "thin-sandwich" formulation does of course presuppose additional topological and metrical restrictions of continuity and differentiability which might not be inherent in the variables themselves.[12] However, they may be justifiable on physical grounds. One further advantage of variational principles is that they overcome Hill's objection that whenever GR introduces boundary conditions, it must introduce particular coordinate systems in which to express them.[13] Instead, it incorporates these boundary conditions into the limits in a still covariant way.

How in fact can we derive information from a variational principle in GMD? If we divide space-time into three-spaces characterized by some arbitrarily chosen time-parameter t, it would seem natural to take two values t' and t'' as our limits and fix the three-geometry on each of these surfaces, the object being to find the four-space in between. However, the quantity R appearing in the variational principle is a combination of components of the four-dimensional Riemann tensor and not just terms from the intrinsic three-geometries alone. Besides, with this three-geometry fixed we need some other quantity to vary in order to find how that three-geometry evolves in time. In addition, a given three-space may be embedded in indefinitely many ways in a surrounding four-space. If one were to draw normals from the three-space in the t-direction they would not necessarily strike the next surface at the same coordinate values as those on the original surface from which they started. We need some quantity that serves to measure these shifts. Along with the *intrinsic* curvature inherent in the surface itself, there is an *extrinsic* curvature that gives its relationship to the higher manifold. To use a standard example, when a flat sheet of paper is rolled up, its two dimensional intrinsic curvature is still zero, since it has not been stretched or internally distorted in any way, but it has acquired an extrinsic curvature in the surrounding three-space.

If we move from x to $x + \delta x$ on the surface t', the normal vector $\mathbf{n}$ will be shifted from that carried to $x + \delta x$ by parallel transport by $\delta\mathbf{n}$, where $-\delta n^a = K^a_b \delta x^b$. The K's form the three-dimensional extrinsic curvature tensor on the surface. Using them, we find that we can write the components of the Riemann tensor (and thus anything derived from it) in terms of three-dimensional quantities by[14]

$$^{(4)}R^{0a}{}_{0b} = \partial K^a_b/\partial t + K^a_c K^c_b$$

$$^{(4)}R^{0a}{}_{cd} = K^a_{c|d} - K^a_{d|c}$$

$$^{(4)}R^{ab}{}_{cd} = {}^{(3)}R^{ab}{}_{cd} + K^a_c K^b_d - K^a_d K^b_c$$

where the vertical slash denotes covariant differentiation in three dimensions (the semicolon for four). We can also write down an expression for the Einstein tensor G^α_β, the left-hand side of the field equations, in terms of the three-dimensional Ricci tensor and scalar and the extrinsic curvature tensor.[15]

Since we have already taken the Riemann tensor as the proper measure of the real gravitational field, we may consider the intrinsic and extrinsic geometries of our surfaces as field variables and formulate the variational principle in these terms. However, it is more convenient to use potentials. Not only is the introduction of the A^α known to be a simplifying convenience in electromagnetism, but in GR the potentials are also metric coefficients and the practical working summary for any concrete problem, since they are the starting point in determining motions by the geodesic law. Arnowitt, Deser, and Misner have introduced a 'lapse factor' N and a 'shift vector' N_i so that the four-metric $\hat{g}_{\alpha\beta}$ is related to the three-metric g_{ik} in the following way: $\hat{g}_{ik} = g_{ik}$; $\hat{g}_{0i} = N_i$; $\hat{g}_{00} = N_i N^i - N^2$. One can raise indices from N_k to N^i by using the three-dimensional g^{ik}, and then calculate the $\hat{g}^{\alpha\beta}$ in terms of them.[16]

If we construct normals from various points on the surface t' pointing to that at t'' (assumed to be very close), the quantity $N(t'' - t') \rightarrow Ndt$ measures the lapse of *proper time* between the two surfaces at the various points. For a given separation, if we parametrize so that $t'' - t'$ is large, then N is small, and conversely. $N^i(t'' - t') \rightarrow N^i dt$ measures the shift between the values of the space coordinates $x^i + \Delta x^i$ at the point on t'' where the normal strikes the surface, and those of its starting point x^i on t'. We can then give the covariant components of the extrinsic curvature as

$$K_{ik} = \frac{1}{2N}\left(N_{i|k} + N_{k|i} - \frac{\partial g_{ik}}{\partial t}\right).$$

This expression shows that (1) the extrinsic curvature is a combination of the *first* derivatives of the potentials, and (2) if we know the intrinsic geometry on one spacelike hypersurface, its rate of change, and the lapse and shift functions there, we can find the extrinsic curvature and thus some information about four-dimensional space-time in the immediate vicinity. Is this all that is needed to get the full four-geometry, and if so, how can we find the lapse and shift functions?[17]

The answer to the first question does in fact turn out to be yes. The general problem of finding compatible initial value data sufficient to determine the whole dynamical history of a system governed by various laws goes back to Cauchy, and the relevant work for GR was done by Cartan, Bruhat and Lichnerowicz, although they required that both the intrinsic and extrinsic geometry of the initial space-like hypersurface must be fully specified, rather than potentials from which these could be derived.[18] The compatibility requirements were embodied in the initial value equations

$$R + (\mathrm{Tr}\ K)^2 - \mathrm{Tr}\ (K^2) = 2\epsilon,$$
$$(K^{ik} - g^{ik}\ \mathrm{Tr}\ K)_{|k} = S^i.$$

Here R is the Riemann curvature invariant for the intrinsic three-geometry of the initial hypersurface, Tr refers to the trace of the appropriate tensor matrix, and ϵ and S^i refer, respectively, to the energy density and momentum or rate of energy flow on the initial surface, multiplied by $8\pi G/c^4$. The energy in question is derived from nongravitational sources, e.g., the electromagnetic field.

Arnowitt, Deser, and Misner approached the initial value problem from a somewhat different direction. They were concerned with the possibility of eventually developing a quantum GMD, and they assumed that any formalism suitable for this purpose would require finding a pair of canonically conjugate variables. The intrinsic g_{ik} were a natural choice for one member of the pair: the problem was to find the other. In ordinary mechanics, position and momentum are conjugate variables, and the Hamiltonian may be expressed as $H = p\dot{q} - L(q,\dot{q})$ where L is the Lagrangian. We have already shown how to express the L for GR in terms of the intrinsic and extrinsic curvatures, but we still need the appropriate 'p' to write down the GR Hamiltonian. Furthermore, if we can find such a canonical tensor P^{ik} conjugate to g_{ik}, we may use it to provide another definition of energy and momentum density in GR. We have seen above in the discussion of integral forms and macroscopic conservation laws for GR that there are difficulties here, and probably several concepts will be needed to cover all the roles played by energy and momentum in classical physics. Nevertheless, the Arnowitt-

Deser-Misner definition seems most appropriate to Hamiltonian approaches, whether quantum or nonquantum.

Following Laplace, we might suspect that K^{ik} can itself be interpreted as the "GMD field momentum" since, along with g_{ik}, it is the quantity needed to determine the history of the system. But in fact Arnowitt, Deser, and Misner have found the quantity P^{ik} to be more suitable, where

$$P^{ik} = g^{1/2}[g^{ik}(\mathrm{Tr}\ K) - K^{ik}].$$

The two tensors are closely related. More precisely,

$$K^{ik} = g^{-1/2}[\tfrac{1}{2}g^{ik}(\mathrm{Tr}\ P) - P^{ik}],$$

so that when either K^{ik} or P^{ik} is known, the other can be calculated.[19] The extrinsic curvature K^{ik} has the more direct geometrical significance, and P^{ik} the more direct dynamical significance. But we really need only one of them to explore the dynamics of geometry.

So far we have been discussing the Hamiltonian or one-surface formulation of the initial value problem. This version is indeed the appropriate analog of Laplace's classical principle of determinism, and it is important to know that we need information on only one initial surface, as long as we have a sufficient amount of it. However, this information is still quite substantial (g_{ik}, K^{ik}, ϵ, and S^i, along with a coordinate system to represent them) and subject to restrictions (the initial-value equations) not necessary for the classical Laplace's principle. In this approach we also make no direct use of the variational principle, with its two limits. The hope is that if we have data on two surfaces, we may need much less on each one, and that such data may be less restricted. In particular, we would like to start with arbitrary geometries on two spacelike surfaces, and show the Einstein's equations determine a unique four-geometry in the sandwich between. This would then be a Lagrangian or two-surface version of the initial value problem.[20]

In considering the most general case of the two-surface version, we may write the action integral as follows:[21]

$$I_1 = \int\left[P^{ij}\frac{\partial g_{ij}}{\partial t} + N_\alpha R^\alpha + (N^i\,\mathrm{Tr}\ P - 2P^{ij}N_j - 2g^{1/2}N^{,i})_{,i}\right]d^4x,$$

where

$$R^0 = g^{1/2}\{R + g^{-1}[\tfrac{1}{2}(\mathrm{Tr}\ P)^2 - \mathrm{Tr}\ (P^2)]\} = g^{1/2}[R + (\mathrm{Tr}\ K)^2 - \mathrm{Tr}\ (K^2)],$$

and

$$R^i = 2P^{ij}{}_{|j} = -2[g^{1/2}(K^{ij} - g^{ij}K)]_{|j}.$$

The last group of terms form an ordinary divergence, and by Stokes's theorem there is a corresponding surface integral which will vanish if the space is closed (since then the boundary shrinks to zero) or is asymptotically flat at infinity. If we restrict ourselves to these cases, we may use the simpler integral

$$I_2 = \int \left(P^{ij} \frac{\partial g_{ij}}{\partial t} + N_\alpha R^\alpha \right) d^4x$$

as our basis for variational calculations. If there are sources or other fields present, we must add a term $-Ng^{1/2}L$, where L is here the Lagrangian for such other fields.[22]

Let us try to apply this Lagrangian approach, where we hold the geometries g'_{ik} and g''_{ik} fixed on the two bounding hypersurfaces at t' and t''. We then try to extremize the integral I_2 with respect to variations in any of the quantities it contains. We have sixteen possibilities: the six g_{ik} in the region between the two surfaces, the six P^{ik}, and the four N_α (not N^α), where N_0 is the lapse function N. If we vary the P^{ik}, we get the equations above which defined them in terms of the K^{ik}, so that the initial assumption that they could be freely variable does not contradict the original definition. If we vary with respect to the N_α, we find that $R^\alpha = 0$, which gives us precisely the initial value equations for the case where ϵ and S^i vanish. But these are in fact four of the ten equations in Einstein's field law $R_{\alpha\beta} = 0$! If then we vary the six g_{ik}, the resulting equations will be the other six of Einstein's equations.[23] From the standpoint of variation, it makes no difference whether we use $P^{ij}(\partial g_{ij}/\partial t)$ or $g_{ij}(\partial P^{ij}/\partial t)$ as the first term in the integral: the result will be the same. Because of this symmetry, we can either hold g_{ij} fixed on both surfaces or P^{ij} fixed on both. In any case Einstein's equations are sufficient to determine the whole intervening geometry $\hat{g}_{\alpha\beta}$.

Now this result implies a conclusion of great philosophical significance. In the original version, $R_{\alpha\beta} = 0$, each of Einstein's ten equations appeared on an equal basis. Each seemed to be saying the same thing, describing one component of the curvature of a static four-dimensional world, forever devoid of matter or fields at all points where the equation held. But now we have achieved the desired separation into two groups of equations. Four of them tell us what must be happening at any given instant of time, in the particular coordinate system and time-axis that we have chosen. The other six say that given such a satisfactory configuration, it must evolve with respect to our time coordinate in a unique and thus deterministic sort of way.

Despite the great freedom we have in being able to specify the geometries (or momenta) on the two surfaces quite arbitrarily with respect to each other as long as the surfaces are well removed, it is mathematically inconvenient to work directly from this "thick-sandwich" formulation, and better to use the one-surface version. (One should remember that the convenience involved has nothing to do with determining the full initial conditions empirically. It is likely that we must deal with considerable idealizations in applying it to any concrete physical situation. We are here concerned with the dynamics of the theory proper, and the behavior of its model in first interpretation, rather than coordination with observables in the second interpretation.) Nevertheless, it is still desirable to show that the two versions of the initial value problem are compatible with each other, and that they reduce to each other in the limiting case of the "thin sandwich," where the bounding surfaces are nearby and the geometries differ by only a small amount. This compatibility has in fact been proved by David Sharp in the course of his investigations of the general variational problem.[24] We can therefore use whichever version is most convenient for the problem at hand.

It is possible to simplify the variational principle still further on a single hypersurface. The terms $N_i R^i$ can be replaced by a surface integral which will vanish if the space is closed. We may also replace the term involving the Lagrangian from other fields with a term with a similar variation expressed only in terms of ϵ and S^i, rather than the full sets of components of the $T_{\alpha\beta}$ tensor. This in turn means that in cases where 'matter' does not vanish, we do not need to know all its hydrodynamical properties in order to solve the dynamics of the system. In fact, once we calculate the complete four-geometry, we can then go back and give all the components of $T_{\alpha\beta}$ at all points of space-time. Using the fact that $Ng^{1/2} = (-\hat{g})^{1/2}$ and the relations among the P's and K's, we get

$$I_3 = \int \{N[2\epsilon - R + (\text{Tr } K)^2 - \text{Tr } (K^2)] + 2N_i S^i\} g^{1/2} d^3x$$

or

$$I_3 = \int \left[\frac{N}{2} (2\epsilon - R) + \frac{\gamma_2}{2N} + N_i S^i \right] g^{1/2} d^3x,$$

where $\gamma_{ik} = NK_{ik}$, $\gamma_2 = (\text{Tr } \gamma)^2 - \text{Tr } (\gamma^2)$, the shift anomaly, as the new integral to be maximized. Once again, extremizing with respect to N and N_i, the unknown lapse and shift functions, gives the initial value equations above. If we look at the first term, it has the form of a typical action integral, and in analogy with the corresponding procedure in classical mechanics, we may think of $R - 2\epsilon$ as the potential term, and $\gamma_2/N^2 = (\text{Tr } K)^2 - \text{Tr } (K^2)$ as the kinetic term. The former describes properties intrinsic to the surface, the latter serves as a scalar measure of its rate of change with respect to the surrounding four-space. The action $\int (T - V)dt$ is then obtained by multiplying each term by $N\Delta t$. If the lapse N is large, as in a thick sandwich, the potential term will dominate, while if it is small, as in a thin sandwich, the kinetic term will dominate.

In either direction the value of the appropriate term can become infinitely large: however, what we seek is a minimum. The only way this can be achieved is by setting the kinetic and potential terms equal to each other, so that $R^0 = 0$. When we do this we find $N = \pm[\gamma_2/(2\epsilon - R)]^{1/2}$. This is an explicit equation for the lapse. It means that if we are given the intrinsic three-geometry (from which we determine R), the energy density ϵ, and the shift vector N_i (from which we determine γ_2), we do not need additional information about the lapse, nor *can* we specify it independently. It is determined by the demand that the integral be an extremum. Finally, we can return to the other three initial value equations, involving the S^i. The lapse factor N appears in these equations, but if we substitute for it according to the equation above, we have only three remaining unknowns, the N_i, and three equations involving them. The last problem is then to solve this set of simultaneous partial differential equations for the N_i in terms of the given g_{ik}, $\partial g_{ik}/\partial t$, ϵ, and S^i.[25] In most practical cases it will be extremely difficult to solve these nonlinear equations, but there is no reason why it should be impossible to do so. Furthermore, it does not follow that a solution for the N_α is the best way to work out the dynamics, or even that there is no other set of initial data which might also be suitable to express dynamical laws. The important point is that at least one such specific way has already been found.[26]

Let us now summarize and evaluate these results. In the first place, the initial value data determines not only the dynamical evolution of the three-geometry: it determines time itself. For we have seen in the description of a given four-geometry that both t—and thus intervals Δt—and N are arbitrary by themselves. However, in combination they give the proper time $N\Delta t$. Consider now the thin-sandwich approach as applied to a given four-geometry, with a given slicing into spacelike hypersurfaces, and a given parametrization of these hypersurfaces by a monotonically increasing time parameter. Two nearby surfaces will be separated by $t'' - t'$. We can then solve the initial value problem to get N and thus learn how far apart in proper time the surfaces really are. If we change the parametrization to increase $t'' - t'$ for these same surfaces, this increase will be perfectly compensated by a decrease in N, so that the product remains unchanged in magnitude. Likewise, given the initial data on one surface, we can calculate the span of time to any other nearby surface. In the thick-sandwich case, there may be a problem if we have to integrate any quantities over t, but in principle we can do the same thing. Thus we may conclude that the three-geometry itself carries information about time, the coordinate dimension normal to it. Finally, since the whole four-geometry is ultimately determinate, we can discover not only separations, but just where our initial surface is located in space-time with respect to any points of symmetry in its evolutionary history, such as a moment of maximum expansion.

The situation is really unprecedented in classical physics. There the choice of a time coordinate was much less arbitrary. It should indeed have direct metrical significance, since there was no quantity to be discovered in the course of a dynamical problem which would relate the chosen 'time' parameter to proper time. The character and flow of time were independent of the physical events taking place in time. According to our equations, if the extrinsic curvature were to vanish, then N, and thus time itself, would also vanish or become indeterminate. The notion of time as an intrinsic measure of change is thus supported by GMD, and in fact embodied much more precisely than in other physical theories. What light does this conclusion shed on Čapek's claims that relativity is based on the dynamization of space, rather than the spatialization of time? To a large extent, the issues remain semantical. One might argue that space is really prior to time, on the ground that information about a spatial manifold tells us everything we can expect to learn about time. On the other hand, one might argue that the extrinsic curvatures, or the lapse and shift functions, are not really properties of a space but rather relations between that space and the evolving space-time. Insofar as a complete characterization of a space in GR requires information about these extrinsic curvatures, which specify its rate of change in time and are thus temporal or dynamical, one might say that space has indeed been dynamized, and if time is not prior to space it is at least a necessary part of it. One position which the above analysis does seem to rule out is that space and time are just alike in GR, or that the relations among them are all symmetrical. The four-dimensional representation by itself shows nothing about the peculiar character of GR. At best we can establish that space and time are quite distinguishable in any space-time allowed by GR, but in fact are connected with each other in complex, diverse, surprising, and unexpected ways. At least the determinism common to all prequantum physics is preserved, but not carried further except in the fact that time itself is determined in GMD. But this claim is far from saying that the character of time must be antecedently assumed; in fact, it cannot be so given unless we can show in advance that the particular specification will never violate the dynamical equations. Most of the debate between defenders of the static and dynamic schools has proceeded in ignorance of the new results of GMD, but one might at least hope that they would refine, if not modify, their position to take account of them.

One curious fact is worth mentioning. We note that the equation for N

is determinate only up to a plus or minus sign. There is no further physical ground for choosing one rather than the other. Now a plus sign (for increasing t) indicates increasing time, or futurity; a minus sign indicates pastness. Thus given two surfaces, we can say exactly how far apart they are in time, but we cannot say within GR which comes first in time. GR therefore can give us no direction of time. The equations are symmetrical with respect to a change in sign, and we can carry our dynamics backward into the past in the same way as forward into the future. In this respect GR is entirely like classical mechanics, and those who interpret the dynamization of space as including directionality cannot support their claims without appeal to something else.

The initial value problem also gives a new plan for using GR, as well as a better understanding of the interrelationship of space and time. Instead of working with specific boundary conditions for especially simple cases, we can now follow this procedure: (1) Select an initial space-like hypersurface. Since we want this to be part of a family of such surfaces, we characterize it by the value t' of some arbitrary parameter t. (2) On this surface we introduce three spatial coordinates normal to t. (3) On this surface t' we then give g_{ik}, $\partial g_{ik}/\partial t$, S_i, and ϵ in the coordinates chosen. At the moment S_i and ϵ appear as foreign, nongeometric entities. However, they will normally come from other known fields, like the electromagnetic field. (4) Thus we must ensure that they satisfy any initial value requirements for such fields. For example, the divergence of the magnetic field must vanish and the divergence of the electric field must equal the charge density. Now the choice of an initial hypersurface is wholly arbitrary. The initial value equations must also hold on every other surface, with a different value of t. To guarantee this, we must know the value of S_i and ϵ on every surface. (5) Therefore we must give the dynamical law governing the time evolution of all the fields responsible for these terms. Of course, if there are no such fields the last two conditions are automatically satisfied and we need only the initial geometry and its rate of change.

From this information we can get everything else about the properties of the four-dimensional space-time in the following sequence. (1) Solve the three initial value equations involving S_i for N_i, substituting for N. (2) Since this solution gives us the needed N_i and γ_2, we can then calculate N, and from this K^{ik} and P^{ik}. (3) Use this result to determine the geometry of a nearby hypersurface. (4) By iteration in accordance with Einstein's equations, or using the latter directly, solve for the metric $\hat{g}_{\alpha\beta}$ of the whole four-space, and any other quantities of interest derived from it. (5) Use the geodesic law to determine the motions of free particles anywhere in this space-time.[27] Whatever the mathematical difficulties may be in actually solving the equations, this plan is applicable in principle to all physical situations, which one could hardly say about the older plan. It is also dynamic in that it enables us to make predictions about what will happen to the spatial geometry in the future when we know it in the present, rather than giving us the whole geometry of space-time in one static hunk.

Let us assume that we have been able to reach a specific solution for the crucial quantities N_i and N. How can we tell whether this solution is unique, or whether there may be many more equally compatible with the initial data? We know that with any set of differential equations, it is normally necessary to add boundary conditions of sufficient strength to rule out all but one solution. Even in a simple case like Poisson's equation in classical mechanics or electromagnetism, where the source is one small sphere, we can still add a set of spherical harmonics compatible with the corresponding Laplace equation. In order to rule them out we normally require certain symmetries because of the assumed isotropy of space, and the requirement that the potential vanish at infinity. In all four of the initial value equations, the absolute value of N_i never appears directly, but only some combination of its derivatives with respect to the space coordinates. It might therefore seem that we could always add a constant to the N_i without affecting anything, just as the potentials of classical physics were determined only by the quite arbitrary convention that they approach zero, rather than some other constant value, as we go to infinity. However, the N_i here are related to components of the full metric tensor $\hat{g}_{\alpha\beta}$, and thus such a change in the N_i values would affect determinations of length. In the variational integral I_3, they appear multiplied by the S^i.

We recall that in the original variational integral I_1 a surface integral term appeared, and it was argued that this would vanish if the space were closed or approached flatness at infinity. The hope of GMD is that closure of the initial surface, or of both surfaces in the thick-sandwich version, can serve as a universal boundary condition capable by itself of ensuring uniqueness. The belief is that for only one choice of the N's can they be integrated over the boundary of this hypersurface without jumps indicating singularities. It may well be that other boundary conditions are needed, or that the closure must be of a specific kind, but there is no clear-cut evidence for such additional requirements. It is also possible to use an open space if we require asymptotic flatness, and perhaps other kinds of open spaces will also guarantee uniqueness. However, it is desirable to emphasize closure for two main reasons: (1) given the orthodox current interpretations of astronomical data, the actual physical universe is normally assumed to be closed; and (2) the restriction to closed spaces is more in keeping with the spirit of Mach's principle, which will be discussed at greater length in Chapter 17.[28]

Why should such cosmological questions play an important role? We must remember that closure and flatness at infinity are both global notions, describing the structure of the space as a whole. It is not enough to give the initial three-geometry in some particular area of interest, or treat some part as an isolated system. In classical physics the latter was possible at least in principle, and certainly the masses in the universe could be considered as confined to some finite region of the surrounding Euclidean space. We needed only the positions and momenta of those particles, wherever they were located. In GR there really are no such things as isolated systems, and local data are not enough.[29] We must know the whole of space before we can trace for all time the dynamics of any one of its parts.

Consider now the choice of our initial hypersurface. According to the principle of general covariance, it must make no difference to the validity of the (tensor) initial value equations how we define such a hypersurface, or what sort of coordinates we use on it, as long as it remains everywhere spacelike. Suppose that on our original choice we have an energy flow, embodied in a nonvanishing S_i. Whatever the source of this energy, the dynamical law governing it cannot allow it to travel faster than light. Thus on a different initial surface, the energy would appear stationary in space. Even in SR, such apparent motion and momentum could be transformed away (locally) by changing the 'orientation' of the surface. We know that on all surfaces the crucial equation $R + K_2 = 2\epsilon$ must be satisfied, where K_2 is the extrinsic curvature invariant, $K_2 = (\mathrm{Tr}\, K)^2 - \mathrm{Tr}\,(K^2) = (K^i_i)^2 - K^i_j K^j_i$. If we were to look at all possible initial surfaces and require only that this equation hold everywhere on them, this law would then include all the requirements in the other three equations involving the S_i. Certainly it is more convenient in any concrete case to choose only one initial surface and use all four equations. However, if our concern is to give physical import to the covariance principle or to find the smallest and simplest set of laws to characterize the content of GR, this fact is essential. Because of it, we may say that the whole content of GR can be summarized in the scalar equation $R + K_2 = 2\epsilon$ *plus* the covariance principle.[30]

This statement, which is equivalent to Einstein's law $G_{\alpha\beta} = T_{\alpha\beta}$, is a new, elegant, and much simpler way of describing the theory. It depends crucially on the possibility of dividing up space-time into infinitely many distinct families of spacelike hypersurfaces, with timelike normals to them. It makes no reference to coordinates; indeed, it could not and still preserve its power. It has the logical character of an $(x)fx$ statement rather than an $f(a)$ statement, where a is an arbitrary substitution instance for x. One could quantify over the tensor statement, but nothing would really be gained. The equation reveals all its interesting features in each particular choice of coordinates, without our having to repeat the process. In the older statement of the law, the only role of covariance appeared to be the decision to use the language of the general tensor calculus to express the laws. Here it is part of the law itself.

In discussing the geometrodynamical interpretation of GR, we saw that the crucial problem was to find the real field law, while $G_{\alpha\beta} = T_{\alpha\beta}$ appeared more like a coordination relation, giving new geometrical names to old phenomenological quantities. This new law does succeed, for it states a relationship between two distinct but equally geometrical quantities, the intrinsic and extrinsic curvatures, as its two distinct field variables. This fact appears even more clearly in the homogeneous equation $R_{\alpha\beta} = 0$. If the matter tensor vanishes in one system, it must vanish in every other at all points where the equations hold. Thus we have $R + K_2 = 0$, or $R = -K_2$, at every point on every member of every family of spacelike surfaces, i.e., throughout all of space-time the intrinsic and extrinsic curvature must cancel each other out. Here there is no reference to the 'foreign' notion ϵ, but only a purely geometrical law. And this law is in no sense trivial. The two quantities are clearly on the same level but have very different meanings and are derived analytically in very different ways. Each function may be very complicated, but it is always perfectly compensated by the other. We have thus achieved our goals of a new dynamical plan and a new intra-level field law, expressed as a relation among the quantities at the geometrical level.

This new plan has indeed had considerable heuristic value, in selecting new research problems within GR and providing the tools to solve them. We can split the work into two problems: finding the conditions on an initial surface, and determining the time evolution of these geometrical conditions, and these will normally be mathematically simpler than trying to get the full four-geometry at once. Much of this work has centered on the time-symmetric initial value problem. If the geometry has a moment of time-symmetry, which we can specify by $x^0 \equiv t = 0$, then the four-geometry must satisfy $g_{00}(x^0,x^i) = g_{00}(-x^0,x^i)$, $g_{ik}(x^0,x^i) = g_{ik}(-x^0,x^i)$, $g_{0i}(x^0,x^i) = -g_{0i}(-x^0,x^i)$. This is equivalent to requiring that on the surface $t = 0$, $K_{ik} = 0$ everywhere. This in turn means that the S_i vanish and the three equations involving them drop out, so that our initial value requirements reduce to $R = 2\epsilon$. If there are no energy sources present, we have the still more simple $R = 0$ as the basic equation to solve. Brill has used this extensively to show that the theory allows the possibility of gravitational waves with no apparent sources, which implode from infinity to a moment of time-symmetry and then explode out to infinity again. By analyzing the behavior at this moment of time symmetry and maximum concentration, and finding a general equation for all such waves, he was able to show that they carry a total energy which is positive definite and approaches the Schwarzschild value at infinity, as one would get from a 'real' mass.[31] This approach is also useful in dealing with cosmologies of the expanding and eventually contracting universe, like those of Friedmann and Tolman, where there is a moment of maximum expansion.

Misner has given the name "geometrostatics" to the time-symmetric problem.[32] Of course, this is not truly a statics since a configuration will not remain in perfect equilibrium forever, and we cannot counter the influence of one force by introducing a force of another kind, as in electrostatics we cancel electric forces by mechanical forces from neutral matter. Nevertheless, it turns out that there are many useful analogies between electrostatics and geometrostatics. The initial value approach has led to a greater understanding of the old Schwarzschild solution, one that reveals its dynamical character despite the usual static metrical form. I will discuss this in Chapter 16. Finally, even in cases where K_{ik} does not vanish on the initial surface, we can still simplify matters if the extrinsic curvature invariant K_2 or the shift anomaly γ_2 vanishes, for then we can use $R = 2\epsilon$ as one of our initial value equations. Wheeler has suggested how such "energy-intrinsic" initial geometries might be investigated.[33]

In attempting to assess the effectiveness of GMD as a unified field theory, it is important to remember just what such a theory is intended to accomplish. In terms of the definitions given above, we require that such a theory be both total and complete. It must exhibit all the supposed 'forces of nature' as connected together in some 'natural' way, as aspects of a single total field, and it must eliminate singular regions where the equations for such a field no longer hold. It is quite possible that the two requirements are not really independent; Einstein always suspected that any satisfactory theory of the total field would automatically be complete.[1] However, the existence of singularities is usually established only with respect to particular solutions, and until a theory has been given definite form, with specific differential equations that can be solved, it is senseless to speak of such particular solutions. Thus most efforts have been devoted to developing a total field theory first, and worrying later about completeness.

One of the great triumphs of prequantum physics was the reduction of all the long-range forces of nature to two separate fields: the gravitational and the electromagnetic. This simplified the program for total field theory immensely, for it meant that only these two forces had to be unified. Insofar as such a theory could be formulated as an extension of GR, it meant that electromagnetism must somehow be incorporated into the metrical structure of space-time and that Maxwell's and Einstein's laws should be derivable from a single principle. Such an accomplishment, of course, would be less spectacular today than in 1920. We now know that the nuclear forces require the introduction of at least two other fields, which manifest themselves in the strong and weak interactions. These incorporate the quantum of action in an essential way, and no attempts have yet been made to combine them with the older forces in a new kind of total field theory. Nevertheless, if we interpret unified field theory as the culmination or ultimate goal of all prequantum physics, the inclusion of electromagnetism will be enough to achieve this. The incorporation of any other fields may be part of the development of a quantum GMD, which then might seem the ultimate goal in the 1970s.

Attempts to develop a unified field theory began almost immediately after the first publication of "classical GR." However, there were two major difficulties blocking progress. (1) As long as GR confined itself to four-dimensional Riemannian geometry, with a symmetrical metric tensor (and, although this was rarely stated explicitly, a Euclidean topology) all the geometrical properties of such manifolds could be interpreted in gravitational terms; there were no extra geometrical features which were clearly non-gravitational and crying for interpretation in terms of other fields. If electromagnetism was to be geometrized in some way, there was no reasonable place to put it. (2) According to the equivalence principle, gravity was a universal force, accelerating all bodies in the same way. This was indeed the motive for interpreting the gravitational force as part of the geometrical structure of the manifold. But electromagnetism was clearly not universal. In the basic covariant law of motion $\frac{D^2x^i}{Ds^2} = \frac{q}{m} F^i_k \frac{dx^k}{ds}$, the charge-to-mass ratio of the test particle, q/m played a crucial role. In gravitation, the ratio of passive gravitational mass to inertial mass, the universal response parameter, was the same for all test particles. There was no analog to the geodesics along which *all* charged particles should move in a given field.

The second difficulty was empirical, but the first seemed to stem solely from the stipulative *choice* of the *kind* of geometry to be used, with only the *particular* Riemannian geometry left to be determined empirically. Therefore it was natural to recommend modifying this assumption. If, for example, we could use a five-dimensional Riemannian manifold, then its metric tensor $g_{\alpha\beta}$ would still be symmetrical, but would now have 15 independent components. Now according to GR we need 10 potentials for a complete description of the gravitational field, and these are identified with the 10 independent components of the four-dimensional metric tensor g_{ik}. In addition, the electromagnetic field can be described by 4 potentials. The scalar electric potential ϕ and the three components of the vector magnetic potential **A** form a four-vector from which the individual field vectors can be derived by differentiation. In a five-dimensional geometry it might be possible to identify the components $g_{i5} = g_{5i}$ (i = 0,1,2,3 or 1,2,3,4) with these potentials or something proportional to them. This was in fact the approach of Kaluza and Klein.[2] In this version all components of the metric tensor are independent of the mysterious fifth coordinate x^5, certain transformations must leave them invariant, and the new five-dimensional coordinates $g_{\alpha\beta}$ are related to the four-dimensional h_{ik} by $g_{ik} = h_{ik} + g_{i5}g_{k5}/g_{55}$ (in close analogy to the relation between the four-dimensional coordinates and those on a three-dimensional spacelike hypersurface), $g_{i5} = 2^{1/2}\phi_i g_{55}$. This leaves the fifteenth potential g_{55} open to interpretation, and here two different schools arose. In the original version, Kaluza made the rather arbitrary (though convenient) assumption that $g_{55} = 1$, thus 'normalizing' the other components. Jordan and Thiry, on the other hand, saw in g_{55} an opportunity for introducing a new scalar potential into the theory of the total field. Although there was no natural interpretation for it, it left a certain flexibility if new gravitational, electromagnetic, or other long-range forces could be derived from it, though from none of the other potentials.

In each case we can find a five-dimensional curvature invariant R_5 for the manifold, identify it with the Lagrangian in a variational principle, and from this derive appropriate field equations for the 14 (or 15) $g_{\alpha\beta}$. In Kaluza's approach these turn out to be identical with the equations of Einstein and Maxwell. Charged particles do indeed move along geodesics in this five-space, whose four-dimensional 'sections' or 'projections', it should be remembered, are not precisely identical with those of GR. The charge-to-mass ratio of a test particle appears as a measure of the angle by which the world line is 'deflected' into the fifth dimension, and enables one to retain the law of motion for charged particles as given above.

A three-space is characterized as cylindrical if there is an axis of symmetry and all quantities are independent of the angle variable measured around this axis of symmetry. By analogy, the requirement here that all the $g_{\alpha\beta}$ be independent of x^5 means that we are dealing not with the most general five-dimensional Riemannian spaces, but only those with some sort of symmetry axis around which the fifth coordinate is measured. Klein attempted a generalization in which complete independence was replaced by the assumption that the $g_{\alpha\beta}$ should have periodic dependence on x^5, now an angle variable with period 2π. This in turn has suggested possibilities for quantization. We know that in quantum mechanics momentum and position, and energy (or mass) and time appear as conjugate variables, such that the product of the uncertainties in measurements of both conjugates must exceed $h/2\pi$. In this respect charge appears in a very different way from mass, since there is no quantity conjugate to charge. Klein has proposed that if we form a five-dimensional Lagrangian for our test particle, the derivative $\partial L/\partial \dot{x}^5$ could be related to its charge-to-mass ratio, making this the conjugate of the fifth coordinate in the same way that energy and momentum were the conjugates of the first four.[3]

If we look at the Jordan version, the scalar which takes the place of g_{55} appears in the Lagrangian as a measure of the relative strength of the gravitational and electromagnetic fields. Since this has been the traditional role of the empirically determined gravitational constant (Newton's G) Jordan suggested that the scalar might be interpreted as a variable gravitational constant, at least in cosmological solutions within the theory. (Presumably it would vary only slightly if at all.) This idea had been proposed earlier by Dirac, and has been followed up in many subsequent publications by Dicke, though without the five-dimensional formalism.[4]

Whatever their various advantages, these theories all seem to be merely formal, nonintuitive, nonpredictive (except possibly Jordan's), and ad hoc. Kaluza's theory is not designed to modify or expand upon the Einstein-Maxwell equations in any way. On the contrary, the interpretation of the other components is chosen so as to reproduce them exactly. The theory does not in bring electromagnetism in a 'natural' way by showing that *only*

Maxwell's equations could be combined with Einstein's into a single formalism. As Pauli expresses it,[5] "It is in no way a 'unification' of the electromagnetic and the gravitational field. On the contrary, every theory which is generally covariant and gauge-invariant can also be formulated in Kaluza's form." Merely identifying the potentials ϕ_i with the mysterious g_{i5} does not establish their geometric character, nor their connection with any possible measurements of space-time intervals. The Lagrangian still appears as the sum of two distinct terms. Since we can give no intuitive, pretheoretic account of even the qualitative features of a possible fifth dimension, we cannot evaluate the theory as agreeing or disagreeing with them.

Aside from these difficulties, the role of a fifth dimension within the theory is unclear. We may, for example, ask whether it is spacelike or timelike, since the 3 + 1 division of the other four dimensions is such a crucial feature of GR. Pauli claims that this decision, which appears as the choice of the sign for g_{55}, is related to the sign of the gravitational constant. In saying that this constant is always positive (whether or not its magnitude changes) so that bodies with positive mass always attract each other, we are saying that the fifth dimension is everywhere spacelike. If the sign were to change, it would become timelike.[6] Nevertheless, the force of such a statement is certainly unclear. Even if it is in some way spacelike, the fifth dimension differs much more from the three ordinary spacelike dimensions than does time. We would need some additional conceptual distinctions, besides that of spacelike versus timelike, to separate it from the other four.

The idea that we might see the significance of the fifth dimension primarily in the quantum domain, especially if it is treated as a constant or angle variable, is extremely intriguing. It might—it just might—provide an effective way of incorporating essential quantum notions into GR. If we do not need to interpret it in the same macroscopic terms as the other four dimensions, we might use other such 'micro-dimensions' to account for those factors responsible for the strong and weak particle interactions. However, this is at best a very long-shot hope. As in the case of singularities in 'classical' GR, elements of mystery are admitted in the hope that they will somehow be clarified once the theory has progressed to a higher stage. Ultimately this procedure may be justified; indeed, it may be that *only* at some higher stage can we give an adequate account of these mysterious factors. However, when the nature and even the existence of such higher stages is quite unknown, it leaves dangerous hostages for the future, and may discourage efforts which could resolve the difficulties with present resources. Somewhere we must draw the line, and even quantum GMD may not be the ultimate panacea.

A second approach to unified field theory retains the four-dimensional world-picture, but gives up the belief that it can be adequately described by a symmetric metric and Ricci tensor alone. This direction was first taken by Weyl, who suggested that we should describe the geometry by a combination of $g_{ik}dx^i dx^k$, and $\phi_i dx^i$. Eddington then introduced the idea of taking the affinity Γ^i_{jk}, rather than the metric tensor, as the basic set of quantities for describing the geometry. We recall that the Riemann (and thus the Ricci) tensor is expressed in terms of the affinity and its first derivatives alone, without explicit use of the metric tensor. However, in orthodox GR the affinity is always identified with the Christoffel symbols, which are *defined* solely in terms of the metric tensor and its first derivatives. The affinity can nevertheless be defined by the behavior of vectors under parallel transport, and in a non-Riemannian space this will not be equal to the Christoffel symbols but an independent quantity in its own right. Eddington still required that the affinity be symmetric in its lower indices, but Schrödinger went a step further by starting with a nonsymmetric affinity, deriving a Ricci tensor from it, finding a Lagrangian expressed in terms of the R_{ik}, and only then defining a metric tensor by the derivatives of this Lagrangian with respect to the R_{ik}.[7]

Einstein devoted much of his own later work to developing a very complex theory in which both a nonsymmetric g_{ik} and a nonsymmetric Γ^i_{jk} appeared on an equal and independent basis. This would presumably be the most general theory of this kind, and further restrictions could later be placed on the symmetry or independence if it seemed desirable. Einstein in fact set only a few requirements on the Γ's to obtain a specific theory. He then chose as a Lagrangian density $\mathfrak{L} = g^{1/2}g^{ik}R_{ik}$. From a variational principle he was then able to derive field equations in two groups, containing 16 and 64 separate equations respectively.[8]

Einstein rested his case on the hoped for ultimate success of this unified field theory.[9] However, this enormous system of simultaneous equations (after the 10 of GR) is so difficult both to interpret and to solve in special cases that a final evaluation may be almost impossible. Unlike the Kaluza theory, which simply aimed to reproduce Maxwell's equations, the nonsymmetric theory could and did hope to account for much more, such as the law of continuity of charge and current. Some of these known laws do seem to emerge from the theory, and there is plenty of opportunity for additional novel predictions. A great deal of work has been done by Tonnelat and other French mathematicians,[10] while Edelen and others have investigated the range and structure of all possible field theories of this sort.[11] Nevertheless, difficulties in earlier and somewhat simpler versions of this approach have aroused considerable skepticism about its worth. Callaway found that one version predicted that a charged test particle would move as if it were uncharged, regardless of how much charge was piled on it,[12] and Johnson claimed that it could not distinguish one solution from another in which all intervals and magnitudes were multiplied by a factor k^n, where n is a positive or negative integer depending on the nature of the quantity.[13] This is in opposition to the fact that masses and charges have definite quantum values and mutual ratios.

In any case, the theory also suffers from serious conceptual objections. Bergmann has emphasized that it is almost impossible to get the 'intuitive feeling' needed to set up an adequate model which would guide us in relating any mathematical solutions to recognizable physical situations.[14] This in turn might prevent us from appreciating important new predictions. There is no central physical principle analogous to equivalence in classical GR which links the formalism with empirical evidence. Formally, we may begin with the Γ^i_{jk} as independent elements and take the g_{ik} as antisymmetric. However, we pay a high price. It is the line element ds, based on the g_{ik}, rather than the affinity which has direct metrical significance in GR, and if g_{ik} is not symmetrical this meaning and empirical content is lost.[15] In addition, the affinity, symmetric or otherwise, is not a tensor and will vanish in an appropriate coordinate system. If we still accept it we must give up a vital part of the covariance principle, that the *basic* quantities of physics should be described by tensors which cannot be made to vanish by mere coordinate transforms. In classical GR the Christoffel symbols were not basic, as they are here. We have modified the 'criterion of reality' for GR. Given any nonsymmetric quantity, we can find its value upon transposing indices, then take one half of the sum and difference of these two to break it up into symmetric and antisymmetric parts. While gravitational theory was entirely symmetric, we know that antisymmetric tensors like the F_{ik}, composed of the components of **E** and **H**, play an important part in electromagnetism. Despite the fact that these refer to fields, rather than potentials, Einstein suggested that they might be identified with the antisymmetric part of the g^{ik}.[16] However, if the symmetric and antisymmetric parts each seem to have a life of their own (and in some cases even transform differently, one as a tensor and the other not) it is hard to see how much is gained by formally putting them together to make a single quantity. Quantum prospects again raise their seductive heads here, since the nonvanishing of such operators as $pq - qp$ suggests that antisymmetric quantities may play an important role, but this again seems only a pious hope offered as a new regulative principle.

Attempts at a unified field theory might of course try to modify still other assumptions of GR. In each of the versions examined so far, we have kept Einstein's original action principle, which required that the Lagrangian be a linear function of the components of the Ricci tensor. Lanczos has proposed that we can keep the original four-dimensional Riemannian geometry of GR and still incorporate electromagnetism and perhaps other things if we give up this physical (as opposed to geometrical) restriction on the Lagrangian. Specifically, Lanczos advocates that we replace Einstein's $L = g^{ik}R_{ik}$ with the quadratic $L = \frac{1}{2}(R_{ik}R^{ik} + \beta R^2)$, where β is a constant.[17] Once again, however, this theory seems to involve various ad hoc assumptions, and Lanczos's actual choice of a Lagrangian, though perhaps more promising than the others considered, is hardly more natural or intuitive.

Geometrodynamics attempts to avoid this artificiality by remaining strictly within the framework of classical GR, the system of field equations as put forward by Einstein in 1916. It accepts the correctness of both Einstein's and Maxwell's equations, but instead of simply coupling them together to form a larger system of field equations it attempts to show that Maxwell's equations appear as *consequences* of Einstein's, plus one additional differential equation and certain algebraic restrictions, expressible solely in terms of geometric quantities like the g_{ik}, R_{ik}, and their derivatives, which establish that the sources of the field are wholly electromagnetic. For this reason GMD can describe GR as an *already unified field theory*. Nothing outside need be added to effect the unification. The electromagnetic field tensor F_{ik} is not an independent element and need not be introduced at all, though in solving particular problems we may find it convenient to introduce it as an 'intervening variable' without ontological significance.

It is well known that given any combination of **E** and **H** fields, we can combine them into a single tensor F_{ik} which is antisymmetric, so that its diagonal elements are all zero, the $F_{\alpha 0} = -F_{0\alpha}$ corresponding to the electric field **E**, and the $F_{\alpha\beta} = -F_{\beta\alpha}$ corresponding to the magnetic field. (Throughout this section we will revert to the convention that Latin indices run over all coordinates 0, 1, 2, 3 while Greek indices run over the spatial coordinates 1, 2, 3.) Also, given any F_{ik}, we can find the corresponding T_{ik} by $T_{ik} = 2(F_{ij}F^j_k - \frac{1}{4}g_{ik}F_{mn}F^{mn})$, where we are again using units in which $c = G = 1$, so that the F_{ik} in ordinary units must be multiplied by $G^{1/2}/c^2$, as is the charge. Having reached this point, we can then solve the field equations to get R_{ik}, and with appropriate boundary conditions, even the corresponding g_{ik}. Matters are further simplified by the fact that the asymmetry of F_{ik} insures that the trace $T \equiv T^i_i = 0$, and therefore R_{ik} is directly proportional to T_{ik}. Thus to any given electromagnetic field there corresponds one and only one Ricci curvature for the geometry.

Well and good so far; but how about the converse? Given a geometry, as embodied in an R_{ik}, can we find a field tensor F_{ik} corresponding to it? Certainly the answer for the general case must be no. There are numerous matter tensors which satisfy the requirement $T^{ik}{}_{;k} = 0$ without being derivable from an F_{ik} by the law above. Mathematically these possible matter tensors are all allowed by the theory. Physically, we have no way to rule them out unless we assume that the only possible extended sources of gravitational fields (regions where the T_{ik} do not vanish) are entirely electromagnetic in character. This might prove to be true, but it would have to be established by careful experiments. At present we must still allow the possibility that there is 'real mass' in nature, irreducible to electromagnetic forces, with a T_{ik} capable of producing gravitational effects. Likewise, in the initial value problem we can deal with the special case where the energy density ϵ on our initial hypersurface corresponds to $E^2 + H^2$ there, and the energy flow S_α to the Poynting vector $(\mathbf{E} \times \mathbf{H})_\alpha$, but this is hardly the only case allowed by the theory.

However, it turns out that there are certain conditions on the Ricci tensor which, if satisfied, will enable one to work in the opposite direction to construct an (almost) unique F_{ik}. It is of course conceivable that there might be some other field characterized by a tensor similar in form to the F_{ik} which was 'really' responsible for the field in the given case. However, we know of no such fields, and would have no reason to postulate their existence unless we found that the conditions were satisfied but we could still not detect an electromagnetic field by any of the standard experimental testing procedures. Specifically, the algebraic conditions on the character of the R_{ik} are as follows: (1) Its trace, i.e., the curvature invariant is zero: $R = R^i_i = 0$. (2) The term $R_{00} \geq 0$. Since R_{00} is proportional to the energy density T_{00} this means that we cannot have negative energy. Since this density is proportional to $E^2 + H^2$, this must be satisfied by any electromagnetic field. In all cases of interest, $R_{00} \neq 0$, which in turn means that some other diagonal term or terms must not vanish, to compensate for R^0_0 in enabling the trace to vanish. Thus we are outside the realm of the homogeneous solution $R_{ik} = 0$, which applies in the Schwarzschild case and which Einstein preferred to use for any concrete problems whenever the T_{ik} could not be fully analyzed. (3) The square of the Ricci tensor (in mixed form) is a multiple of the unit matrix, $R^i_j R^j_k = \delta^i_k (R_{mn} R^{mn}/4) = z^4 \delta^i_k$. Here z is a scalar and δ^i_k is the unit matrix. The actual T^i_k will then appear as a sort of square root of this matrix. In a suitable coordinate system which puts it in diagonal form, it will appear as

$$T^i_k = z^2 \begin{pmatrix} -1 & 0 & 0 & 0 \\ 0 & -1 & 0 & 0 \\ 0 & 0 & +1 & 0 \\ 0 & 0 & 0 & +1 \end{pmatrix}.$$

Now we can take a further step toward our goal of calculating F_{ik}. We can write the eigenvalue equation $T_i^k b_k = ab_i$, where the b_i are eigenvectors for the matrix. Here the four eigenvectors are $p_i = (1, 0, 0, 0)$, $q_i = (0, 1, 0, 0)$, $r_i = (0, 0, 1, 0)$, $s_i = (0, 0, 0, 1)$. The first two have $-z^2$ as eigenvalue; the last two, $+z^2$. We can define bivectors like $(pq)_{ik}$ as $\frac{1}{2}(p_i q_k - p_k q_i)$. We may also introduce the notion of the dual of a tensor: $^*A_{ik} = \frac{1}{2}(-g)^{1/2}\epsilon_{ikmn}A^{mn}$, where $^*A_{ik}$ is called the dual of A_{ik} and ϵ_{ikmn} is 1 for ϵ_{0123} and all even permutations of these indices, -1 for all odd permutations, and 0 when any two indices are equal. We then find that $(pq)_{ik}$ and $(rs)_{ik}$ are the duals of each other, and we make the following choices as a further sort of square root: $F'_{ik} = 2z(pq)_{ik}$, $^*F'_{ik} = 2z(rs)_{ik}$. In terms of the duals we can write $T_k^i = F_j^i F_k^j + {}^*F_j^i {}^*F_k^j$, and if we substitute the primed F's we find that all our requirements on the T's, and thus the R's, are satisfied. We have one particular solution to the problem of finding an F_{ik} given an R_{ik}.

In terms of the usual field vectors, the operation of taking the dual of F_{ik} has the effect of converting **E** into **H**, and **H** into $-$**E**. We can also identify the invariants formed out of these components by $\frac{1}{2}F_{mn}F^{mn} = H^2 - E^2$, $\frac{1}{2}F_{mn}{}^*F^{mn} = 2\mathbf{E}\cdot\mathbf{H}$, $z^4 = \frac{1}{4}R_{mn}R^{mn} = (H^2 - E^2)^2 + (2\mathbf{E}\cdot\mathbf{H})^2 = (H^2 + E^2)^2 - (2(\mathbf{E}\times\mathbf{H}))^2$, all taken in a suitable Minkowski frame of reference. Yet our F' is hardly the most general solution possible. In this given frame F_{ik} (as opposed to $^*F_{ik}$) corresponds to the case where **H** vanishes and **E** is parallel to the x-axis, with magnitude z. However, if we now make what is called a 'duality rotation' $F = e^{*\alpha}F'$, which converts F'_{ik} into $F_{ik} = F'_{ik}\cos\alpha + {}^*F'_{ik}\sin\alpha$, $^*F'_{ik}$ into $^*F_{ik} = -F'_{ik}\sin\alpha + {}^*F'_{ik}\cos\alpha$, we have the most general solution. The scalar quantity α is called the complexion of the field, and represents the division of the total electromagnetic field into electric and magnetic parts. We already know from SR that the electric and magnetic fields are not independent of each other, and that the breakdown into these separate parts depends on the coordinate system, and can be altered by a Lorentz transformation. As long as the complexion α remains undetermined, we seem to have a further degree of arbitrariness. Without changing coordinates, we can formally convert a pure electric or magnetic field (as embodied in F' and $^*F'$) into any sort of mixture by the device of a duality rotation. Since the T_k^i remain invariant under such rotations, we apparently cannot determine the complexion from the R_{ik} alone. Since some sort of distinction is important within any particular coordinate frame, this incomplete determination of the field from the geometry might appear to be a serious weakness in "already unified field theory."

But we have not yet exhausted all the geometrical or physical information available to us. We have treated F_{ik} so far only as an object capable of generating a T_{ik}, without regard for the fact that it obeys an important set of dynamical laws in its own right, i.e., Maxwell's equations. Taking account of individual components, there are precisely eight of these. In the usual covariant form they appear as $\epsilon^{ijkl}F_{kl,j} = 0$, corresponding to div $\mathbf{H} = 0$ and curl $\mathbf{E} + \partial\mathbf{H}/\partial t = 0$, and $(-g)^{-1/2}((-g)^{1/2}F^{ik})_{,k} = 4\pi J^i$, corresponding to div $\mathbf{E} = 4\pi q$ and curl $\mathbf{H} - \partial\mathbf{E}/\partial t = 4\pi\mathbf{J}$. In the case where there are no sources, $J^i = 0$, we can rewrite these equations as $F^{ik}{}_{;k} = 0$, $*F^{ik}{}_{;k} = 0$. Upon subsequent recombination, we find that four of these are equivalent to the requirement that $G^{ik}{}_{;k} = T^{ik}{}_{;k} = 0$, the Bianchi identities. The remaining problem is then to find some law expressed in geometrical terms alone which is equivalent to the other four Maxwell equations.

Fortunately this can be done. We define a new vector α_i as follows: $\alpha_i = (-g)^{1/2}\epsilon_{ijkl}R^{jm;k}R^l_m/R_{st}R^{st}$. This is expressed entirely in terms of geometrical quantities, the Ricci tensor *and* its first derivatives. Despite its complex and rather odd form, this vector is by no means arbitrary. In fact, when the Maxwell conditions are satisfied it appears as the gradient of the complexion α. For we find that its curl then vanishes, $\alpha_{i;k} - \alpha_{k;i} = \alpha_{i,k} - \alpha_{k,i} = 0$ identically, which is the condition for its being a gradient. This in turn provides another advantage. Whenever the curl of a vector field vanishes, the integral of that field from one point to another is independent of the path of integration as long as the space is simply connected. It therefore follows that the complexion of the electromagnetic field is not wholly arbitrary. We can define it as $\alpha = \int_0^x \alpha_i dx^i + K$, where K is its value at the lower limit of integration, $x = 0$. To be sure, the value of the complexion at this initial point is wholly arbitrary and a matter of free choice. Nothing in the geometry can be used to specify it. Nevertheless, once K is fixed the complexion is fully determined at all other points in space-time. We cannot freely perform duality rotations to satisfy our preferences for electric or magnetic field vectors, any more than we could in classical theory; we use the rotation law only to discover the actual field F_{ik} in terms of the extremal field F'_{ik}. Furthermore, the requirement that the curl of α_i should vanish is precisely equivalent to the remaining four of Maxwell's equations. If we accept this as requirement (4) on the Ricci geometry and its rate of change, we have then succeeded in reproducing the entire content of source-free classical electromagnetism and deriving Maxwell's laws, even though in none of the four requirements was there any explicit mention of F_{ik}![18]

There is one slight flaw in the elegance of this system. In the case of null fields, where $H^2 - E^2 = 0$ and $\mathbf{E} \cdot \mathbf{H} = 0$, so $\mathbf{E}$ and $\mathbf{H}$ are equal in magnitude and perpendicular to each other, the equations break down, since $R_{mn}R^{mn} = 0$ here, and the α_i cannot be defined by the expression given above. It is still possible to write down a suitable R_{ik} by introducing a null vector p_i such that $p_ip^i = 0$, $R_{ik} = 2p_ip_k$. If we then take a unit vector g_i orthogonal to p_i, we can then use $F'_{ik} = 2(pg)_{ik} = p_ig_k - p_kq_i$ and use a duality rotation to get F_{ik}.[19] Here the problem of finding the relative complexion is more complicated, and requires additional conditions on the Ricci tensor, but still appears to be solvable.[20] Peres, on the other hand, has taken a more pessimistic view. He claims that although this procedure works for most null fields, it breaks down in some exceptional cases, with an unwarranted degree or arbitrariness that makes the field not fully reducible to geometrical concepts. Since null fields also do not appear as the limit of non-null fields in any straightforward way within the theory, he suggests that they may have to be given a special conceptual status in GMD, despite the fact that they seemed perfectly ordinary in classical theory and there is no clear physical reason why they should not occur (at least within limited regions of space-time) in Nature.[21]

Much effort has been devoted to examining the character of the geometries that satisfy the Einstein-Maxwell requirements. One striking feature is the fact that the T^i_k has two positive and two negative components. This divides up the four-geometry in the vicinity of any given point into two orthogonal surfaces or 'blades', one in the xt-plane and the other in the yz-plane. Any vector in one of the blades will be perpendicular to any vector in the other. If they join together neatly, or 'mesh' at any one point they will in general not mesh at other points. According to Rosen, they will mesh if and only if the complexion α is everywhere constant. By adjusting the constant K, this constancy of α can be interpreted as a field which is everywhere purely electric, as with the field of a static point charge, and provides a convenient geometrical way for characterizing such fields.[22] From the algebraic standpoint, it is important to note that the equation involving the curl of α_i, which incorporates Maxwell's remaining equations, does violate one of the basic assumptions which Einstein used in setting up his field law for classical GR. The R_{ik} are based on the second derivatives of the g_{ik}; however, α_i, which uses a (covariant) derivative of R_{ik}, is based on the third; and thus the curl of α_i must introduce the fourth. This is in opposition to Einstein's belief, presumably based on his general knowledge of physics, that the basic laws should involve only second-order differential equations. This assumption had been criticized on occasion, and Lanczos in fact had previously tried to replace Einstein's law with a fourth-order equation.[23] Certainly a fourth-order equation, particularly one with the nonlinearity and complexity of this one, would be hard to solve and primarily of theoretical interest, but it still appears a legitimate price to pay for the ontological advantages of the unification.

It is usually considered advantageous, if only from an esthetic standpoint, to be able to derive all the basic field equations of a theory from a single variational principle. In a generally covariant theory, this requires that we should find some scalar L such that $\delta\int L(-g)^{1/2}d^4x = 0$ when the variation is performed with respect to the path or any of the independent variables in L. In GR this was particularly convenient because $L = R$, the scalar curvature invariant, in the source-free case, leading to the equations $G_{ik} = R_{ik} = 0$. If our sole concern is to couple together the equations of Einstein and Maxwell, without really unifying them or geometrizing electromagnetism, we can simply add a term to R to get the total L, i.e., $L = R + \frac{1}{2}F_{mn}F^{mn}$. The field components of which the second term is a function are considered quite distinct from the metrical components in the first term, so that corresponding variations are independent of each other. What we seek for already unified field theory is a Lagrangian whose variables are all geometrical, rather than one with 'foreign' terms.

Sharp has proposed that we can at least derive the curl equation if we use $L = R - \frac{1}{2}(R_{mn}R^{mn})^{1/2} \cos 2\int_0^x \alpha_i dx^i$, and vary this with respect to both the path of integration of the α_i and the metric. The first variation requires that the integral be path-independent, which is equivalent to the vanishing of the curl of α_i, and allows one to use the more simple $\cos 2\alpha$ for the last factor. *If* we can accept the idea that the complexion α is a purely geometrical notion, we *may* have what we seek: a single geometrical quantity from which both Einstein's and Maxwell's equations can be derived. However, it is not clear whether this principle is strong enough to enable us to deduce the other algebraic conditions on R_{ik} (as opposed to α) which must also be satisfied if we are to regard the whole field as electromagnetic in origin.[24]

An interesting new analogy suggests itself here. In terms of dimension and physical interpretation, the field components F_{ik} are comparable to combinations of the first derivatives of the metric tensor (e.g., the Christoffel symbols) rather than to the metric itself. Nevertheless, in the current variational principle the F_{ik} and g_{ik} appear on the same level. The electromagnetic field components must satisfy certain requirements, both on an initial surface and generally (Maxwell's equations), and these can be guaranteed if we introduce the vector and scalar potentials from which they can be derived by differentiation. In the variational principle it proves more useful to have these potentials as the quantities fixed on the limiting surfaces. Sharp has therefore proposed that we search for a set of 'superpotentials', of which the ordinary metric potentials g_{ik} would appear as combinations of first derivatives. These in turn could guarantee that the initial value equations discussed above would automatically be satisfied on the initial surfaces, and that the Ricci tensor would have the appropriate electromagnetic form.[25] But as yet these wonderful superpotentials (if indeed they exist) have not been found.

There are also problems in regarding the complexion as a completely geometrical notion. We know that in any case α is determined only up to an additive constant K, so that it makes no sense to speak of the absolute complexion of the field at a given point. As long as the *relative* complexion between any two points is unambiguously defined, this is no great loss, since no physical observations would be affected by adding the same constant to the complexion at all points of space-time.[26] However, we can define this relative complexion only if the two points can be connected by a path along which the electromagnetic field does not vanish. If, on the other hand, we have two regions, each containing an electromagnetic field but separated by a region in which this field vanishes, we could choose a different constant K in each region. There is no natural way to make the regions parts of one compatible system. Furthermore, this is not just a hypothetical situation. We know how to construct various screening devices which can confine an electromagnetic field and make it effectively vanish outside, though this cannot be done with gravity.

Penrose has therefore emphasized the following problem for already unified field theory. At some initial time t_0 region A might either be wholly electric or wholly magnetic. From information concerning the complexion in region B, separated from A by empty space, we could never tell which. On the other hand, if we started off with the field tensor F_{ik} as an irreducible quantity, we could certainly distinguish the two situations. Now let the regions A and B evolve so that they coalesce at some later time t_1. Now we have one single region in which the relative complexion must be univocal everywhere. But the two possible initial configurations in A will lead to measurably different results at t_1. If we stick to F_{ik}, we can predict the outcome then exactly; if we try to rely on α, we can only guess. One might therefore take the pessimistic view that the program of GMD for a complete geometrization of physics has failed. There is an observable physical distinction which GMD cannot make, but which can be made by other means.[27]

Of course, once the regions do coalesce we can then work backward to determine the correct relative initial complexions ex post facto, but it may not be convenient to wait so long, especially if we are interested in prediction rather than postdiction. Wheeler and Sharp have thus raised the interesting suggestion that the fault lies not in the ideal of physics as geometry, but in the choice of the particular *language* for expressing this ideal, that of local differential geometry based solely on the metric tensor. The moral of the Penrose problem is that we should search for other means of expressing geometrical concepts, rather than taking the historically first as an absolute. In particular, we may require more use of global or topological ideas.[28] If Nature really is geometrical, this may well be true. However, to assume this is to beg the question, and once again we may be elevating a pious hope into a regulative principle. The old semantical danger returns: in allowing for these very vague possible developments, we may lose much of the meaning of 'geometry' and the force of the ideal of geometrization.

It is a curious fact for the history of science that the mathematical formalism for already unified field theory, as presented above, was first worked out by Rainich as early as 1924.[1] Since it does remain wholly within Einstein's own framework, and is simpler by almost any standard than the asymmetrical or five-dimensional approaches, one might suspect that the problem of a unified field theory had already been solved, that all the subsequent work was quite superfluous, and that Rainich would have been recognized for a long time as one of the major contributors to the development of relativity theory. However, this did not happen. On the contrary, the equations had to be discovered independently thirty years later, by Misner in 1955. Only later did Misner and Wheeler learn that Rainich had anticipated their work and give him belated credit as a precursor.[2]

In view of the intense interest in unified field theory around 1925, though swamped by the new quantum mechanics almost immediately afterward, it seems incredible that such an important paper could have remained unknown. Some of the reasons for this may of course be psychological or sociological. It is undoubtedly true that Rainich was an unknown in the field at that time, and he may have been a rather retiring person who did not push hard to publicize his results, or reiterate it to meetings and conventions of physicists in the field. Since Rainich was more a mathematician and wrote his major article in a mathemetics rather than a physics journal, other physicists may have suspected that his was merely a formal result, with no significance either for ontology or for a major program within theoretical physics. Rainich himself may have failed to appreciate its importance, or even that of the question which it seeks to answer. It is noteworthy that in a book on relativity written in 1950, Rainich does not even mention his earlier incorporation of electrodynamics at all![3] In any case some physicists may have believed that his use of a fourth-order equation was against the 'rules' for developing a legitimate theory.

Nevertheless, there is an important conceptual weakness in Rainich's original scheme, which (one might hope) was the biggest stumbling-block for getting his results accepted. For the theory above works only in the source-free case; the Maxwell equations which come out of it are only correct when charges and currents all vanish. It is to Maxwell's credit that he recognized that electric and magnetic fields could exist without always being reducible to particular sources. Nevertheless, it is clear that most of the fields in ordinary experience do have sources, and that charges and currents do exist and must somehow be comprehended in a satisfactory unified field theory. At least Einstein's elaborate asymmetric theory had tried to take acount of them and derive, among other things, the law of conservation of charge. Rainich, some physicists might have argued, had pushed four-dimensional Riemannian geometry about as far as it could go. If the notion of discrete charged particles still had no home in it, then it seemed that Einstein's original GR was indeed a blind alley here, and that one should turn to more flexible approaches.

The great accomplishment of Misner was to give such a new account of charge, which treated charge as a characteristic but natural feature of the geometry, rather than as a nongeometric foreign element or as a singularity in the otherwise continuous geometric structure of space-time. His approach did not involve changing the dimensionality or Riemannian character of space-time, but it did allow that its topology might be more complicated than had been hitherto assumed, usually implicitly. This idea had not appeared previously in electromagnetic theory. However, its germ was contained in the Einstein-Rosen attempt to eliminate source masses as singularities, as was discussed above. The achievement of GMD was thus one of synthesis; it took the mathematical formalism derived by Rainich (and later, Misner) for source-free electromagnetism with a model based on multiply connected topologies in gravitational theory and combined them into a single theory. This new model went far toward eliminating both mass and charge, and it helped approach the ideal of a field theory that was complete as well as total.

In order to examine the importance of this achievement more closely, we must consider the GR solution of the field equations corresponding to a single spherically symmetric source particle carrying mass m and charge q. This case, which is the customary first generalization of the Schwarzschild solution, was first worked out in 1918 by Reissner and Nordström. If we choose a spherical coordinate system in which the particle is at rest, the field is purely electric, with the same complexion everywhere. The only nonvanishing components are $F_{01} = -F_{10} = q/r^2$ (Coulomb's law), $T_{00} = T_{11} = -T_{22} = -T_{33} = q^2/r^4$, neglecting constants of proportionality. The Reissner-Nordström solution turns out to be one of the general class $ds^2 = -\phi dt^2 + \phi^{-1}dr^2 + r^2(d\theta^2 + \sin^2\theta d\psi^2)$. In the Schwarzschild case $\phi = 1 - 2m/r$, as discussed above; for the Reissner-Nordström, $\phi = 1 - 2m/r + q^2/r^2$, where m and q are of course measured in units of length.[4]

If we compare the two solutions, we notice that simply charging a source particle, without changing its mass, will affect the metric tensor, and, since the coordinate system remains the same, the underlying Riemann and Ricci tensors. In this new geometry we will have a quite different set of geodesics. But these are just the paths of *neutral* test particles! In classical physics it was always assumed that gravitation and electromagnetism were quite independent of each other. If a test particle were uncharged, and there was no charge induced on it by passage through the field of a charged source particle, it could respond only to the *gravitational* field of the source. Since this gravitational field derived solely from its mass, charging it should have no effect whatsoever. The tracks of neutral test particles could indeed serve to measure this gravitational field. Of course, if the test particle were also charged, it would respond to the Coulomb force from the source, but this was a separate effect which could be added to the gravitational to get the total force. Here this separation is no longer possible. We must say either (1) charging a source particle changes its gravitational field, or (2) a neutral test particle also responds to an electromagnetic field. (1) is usually the more convenient version, since otherwise we cannot identify gravitation with geometry. However, this suggests that any breakdown is to some extent conventional. Qualitatively, this effect could have been predicted on the basis of SR and the equivalence principle. Since the electromagnetic field carries energy, proportional here to $E^2 = q^2/r^4$, there must be a corresponding inertial mass density proportional to q^2/c^2r^4, and by the equivalence principle this must have gravitational effects.

Let us consider just how the interactions predicted by GMD differ from those of classical physics. In comparing the Schwarzschild solution to the ordinary Newtonian prediction, we noted that in these coordinates the relativistic correction for a particle with angular momentum amounted to an additional attraction proportional to r^{-4}, though the r^{-2} term remained the same. If we write out the equations of motion for a neutral particle of mass μ in the Reissner-Nordström metric, we find that we must add to the usual r^{-2} attraction a repulsive force proportional to $\mu q^2/r^3$ even if the particle is moving along a radial line. Since this will eventually dominate the $\mu m/r^2$ attraction as $r \to 0$, we note an observable difference in gravitational effects: if the source is uncharged, a radial test particle with mass μ will eventually strike it, while if it is charged the particle will be repelled before it reaches $r = 0$. If the test particle has angular momentum proportional to h, it will also be subject to a repulsion proportional to $\mu q^2 h^2/r^5$, which will eventually dominate even the r^{-4} attraction. Finally, if we give the test particle a charge ϵ, so that both test and source particles have both mass and charge, we have in addition to the ordinary Coulomb force an electromagnetic attraction (never a repulsion!) proportional to $(\epsilon^2/\mu^2)(\mu q^2/r^3)$ working against the $\mu q^2/r^3$ repulsion.[5] Since this force does not affect neutral test particles (for which $\epsilon = 0$), we cannot call it gravitational by our convention above. However, it depends explicitly on the *mass* of the test particle, as the Coulomb force did not. This new force is proportional to μ^{-1}, the acceleration to μ^{-2}.

Thus our conclusion about the interconnection of the two forces is symmetric: not only does the gravitational force depend on charge, but the electric force depends on mass, both of which contradict classical physics. This result, which derives from the earliest days of "classical GR," before any of the attempts to develop a unified field theory, further strengthens the conclusion that any adequate field theory must deal with the total field, if any attempt to break it down into its parts is arbitrary. The apparent asymmetry resulting from the fact that the charge in question pertains to the source particle, and the mass to the test particle, so that a charged and uncharged test particle will respond in the same way to an uncharged source, is of no importance. It simply reflects the fact that we are not dealing with a genuine two-body problem (where one should indeed expect symmetry) but rather one in which the test particle is so small in both mass and charge that the self-fields which it generates are a negligible part of the overall field. Without such an approximation neither the Schwarzschild nor the Reissner-Nordström can be considered a rigorous solution. In addition, the 'pseudo-classical' attempt to get the Schwarzschild correction without using the conceptual framework of GR likewise loses its value, since we would undoubtedly have to postulate ad hoc a great number of distinct new forces to account for all these relativistic effects. No one has ever put forward a theory capable of doing so effectively.

Let us return to the ϕ factor, which measures the deviation from the flat Lorentz metric in both cases. Two important facts emerge. (1) The charge q never enters by itself, but only in the factor q^2, which is always positive, regardless of the sign of q. Thus we may conclude that positive and negative charges produce identical gravitational effects. By observing the behavior of neutral test particles, we could discover the existence and magnitude of a charge at the source, but we could never determine its sign. (This is also true of the 'electromagnetic' correction, involving $\epsilon^2 q^2$.) The mass m, on the other hand, does enter directly. (2) If it is really true that mass always has positive sign, then the mass and charge parameters always tend to bend space-time in opposite directions. The $2m/r$ effect of mass is attractive, tending to close the space; the q^2/r^2 effect of charge is repulsive, tending to open it. As far as their gravitational effects are concerned, mass and charge work in opposite directions and tend to cancel each other out. On the surface of the sphere at $r = q^2/2m$, $\phi = 1$, and a neutral test particle would receive no gravitational acceleration.

If $m < q$, the ϕ factor is always positive, and the metric becomes singular only at the source point $r = 0$. If $q < m$, the metric does have singularities at $r = m \pm (m^2 - q^2)^{1/2}$ in this coordinate system, but once again we can use a general method to find a new coordinate system in which these singularities no longer appear.[6] With elementary particles like the proton, the charge is much greater than the mass, as measured in our length units, and we might expect it to play the dominant role in determining the gravitational field, even at some distance from the source proton. However, the gravitational field of a proton is so tiny in any case that it is still impossible to measure. On the other hand, the total bulk of the sun, earth, or any other body large enough to make the relativistic correction observable, certainly has a greater mass than net charge, and this charge is not known to a very high accuracy. Thus we have as yet no direct test of the Reissner-Nordström solution or the gravitational effects of charge, despite the fact that in principle it allows many new predictions.

It is very tempting to do some speculating here. Throughout classical physics it was assumed that a particle was characterized by only two intrinsic properties, the parameters m and q. There was never an explanation for (1) why there should be only two, or (2) why mass always had the same sign, while charge could be positive or negative. Having seen that both of these parameters must appear in the expression for the full gravitational field, we may consider the following possibility: Suppose ϕ really has the form of a series, $\phi = 1 + \Sigma_n a_n (\alpha_n / r)^n$. Here the α_n represent the characteristic properties of the particle, so that $\alpha_1 = m$, $\alpha_2 = q$. The a_n are numerical factors, with $a_1 = -2$, $a_2 = +1$. The α's of course must be measured in length units, and we have deliberately left open the upper limit of the series. Such a series is perfectly permissible according to the field equations of GR. In general, the matter tensor will not vanish, but by adjusting the a's we can satisfy the requirement that $T^{ik}{}_{;k} = 0$.[7] Furthermore, if the series does break off after a finite number of terms, we can eliminate any apparent singularities which may appear in this coordinate system.[8]

If one follows this line of argument it would seem that there is no reason why there should be only two parameters; indefinitely many more might be included. Nevertheless, if we assume that the α_n are all the same order of magnitude (abbreviated α) as well as the a_n, each term in the series is approximately α / r times its immediate predecessor. Certainly for ordinary m and q, this α / r quantity will usually be a very small number. When raised to higher powers, it soon becomes far too small for detection in gravitational experiments. Even the q^2/r^2 term has not been observed yet. Thus our limitation to two parameters might simply be a well-justified approximation. However, we can always find the contribution the higher terms make to the T_{ik}. We may then be able to calculate the sort of field that would give rise to it, just as we could calculate F_{ik} when given the T_{ik} for the electromagnetic field. If such a procedure were really successful, it would go far toward showing the interconnection of all the forces in nature, and also the role of gravitation as a master force, laying down the geometry for which the others appear as special but ultimately eliminable aspects. It is at least possible that the various parameters now used to characterize elementary particles with respect to the strong and weak interactions might be related to, or replaceable by the higher α's.

We may also note that if n is even, the sign of α_n will make no difference with respect to gravitational effects, as in the case of charge ($n = 2$). However, if n is odd the sign and direction of the nth term will depend on the sign of α_n. If we can give a reason why each term in the series should be a particular sign (alternating in the first two terms) we might be able to explain why mass and the other odd parameters must have the same sign for all particles, while charge and the other evens may have either sign. Certainly all this is very speculative. However, if the program of already unified field theory were to be carried still further, it would be reasonable to hope that all the forces of nature were geometrical, in the sense of Einstein's four-dimensional Riemannian geometry; that they should therefore exhibit characteristic gravitational effects; and that their gravitational role should account for some of their other puzzling features, or at least be consistent with them.

Throughout this whole discussion we have treated the source particle as a true geometrical singularity, however many parameters might be needed to characterize it, or even if the series in which they appear might be replaced by a single analytic function. But this would at best give us a total, and still incomplete, field theory. Our geometrodynamical xenophobia calls us to eliminate the foreign bearer of these properties, and to absorb them into the singularity-free structure of space-time. Fortunately, we have some precedents. Let us recall Einstein and Rosen's attempt to remove the singularity caused by the uncharged Schwarzschild mass. They interpreted the particle not as a singularity but as the 'bridge' between two identical spaces separate from each other but each approaching Euclidean values far from the source.[9] This bridge effect shows up in several ways. If we use the isotropic radial coordinate ρ, r is related to it by $r = \rho + m + m^2/4\rho$. If we follow ρ below the symmetry point $\rho = \frac{1}{2}m$ ($r = 2m$), we find that the values of r increase out to infinity again, instead of decreasing. Flamm had also shown that the shape of this bridge was that exterior to a paraboloid of revolution.

This interpretation in terms of two spaces was fine as long as we were dealing with only one source mass. However, if we were forced to introduce a separate 'mirror space' for each particle, we would be stuck with a vast number of unrelated and uninterpreted spaces, a situation that parsimony would dictate ruling out. Wheeler and Misner then made the crucial suggestion that the bridge should not provide each particle with a separate space of its own, but rather re-emerge somewhere in a distant region of the main space. Having been brought back from splendid isolation to join the 'real world', the bridge acquired its married name: "wormhole." Call it what you will, it implies that the space, while free of singularity, is now multiply connected. There are topologically irreducible paths between two points, one through the main space and one through the handle or wormhole.[10]

One might assume that the use of such complex topologies is either ruled out by Einstein's equations or else represents a significant increase in the conceptual resources allowed by Einstein in the "classical GR" of 1915. However, this is not so. Weyl has emphasized that the field equations are entirely local differential equations, and cannot rule out multiply connected or even nonorientable spaces, like that of a Klein bottle.[11] Bass and Witten have even claimed that a space which is finite in all directions, temporal as well as spatial, *cannot* be simply connected, as some sort of four-dimensional sphere.[12] The restriction to simply connected manifolds in earlier discussions reflected either uncriticized and often unconscious habit, or else a lack of any previous conceptual interpretation of these new topological features. Finally, the notion of complexion for already unified field theory is compatible with multiply connected topologies as long as we impose the requirement that the integral of the gradient of the complexion over every irreducible closed path be $\oint \alpha_i dx^i = 2\pi n$, where n is any integer. This is a nonlocal requirement, but still a geometrical one, and the program of GMD does not restrict us to local differential geometry.[13] It appears that wormholes are mathematically permitted by GR; whether they are also conceptually permitted must be considered in greater detail.

In trying to interpret charge by the wormhole or bridge notion originally developed for mass, we are helped by the similarities between the Schwarzschild and Reissner-Nordström solutions. In the latter case we can also introduce an isotropic coordinate ρ, which is related to the original r by $r = \rho + m + (m^2 - q^2)/4\rho$. If we follow ρ in toward 0, we again find a point of symmetry, at $\rho = \frac{1}{2}(m^2 - q^2)^{1/2}$, $r = r_1 = m + (m^2 - q^2)^{1/2}$. The coordinate r reaches a minimum here and then starts out to infinity again. The shape of the resulting bridge is not precisely the exterior of a paraboloid of revolution, as it was before, but a figure very similar to it. It extends out to infinity at both ends, and reaches a minimum at $r = r_1$, a value somewhere between m and $2m$. In the limit of all these expressions as $q \to 0$, the result is identical with the pure Schwarzschild case. As long as we satisfy the requirement that $q < m$, the effect of charging the source mass is not to change the topology or the general shape of the figure in any way, but simply to narrow the width of the bridge or wormhole. r_1 may be taken as an appropriate measure of this width.[14]

But there is a new special advantage in the electromagnetic case. There appears to be a basic difference in classical physics between electric and magnetic fields. For the latter, div $\mathbf{H} = 0$ everywhere. If we use Faraday's language of lines of force, the lines of magnetic force are always continuous; they never start or stop anywhere. On the other hand, div $\mathbf{E} = q$, where q here represents charge density. Lines of electric force do have sources and sinks; they start on positive charges and stop on negative charges. The existence of such charges was the stumbling block for the original Rainich theory. But with a multiply connected topology, lines of electric force need never start or stop; they simply thread through the wormhole connecting a positive and negative charge. If we use a coordinate system that can cover both sides of the bridge or wormhole, we can show that the flux lines are not only continuous, but have the same direction of flow throughout their circuits. However, this will carry them in opposite directions in the main space and the wormhole. In the main space they will flow from positive to negative charges; through the wormhole they will flow from negative to positive. But a wormhole mouth is not something foreign to the geometry, nor is it something localized at a single point. It is rather the name given to the general region where the topologically distinct paths branch off. The flux through each wormhole remains a constant whatever the detailed shape of the mouth may be, or whatever its motion in the main space. The only thing that can change it is a change in the topology, as when two wormhole mouths coalesce into one, or a handle is somehow raised up from an 'eruption' whose interior splits away from the main space. Thus a wormhole mouth plays all the roles of the classical notion of charge, and the flux through it satisfies the same sort of conservation law. Finally, insofar as a wormhole must have two mouths (or a bridge two sides, for those who prefer that model), we can see why individual charges are never created or annihilated separately, but only in positive-negative pairs. Since we now do not need to postulate charges as regions where the electric field diverges, we can use the Rainich equations (including div $\mathbf{E} = 0$) everywhere. And we do have a complete theory of electromagnetism, which requires no singular sources, and which reproduces all the features of the classical theory.[15]

Nevertheless, there are still problems in this wonderful scheme. It is easy enough to visualize a multiply connected two-dimensional manifold: we simply consider it embedded in a three-dimensional one, with the handle sticking out into this additional dimension. Likewise, if we are dealing with a finite three-dimensional object (like a cup) rather than the whole three-space, we see the handle as occupying a different region in the overall three-space. But here we are demanding that the four-dimensional space-time as a whole (or any unbounded spacelike hypersurface within it) be multiply connected. If we try to introduce higher dimensions to take care of the wormholes, we run into all the difficulties discussed in relation to the embedding problem. And since our space is all-inclusive, it leaves open no regions of the same dimensionality. Here the difficulties stem from an excessive reliance on visualization and a confusion of the range of space-time and the range of coordinate systems. We do visualize in Euclidean terms, and tend to suspect that there is really something outside a closed space-time corresponding to the infinite stretches of Euclidean space. However, this is quite unwarranted and would not cause dilemmas if we could use an intrinsic, rather than such an extrinsic, characterization of our space. We can imagine spaces which do not 'use up' a coordinate system, in that certain values of the coordinates either do not name real points at all or else refer to points that have already been named by other values of that coordinate. (Consider Euclidean space with polar coordinates: the values of the angle variable $\theta \pm 2\pi n$ all name the same points.) It is less obvious that a given coordinate system may not be enough to cover a space, even when its coordinates run through their entire (perhaps infinite) ranges, or that one set of coordinate values might have to refer to several points. Yet this situation is equally possible. Our coordinate system is designed to cover all the points in the 'main space', even if some of these represent singularities and/or can never be reached by any allowed trajectories. The points in the wormhole or handle may be just as real, but we must change to a different coordinate system to refer to them. It is only the coordinate values that are used up in the main space, not the totality of possible spatial positions.

This relation between physical points and coordinate values might be reasonably clear in the simple case of an Einstein-Rosen bridge for a single particle, whether in the Schwarzschild or Reissner-Nordström solutions. Here we have one main space and one mirror space, with a perfectly determinate metric on each. But in using the wormhole model we demand that the wormhole emerge somewhere else in the main space, rather than a mirror space of its own. No indication is made of where or how far away this should happen. And if we do specify that two particular 'apparent charges' are to be considered the mouths of one and the same wormhole, we will want to compare the distances between them within the main space and within the wormhole. For the former distance, there may be some difficulty in determining the end points, since the space curves into the wormhole instead of reaching the assumed point charge. However, we can at least measure along the geodesic line connecting them and arrive at a reasonable measure of distance. The latter, the distance through the wormhole connection, presents the greater problem. It is hard to see why this should not have a definite value, which may be smaller or larger than in the main space. For example, the handle on a cup is normally more sharply curved than the cup itself in the region in question, so a path through the handle should be longer. On the other hand, the airline distance between two points on the earth's surface, measured along a great circle there, will be longer than that along a wormhole following a three-dimensional Euclidean straight line through the interior of the earth. However, in GMD not only is there no criterion for measuring distance through a wormhole, but this is even treated as an illegitimate or meaningless question. This seems to me an attempt to have one's case and eat it too. We know that the main space is fully metric: as long as we measure along geodesics, we can find the distance between any two points. We cannot have it both ways. If we are interested only in topology, we must give up talk about metrical features in the main space; if we keep the latter, the wormhole can't be *just* a topological feature. It must share the same sorts of metrical properties as the main space. Without this we must make an arbitrary and unwarranted separation of the points of space into two classes, depending on whether or not they allow metrics. If the wormhole cannot be fitted more neatly, we should cease talking about it altogether. Misner has in fact suggested that certain solutions can be fully analyzed in Einstein-Rosen bridge, but not in wormhole, terms.[16]

In classical theory, it was perfectly possible to have, say, a large positive charge sending out lines of force which ended on a great number of different negative charges, which might be widely dispersed through the region. If we are to model this situation in GMD, we must assume that a large wormhole representing the positive charge branches off, with individual tributaries connecting it with the various negative charges. The flux flows toward the large mouthlike streams which flow into a river, which in turn finally reaches a delta and flows into the sea. This suggests an elaborate capillary network of wormholes under the main space, joined together in myriad ways which change constantly in response to the motion of the charges in the main space, whereby the lines of force shift to new charges that become nearer or stronger during its passage. This interpretation is, I suppose, possible, but it certainly complicates the original picture. Somehow the total flux entering and leaving each of these wormholes remains constant, despite the fact that there is constant juggling back and forth in the network, perfectly parallel (though opposite in direction) to that in the main space.

Alternatively, we may suppose that there is no interlocking network, but that each wormhole mouth is connected to another equal and opposite in sign. One might be tempted to think of proton-electron pairs, and argue on this basis. However, this would seem to introduce quantum notions into a theory specifically designed to avoid them at this point; it is the classical notion of charge—which, like mass, could take on all possible values—that the wormhole model seeks to reproduce. Furthermore, it introduces an affinity between a particular proton and electron which does not agree with experimental results in particle physics. In either case, the theory assumes that the net charge of the universe as a whole is zero. If this were not so, we would have wormholes carrying flux to or from infinity, there to disappear in the unknown. It may be true that the positive and negative charges in the universe do ultimately cancel each other out, and if this is so, it would be desirable for a satisfactory electromagnetic theory to explain it. But as yet there is no clear evidence for it, and it is risky to assume this a priori.

Consider now a large body. It contains a great number of positive and negative charges. Most of these cancel each other out, but there is still an unbalanced net charge, which causes it to exert an electric force on other bodies. A macro-observer, not knowing the internal constitution of the body, would interpret it as a single large wormhole mouth, with its size and shape determined by the mass and charge of the body. However, there are 'really' many tiny wormholes inside it. Shall we say that there is not one big wormhole, macroscopic appearances to the contrary notwithstanding, but only a series of small parallel wormholes from the unbalanced charges in the body? or shall we say that there is still one big wormhole, however the smaller ones are related to it? If we choose the latter, we must give some notion of 'topological addition' which explains how many little wormholes can unite to function as one big one. And the Betti number, or whatever other mathematical concept is used to characterize the degree of connectivity of the space, then becomes ambiguous or context-dependent. Or shall we say that the wormhole notion is appropriate on different levels of size, which must however be kept separate, i.e., there is either one or many wormholes depending on the level, but not both at once? Then we must give some arbitrary boundary between these size levels, and give up the idea that the geometry has intrinsic properties, independent of the observer and the questions he is asking.

Finally, classical theory allows charge to appear on individual point particles, or as an extended distribution through space. The Reissner-Nordström solution, on which the wormhole model is based, assumed the charge to be concentrated at a point, which originally appeared to be singular. It is possible that a continuously extended, but still well localized, distribution of charge could also give rise to a wormhole mouth, though perhaps one with a rather different size and shape. Perhaps this distribution must be conceived instead as a set of tiny wormhole mouths close together. If charge were to be distributed continuously throughout all of space, giving rise to densities but no individual charges, it would be still harder to use the wormhole model. We may try to rule out the latter cases on empirical grounds, but this would be almost impossible for mass, to which a similar analysis applies. Otherwise we might have to allow that there could be two different kinds of mass or charge, one of which manifests itself in wormholes while the other does not.

Even if one were to consider these mere points of detail, there remains one serious difficulty. We have insisted throughout that GR is essentially a field theory, and the interpretation of GMD certainly wants to preserve this feature. But in a field theory a body is influenced only by the fields in its immediate neighborhood, and any causal field influence must be propagated with a finite speed. There can be no signal, force, or influence instantaneously connecting two points separated by a finite region of space. GMD also accepts the further restriction that not only are propagation velocities finite, but that they cannot exceed the speed of light. We can always draw a light cone at any point in space-time to determine the range of possible causal connections with that point, though in a general coordinate system the boundaries of such a cone will not normally be perpendicular straight lines.

Now suppose that we have a wormhole connecting two points. Their separation in the main space is l; through the wormhole it is l'. A light signal sent through the main space will require a time $t = l/c$. Next suppose that the main space is bent so that it curves back on itself, and the wormhole is the shorter route, with $l' < l$. It would then seem that the time required to pass through the wormhole, $t' < t$ as long as v, the velocity of the signal sent through the wormhole, exceeds cl'/l. This requirement can be satisfied for $v < c$ as long as l'/l is small enough. If so, we have violated our causality postulate. It *is* possible for a signal, sent through the wormhole at v, to reach the other point faster than the light signal. By making a suitable Lorentz transformation, we can then find a coordinate system in which the signal is received at the second point before it is transmitted through the main space from the first. This paradox, first suggested by Niels Bohr, could prove the coup de grace for the wormhole theory unless it can be avoided.[17]

It does seem possible to overcome this disastrous result at least in the Schwarzschild case, the wormhole corresponding to mass without charge. However, this is by no means the only possible type of multiple connection, and there is no general way of solving the problem for all other possible cases. It may be that we can rule out other kinds of wormhole on empirical grounds, as corresponding to no observable physical situation. Without either of these general guidelines, we must look at each case individually and hope for the best.

Fuller and Wheeler have pointed out that the Einstein-Rosen bridge is a perfect representation of this sort of wormhole situation. Here the wormhole mouths open onto the two 'parallel' pseudo-Euclidean sheets. The sheets would be connected again only at infinity, so that the mouths would be infinitely far apart if considered as belonging to a single main space. On the other hand, the throats or necks (points of minimum radius) corresponding to the two mouths are identified so that the separation vanishes. However, this situation is not static in time. What we are describing is really a wholly spacelike initial surface, and indeed one at a moment of time-symmetry, or (for so it turns out) maximum expansion of the throat. I have shown that in the finite time πm after (or before) this moment of time-symmetry this throat will contract and finally 'pinch off', reaching the singularity corresponding to $r = 0$. An observer stationed at the throat will be caught in this pinch-off; a particle moving along the corresponding timelike geodesic will hit the singularity, beyond which the analysis cannot be carried further, rather than passing through to the other side to announce its triumph over light.[18] Fuller and Wheeler have calculated the trajectories of a variety of radial geodesics, all leading to the same result. This result shows up even more clearly in the Kruskal coordinates, which enable one to write down the whole metric for $0 < r < \infty$ without singularity. These have the advantage of being isothermal, so that the paths of all radial light rays are straight lines at 45° angles. In this system we do indeed have two symmetrical regions corresponding to $r > 2m$ on each of the two sheets, but it is impossible to draw a timelike or lightlike geodesic that passes through both regions and thus might be considered a signal connecting the two. There are also two regions corresponding to $r < 2m$ on the two sheets, and a particle here can reach either of the regions beyond $r = 2m$. But such a particle must soon either come from or arrive at a region of infinitely high curvature corresponding to $r = 0$. Fuller and Wheeler postulate that laboratory physics, which would include the sending and receiving of any real signals, cannot be done in such regions.[19] Not only is the particle being sent as a signal moving, but also in effect the spatial framework itself, and this latter motion proves faster.

We have said that our initial time is a moment of maximum expansion, and that the throat never simply remains there; it reaches that radius and immediately begins its contraction. Yet the initial time, the choice of our zero-point for the time coordinate, should be arbitrary. We would then have the paradoxical situation that whether the throat appeared open or closed at a given time would depend on whether or not that observer happened to choose that particular time as his zero-point. This seems wholly unreasonable. At a given moment the wormhole should appear unequivocally open or shut, whatever name or number the observer gives to that moment. Furthermore, once the throat pinches off, we have no way of telling (in any coordinate system) how long it will remain shut, or when and why it will suddenly start to reopen. In a metric which is supposed to represent a single material source particle fixed at rest, this nonstatic behavior is surprising and hard to interpret or explain.

If we consider all possible timelike geodesics from a given point in space at our initial time, we find that if they keep going across $r = 2m$ they must all eventually reach the singularity at $r = 0$. This agrees with our earlier 'pseudo-classical' interpretation of the Schwarzschild metric, in which it was claimed that an attractive force would always dominate at small distances, pulling a test particle into the center if it approached close enough. Now let us calculate the spacelike geodesics passing through this point, as possible points of simultaneity. A further surprising result emerges: while these geodesics can pass through some of the points with $r < 2m$ and then emerge on the other side of the bridge, there are some points in this region which cannot be reached by any spacelike geodesics: in particular, $r = 0$ is excluded. Such points could therefore never be part of the world, or initial space, of our geodesic-constructing observer. Any signals sent from these points would never be received, no matter how long he waited and no matter how his initial space evolved.[20] This result may have implications for the unity of the four-dimensional world, or the extent to which it can be represented as a sequence of three-dimensional spacelike hypersurfaces.

In the Reissner-Nordström case, the situation is more complicated. Here we have the lines of electric force trapped in the wormhole as our model for charge. If the wormhole starts to pinch off, these lines (which cannot change in number) will in turn be squeezed together. However, the normal effect in such cases is that they repel each other, causing a subsequent expansion. Arguing in these qualitative terms, we may suspect that there will be some sort of "elastic electromagnetic cushion" to resist the pinch-off. When one makes the calculations, this turns out to be perfectly correct. Whatever 'force' is responsible for the gravitational pinch-off contraction, it is perfectly balanced by the electromagnetic repulsion. Indeed, an observer stationed at the throat will see a harmonic oscillation between the limits $r = m \pm (m^2 - q^2)^{1/2}$. Furthermore, although the amplitude of the oscillation is clearly a function of q, the period or frequency is not: it turns out that the time for passage from one limit to the other is πm, the same as the total time required for pinch-off in the pure Schwarzschild case![21]

Now this result at least agrees with our pseudoclassical analysis, according to which a repulsive force (r^{-3} or r^{-5}) will always dominate eventually and prevent any neutral test particle from reaching $r = 0$. Furthermore, it does reduce to the Schwarzschild complete pinch-off in the limit where $q = 0$. We might in fact try to work the analogy in the opposite direction and assume that after pinching off the Schwarzschild wormhole begins to reopen immediately, so that in both cases the time average of the radius of the wormhole remains a constant, $r = m$. This might help remove some of the asymmetries involved in using a moment of maximum expansion as our initial time. But the price is then to bring back the causality paradoxes in full force. It is still true that the topology is somewhat more complicated in the Reissner-Nordström case, and it is still true that the metric will eventually become singular, though perhaps not in the same way as the Schwarzschild. However, it certainly looks as if a wormhole containing charge will always remain open to some finite degree, and if so it should be possible to send a signal through it that will arrive before a light ray in the main space. Perhaps a more detailed analysis of possible geodesics here might show how the problem can be avoided, but otherwise we are faced with a real dilemma of interpretation: either we must face the mysterious singularity of the pinch-off, which seems to limit the completeness of GMD as a field theory; or we must allow possible violations of the causality postulate.

At this point the skeptical reader (and writer) may well wonder whether anyone making these statements about the dynamics of wormholes can possibly know what he is talking about. To be sure, we have a conceptual model, a material counterpart to the Rainich formalism of GMD. This model in turn suggests certain kinds of questions that we ought to ask, along with a range of possible answers. Nevertheless, these questions seem very remote, not only from empirical tests, but from possible analogies within other models. We run the danger of becoming wrapped up in a closed, self-contained conceptual system and mistaking it for some level of reality, or even the whole of reality. There is yet no assurance that it mirrors nature, and our paradoxes might better be taken for inadequacies of the model, rather than mysteries of the world.

Suppose we are comparing two models A and B. If every statement S_A within A has a natural counterpart or analog S_B within B, we may say that A and B are fully comparable. If in addition S_A and S_B not only refer to the same sort of situation but say the same thing about it (i.e., make the same predictions) the models are equivalent, and the decision between them must be made with respect to 'inner' characteristics like conceptual coherence or intelligibility. Now suppose that A and B are not fully comparable, so that there is at least one S_A which has no counterpart in B (and for simplicity, we will assume the opposite does not hold). Then A may be called richer or more comprehensive than B. If S_A can be tested empirically and shown to be true, then clearly A will be a better or more adequate model than B, which simply cannot deal with such an observable situation. Of course, A also takes the greater risk. If S_A disagrees with the results of observations, then we may be forced to reject A. But what if S_A does not appear to be testable? At a given historical stage of a science, it is in general impossible to predict what sorts of observations might be interpreted in the future as confirming or disconfirming evidence for S_A, even if they are not available now. Only blind foolishness or positivism could lead one to decree now that such statements must always be meaningless. Maybe S_A will prove a useless blind alley; but maybe it will eventually lead to new coordination relations which will connect the model to a new and often unsuspected realm of empirical data.

Let us apply these considerations to the case at hand. Suppose that some of our concepts do have observational analogs, e.g., 'mouth of wormhole' $\leftrightarrow$ 'classical charge', and statements in terms of them can readily be tested, e.g., 'evolution of a configuration of wormhole mouths' $\leftrightarrow$ 'motion of a system of charges'. Other concepts seem to have no observational analogs. Whatever corresponds to the mouth of a wormhole, there is no concept in classical physics or obvious observable feature corresponding to the interior of a wormhole, despite the fact that this seems to be an equally important element of the model. In trying to describe what happens to it, we cannot be guided by analogs in other models; we must work from within, considering how it should behave with respect to the other elements of the wormhole model. The latter procedure is heuristically valuable, but inherently risky when the model is considered as a candidate for an ultimate ontology.

We can of course find a coordinate system suitable for describing the wormhole geometry, and we can construct geodesics in such a system, including some which appear to go through the interior. However, it is only an assumption, albeit central to the present interpretation of GR, that every such timelike or lightlike geodesic is a possible path of a real signal, light ray or free particle. There may be yet undiscovered grounds for ruling out some of these as being physically impossible, just as the causality postulate of SR ruled out spacelike geodesics. It may be that the wormhole geodesics will be thus forbidden. Furthermore, when we move from the language of geodesics to the language of observers performing experiments, we must make further assumptions. We certainly do not *see* a wormhole mouth as such. No one has any idea how to send anything through a wormhole, or even to recognize the event if by chance they succeeded in doing so. Under the circumstances, it is not clear just what sort or how great a modification of the causality principle would be required, even if one tried to modify it in some ad hoc way. We should want to be able to describe this modification in nonwormhole terms, since it would presumably apply to other theories and physical situations.

Wheeler has attempted to argue that the empirical existence of electric charge in Nature is evidence for the claim that space-time really is multiply connected, with whatever else that may imply, on the ground that in merely following Maxwell's equations the fields will eventually evolve to crests which can be interpreted as flux-trapping wormholes.[22] However, his arguments are not really convincing. One can defend the alternative position that these crests correspond to singular points where by some mysterious process an 'electric jelly' is created or destroyed. It is certainly true that this interpretation also runs into many problems, though whether these are more or less than those for the wormhole model may be a matter of personal judgment. The choice is not between two hypotheses h_1 and h_2 within a single model, where experimental results should give a clear-cut answer. The choice is rather between two distinct models, each of which has advantages and disadvantages. We may try either an everywhere regular but multiply connected space, or an everywhere singly connected but sometimes singular space. These models may resemble each other closely in certain features, but they are still distinct. Our decision must really be made on a hope for the future, as Wheeler indeed seems to recognize: we ask, which model will be richer? Which model will be more effective in revealing the vast potential richness hidden in Einstein's equations? On this basis Wheeler does have a strong case. Until and unless we can prove that multiply connected topologies are conceptually impossible as candidates for physical space, only a dangerous prejudice will prevent us from at least examining them. And a singularity is an inherent *ignorabimus.* It places an ultimate limit on possible analyses, rather than suggesting where they might be carried further. These new lines may be hard to fathom, and they may eventually fail, but at least they are better than nothing. We need not *accept* the wormhole model as yet, but we should at least wish it well.

Having decided to countenance wormholes, we can then use them to shed light on other unresolved problems within GR. In discussing the problem of motion and the two-body problem in the previous section, we remarked that there were difficulties if we treated the moving masses either as singularities (the Einstein-Infeld-Hoffman approach) or as extended bodies with nonvanishing T_{ik} (the Fock-Papapetrou approach). We now have a third alternative. We may represent each of our massive bodies, with neither one singled out as being the source or test particle, by distinct Schwarzschild wormholes. At an initial time they will appear together on a single spacelike hypersurface which satisfies the initial-value equations. Then we know that this initial two-wormhole configuration will evolve in time according to Einstein's equations. Since gravitation is always attractive, we would expect the wormhole mouths to approach each other. Since the space curves smoothly into each mouth, we would also expect that the shape of the throat, spherical in a one-body problem, would become increasingly distorted along the line of centers as the mouths approach, and perhaps eventually coalesce.

The relative motion of the mouths has been investigated in detail by Dubman and Misner, both starting from a moment of time-symmetry when the metric has a particularly simple general form. Starting with the simplest possible case, Dubman was able to find a relationship between the supposedly intrinsic masses, the m's, and the effective gravitational parameters, the α's discussed above in connection with the problem of adding effects from several sources. He was then able to find a reasonable measure for L, the effective distance in the main space between the throats of the wormholes, corresponding to Newton's Euclidean separation r. By expanding $L(t)$ as a Taylor series in t, and taking the coefficient of t^2 in this expansion as a measure of the acceleration, he found that it did agree to the first order with Newton's r^{-2} acceleration, as would be expected. He also found that the minimum radii at the throats were affected by the interaction.[23]

Misner has attempted to deal with a more general case, one which takes account of the fact that there are many non-Newtonian features present in the two-wormhole problem. In particular, the accelerating masses can radiate gravitational waves and respond to any already present in a variety of ways, which means that there is no unique metric corresponding to the evolution of the wormholes. Furthermore, the distance L is not unique but must be fixed by additional (reasonable) conditions. What he specifies is rather the entire path length through the main space and wormhole, requiring that this and the total mass of the system have some definite value in order to fix the various constants of integration. Qualitatively, the same sorts of motions again appear, with the additional conclusion that the shape as well as the size of the wormhole throat is distorted by the presence of the other mass, taking on an increasingly cigarlike elliptical shape.[24]

In any case we have a legitimate two-body problem, along with the means for its solution; though the apparent masses involved are part of a single configuration rather than a superposition of two distinct centers of interaction as in the Cartesian vortex theory, which, however, seemed incapable of accounting for gravitational attraction.[25]

The existence of so many symmetries or resemblances between electric and magnetic fields, and between electromagnetic and gravitational fields (in the initial value problem) has led to renewed interest in the questions of whether there might be magnetic charges, or poles, as well as electric charges, and whether there might be some geometrodynamical analog of charge. If we look at Maxwell's equations in the presence of sources, it would seem that symmetry would be increased if we could allow magnetic charges, responding directly to magnetic fields, and magnetic currents or charges in motion responding to electric fields. This apparent asymmetry shows up whether the equations are written in ordinary three-dimensional vector notation or in covariant four-dimensional form. In the completely source-free equations of GMD, this asymmetry disappears, but we still must explain why electric but not magnetic lines of force can get caught in a multiply connected topology. According to Dirac, quantum mechanics as well as Maxwell's electrodynamics allows the existence of magnetic poles, and indeed appears incapable of explaining their nonexistence beyond postulating it as a brute fact. Such magnetic monopoles would necessarily have a mass considerably greater than that of any known elementary particle, and would be bound together by a much stronger force. But they still are permitted, and it is remarkable that the many experiments designed to detect them have all failed to reveal any.[26]

In the vicinity of a wormhole mouth the complexion should be almost constant, though Maxwell's equations do not specify what this value should be. Since the absolute complexion always contains an arbitrary constant of integration, we can give the field a constant duality rotation at all points of space without changing any physical situation. Thus if we started with a solution in which all the wormhole mouths were purely magnetic, or a mixture of electric and magnetic in the same proportion on each, we could convert them all to purely electric by means of a duality rotation. A problem would arise only if we started with a mixture of electric and magnetic charges, or different proportions at the different mouths. In such a case we would be forced to postulate either (1) the existence in our solution of irreducible magnetic poles, or (2) that wormholes can also capture lines of magnetic force.

If we wish to make GMD agree with the 'negative observation' that there are no magnetic poles in Nature, we must require that the characteristic complexion of each of the wormhole mouths differ from any other by

$2\pi n$, where n is any integer, including 0. It may be possible to derive this either from Maxwell's equations alone, which seems very unlikely, or from the fully coupled Einstein-Maxwell equation of GMD. If not, we must add this as an additional postulate, which would then preclude any deeper explanation. This requirement is certainly nonlocal, since it relates the complexions at points of space which may be separated from each other by great distances. It looks reminiscent of the other nonlocal requirement, that the line integral of α_i over any closed path should be an integral multiple of 2π, but the two do not seem to be mutually derivable.[27]

One might try the following argument: according to de Rham's theorem in topology, the existence of a magnetic vector potential is equivalent to the nonexistence of any magnetic poles, or wormhole mouths whose flux is wholly or partly magnetic. And the magnetic field **H** can always be derived from a vector potential **A** by $\mathbf{H} = \text{curl } \mathbf{A}$. Therefore, there are no magnetic poles. However, this potential **A** is by no means a basic part of electromagnetic theory. It was introduced only because it appeared that there were in fact no magnetic poles, and the div $\mathbf{H} = 0$ everywhere. The effect of discovering such a pole would be to give up the vector potential, so our argument really puts the cart before the horse. In any case, we would still have to explain why there is an asymmetry such that **E** can *not* be derived from a vector potential.[28] It may turn out that we could derive the existence of the vector potential from the general Einstein-Maxwell equations, but this seems doubtful. Witten has indeed found a solution containing a mixture of poles and charges by relaxing this requirement, though the solution is not static and the wormhole mouths violently repel each other.[29]

In the initial value problem for electrodynamics, if we specify **H** and $\partial\mathbf{H}/\partial t$ on the initial surface, we cannot determine **E** uniquely, but only up to a parameter representing the charge on that surface. This result is closely bound up with the fact that the absolute value of the scalar electric potential has no significance. It can jump discontinuously across the initial surface if there is charge there, and still satisfy the initial value equations for electrodynamics. In GMD we have seen that the shift vector N_i plays the corresponding role of a potential for the extrinsic curvature K_{ik} or the 'field momentum' P_{ik}. If only the derivatives of N_i had any physical significance, we could also allow discontinuous jumps in the N_i. These in turn would be governed by parameters that might serve as the geometrodynamical analog of electric charge. However, the situation is quite different in GMD. The absolute values of the N_i, and the requirement that they be continuous, stems from their other role as coefficients in the four-dimensional metric tensor. If they could change discontinuously, we might have paradoxical results of measurements and unwarranted singularities in the geometry. Thus Wheeler feels confident in concluding that GMD exhibits no analog for electric charge. Both K_{ik} and P_{ik} are uniquely determined by the initial value equations, with no room for further specification.[30]

One slight possibility remains: there might be a GMD analog to magnetic poles rather than electric charge, as a feature of the intrinsic, rather than extrinsic, geometry. This would require solving the 'conjugate representation' of the initial value problem, wherein one starts with the field momentum and its rate of change and tries to derive the intrinsic geometry g_{ik}. But this has not been done as yet. It is also questionable whether the apparent nonexistence of magnetic poles makes this intrinsic geometry kind of charge any more or less likely. At least we have not exhausted all the varieties of charge which derive their existence from topology.[31]

Indeed, the geometry of space-time may exhibit many other peculiar features besides multiple connectedness. Finkelstein and Misner have shown that a simply connected manifold may have 'twists' which prevent it from being flattened out to Euclidean character by continuous topological transformations. These global features of the geometry satisfy new conservation laws of their own, and may eventually be used as a model for other physically interesting possibilities, though as yet the scheme has only mathematical interest. Like Betti numbers for characterizing the degree of connectivity, these conserved quantities can take on only integral values, and may correspond to something like the number of particles represented by the geometry.[32] The range of possibilities here may be limited only by the mathematical imagination, and we can hardly know in advance the number of physical possibilities which we may seek to comprehend in our theory.

The foregoing analysis has dealt with mass entirely in the sense of particular masses, concentrations that in their classical description can be localized at a single point, at least to a sufficiently good approximation. This is equally true of both the Schwarzschild and the Reissner-Nordström solutions. We begin by asserting, in the language appropriate to a more observable level, that there are such concentrations of 'real mass', a fact which seems clearly borne out by experience. We then try to ground this fact in the level of GMD by coordinating such masses with the mouths of wormholes, which are assumed to provide an adequate model for the properties of these masses. However, the theory does not really explain where and why such wormholes should arise, except in rather vague qualitative outlines. It simply accepts their existence as a brute fact, without relating it closely to other nonlocalized things (like waves) which also affect the structure of space-time. In addition, it makes mass and charge appear in almost too similar a way. Multiple connectivity was first proposed as a means of eliminating charges as sources and sinks of the electric field, which could then be everywhere continuous like the magnetic field. That it happened to take care of mass and eliminate mass singularities at the same time was more a convenient dividend than anything else. This is certainly important, since it does seem empirically true that all charge in nature is associated with mass, and a theory which eliminated the charge-part of a singularity but left the mass-part would not be very helpful. Nevertheless, while this may be a complete and comprehensive account of all charge, it certainly is not a full account of mass. All charge may be localized; but any kind of wave that carries energy also represents a distribution of mass, continuous throughout an extended region of space-time.

Maxwell's theory allows source-free electromagnetic waves, which are one possible source of such mass. Are there also gravitational waves, both radiated from massive bodies and source-free? The theory does in fact allow both kinds of waves. Taub has established that there can be no perfectly spherically symmetric or plane-symmetric gravitational waves (in contrast with the electromagnetic case), but there can be others, particularly of the axially or cylindrically symmetric variety. In fact, it is possible to construct a model universe out of such gravitational radiation alone, and if there were enough of this radiation the mutual gravitational attraction of its parts would be sufficient to create a closed universe.[1] Although there had originally been considerable skepticism that the formal cylindrically symmetric source-free solution first developed by Einstein and Rosen could really be taken to represent physical waves of any sort,[2] Brill was able to show that such solutions must always have positive-definite mass, and thus close in the metric in the same way as ordinary mass.[3] Here he was aided by the powerful new techniques of the time-symmetric initial-value problem, since such waves could be assumed to implode from great distances, reach a moment of time-symmetry or minimal extension, and then explode out again in a similar fashion.

Weber has been especially concerned with the problem of gravitational radiation from massive bodies, or from combinations of bodies moving like the ends of a dumbbell about an axis of rotation.[4] The general theoretical procedure here is well known, and goes back to Einstein.[5] Since such waves are assumed to be very weak, we take a Minkowskian background space and construct a tensor h_{ik} to represent the deviation of the actual g_{ik} from this background. The wave equation thus appears as a proportionality between the d'Alembertian of $h^i_k - \frac{1}{2}\delta^i_k h$ and the integral of T^i_k/r, evaluated over all space as a retarded potential, with r the Euclidean distance from the point where the field strength is presently being measured to the spatial point where the signal originated at a time t before now. Since the waves are assumed to travel at the speed of light, $r = ct$.

Aside from recognizing the theoretical possibility, it would of course be desirable to know if such radiation really exists. A variety of attempts have been made to set up apparatus capable of detecting, or even generating, gravitational radiation.[6] This could in no case be observed by its effect on a single particle, but by its effect on the mutual distances of pairs of particles, such as the ends of a rigid rod. To determine polarities, one should have an arrangement of four particles capable of exhibiting quadrupole effects. As yet, no such test has given any positive results. This is hardly surprising, for gravitational waves, like other gravitational phenomena, are far weaker than electromagnetic and nuclear interactions, as reflected in the relative smallness of the gravitational coupling constant G. In fact, if such radiation were really strong enough to be observed, it might close the universe much more strongly and make it a smaller place than present astronomical data seem to suggest. However, there does seem to be a slight discrepancy in the other direction. The universe seems larger than would be expected if ponderable masses were the sole sources of its overall curvature. A small amount of gravitational radiation, whether coming from present massive sources or left over from the "big bang" at the beginning of the universe, might just be enough to account for the difference. Such a hypothesis is testable in principle, but not in practice, given present experimental capabilities. It is too convenient, as a fudge factor to make the astronomical data consistent, but it cannot be ruled out as a possible correct account. In any case there are several other hypotheses, at least equally plausible, under consideration at present.

A further problem in setting up detection experiments is that we have no idea where in the spectrum to look. In the electromagnetic case, we were already very familiar with a particular range (visible light) and could extrapolate from our experience here to make predictions and set up experiments appropriate to other spectral regions. Liebes has suggested that radiation of very long wavelength, while undetectable in ordinary terrestrial laboratories, would have a focussing effect, making distant galaxies appear brighter by a significant factor than their counterparts in other areas of the sky.[7] While it is perfectly clear that universes filled entirely or even primarily with source-free radiation bear little resemblance to our own, they are still philosophically interesting, for they show that it is possible to satisfy the demand of the initial-value problem, that the initial three-space be closed, without introducing wormholes, singularities, or any other device to represent mass.

Although we have analyzed the gravitational effects of extended radiation, we have still not dealt with the concept of body in GR. Wormholes were appropriate only for (practically) point particles. They did account for the problem of interpreting masses and charges as characteristic lengths within a four-dimensional manifold by taking them to be parameters measuring the sharpness or curvature into the wormhole, or the radius of its mouth at a moment of time-symmetry. As long as the objects were simply points, no other lengths had to be considered, e.g., we did not have to ask whether the body was small enough in extension to fit into its own wormhole mouth. However, bodies massive enough to show interesting and observable gravitational effects are in fact extended; indeed, their observed space-time dimensions are much larger than their dynamical ones, given the smallness of the G/c^2 factor. Radiation, on the other hand, is diffuse: it has no identity—in the sense of size and shape—that it can maintain through time. To 'save' the observed phenomena more completely, we would like to be able to show that GR does allow the construction of extended bodies of definite size and shape, which maintain an identity and behave as a unit through long periods of time, and which exhibit the gravitational behavior of ordinary mass.

It turns out that the construction within GR of such bodies is theoretically possible. In classical physics, thanks to linearity and superposition, we could pile up as much radiation as we wanted in a given region without generating any new gravitational effects. However, we know in GR that electromagnetic waves, even if they are originally propagated in straight lines at the speed of light, can both exert and be deflected by gravitational attractions. If we put enough of these waves together, their mutual gravitational attractions can bind them together and confine them to a quite definite region of space, rather than shooting off to infinity, even in the absence of other 'real masses', reflecting surfaces, etc. Such localized concentrations, which exhibit the same sort of mass-equivalents as ordinary radiation, are called "geons." Since the 'stuff' from which these geons are made is only the sort of curvature well known and well analyzed in GMD as corresponding to ordinary radiation, we do not need to bring in from outside any non-GMD assumptions about the hydrodynamical state and constitution of the mass, such as are made in the Schwarzschild interior solution and the Fock-Papapetrou approach to the problem of motion. Furthermore, the metric and other relevant quantities are entirely free from singularity, both at the initial time and throughout the lifetime of the geon as a recognizable unit.[8]

The simplest sort of geon is spherically symmetric, where the outward radiation pressure is perfectly balanced by the inward gravitational pull to confine the electromagnetic radiation in a spherical shell of small thickness. The radius at which this occurs turns out to be related to the mass (measured in length units) by $r = 9m/4$. Outside the geon the metric has exactly the same form as the Schwarzschild solution and thus will look to a distant observer just like an ordinary mass—the only difference is that the mass m will come entirely from the radiation, rather than any 'stuff'. Inside the shell the metric tensor will be constant, as in flat space, though the g_{00} component will be only $1/9$ its Minkowski value, and in the narrow 'active region' there will be a continuous transition of the metric values.

Aside from its behavior as a source particle, a geon will respond like an ordinary test particle to weak and slowly varying fields, i.e., the evolution of the geon through time in response to such fields will correspond to the geodesic law of motion. However, there are further complications which make it less than ideal as a GMD model for the classical notion of mass. In the first place, its mass cannot remain constant through time. There will always be some leakage of the trapped radiation into the surrounding space, even in the absence of disruptive forces. Since the leakage rate is governed by a negative exponential of the ratio of the radius of the active zone to the wavelength, the rate is small for short-wave radiation, but it never vanishes entirely. In addition, if the leakage has a preferred direction, as is quite possible, this will accelerate the geon in the opposite direction by a rocket effect. This acceleration would of course cause the geon to deviate from the geodesic path set up by the background fields. The very strong fields internal to the geon will also strongly accelerate particles in its path, increasing their energies at the expense of that of the geon. On the other hand strong or variable fields will affect different parts of the geon in different ways, breaking down its unity and coherence. This would certainly be the case if two geons were to approach each other, so that we would have something very different from the situation of a classical billiard-ball collision.[9] One such outcome might be the 'pair annihilation' of both geons, i.e., the releasing of their energies in the form of free radiation. In any case, it suggests that the conceptual separation of a body with its constant mass from the fields surrounding it should not be made too sharp, and cannot be if geons are used to represent masses. Transmutations are always possible, and there may in fact be a continuous interchange between body and field.

Since we wish to confine ourselves at this stage to classical or prequantum GMD, it is important that our geons be large enough so that quantum effects do not occur. Wheeler has investigated the minimum size possible for fully classical geons. Concluding that the critical field strength beyond which creation of electron-positron pairs takes place is the crucial factor in determining this size, he finds that for sufficiently weak fields the mass must be of the order of 10^{39} grams—far greater than that of the sun![10] Fully classical geons must lie in the range from 10^{39} to 10^{57} grams. It may of course be possible to have smaller geons with much stronger internal fields, but these would be specifically quantum objects, and we as yet do not have a suitable quantum GMD to deal with them. There is no indication that any of these enormous geons do exist. They might, of course, exist in the distant reaches of space, but it would be hard to determine (certainly from gravitational behavior alone) whether a massive and distant astronomical object was really a geon or some other kind of star, and their interest at present is primarily theoretical, in showing that GMD does allow a self-consistent notion of extended body.

Geons need not be spherical, though the latter have the value of being more directly comparable to the Schwarzschild solution and the classical notion of mass. Ernst has shown that it is possible to confine radiation in a long linear tube or cylinder, and this in turn can be bent around itself to form a toroidal geon.[11] Power and Wheeler have investigated other special cases, and it seems likely that geons of many other shapes can be constructed, whether or not they have any physical existence.[12] One can also construct a geon out of pure gravitational radiation, as well as electromagnetic waves.[13] Given the weakness of gravitational radiation, it would seem even more unlikely that one could put together a sufficiently large concentration to form a real gravitational geon. However, such geons have the conceptual advantage of being constructed from purely gravitational quantities, without using the geometrical properties that correspond to the electromagnetic field.

Aside from the interaction of a geon with its surrounding field, another kind of transmutation occurs in the phenomenon of gravitational collapse. A large massive object, whether a star composed of ordinary mass, a geon, or perhaps even the universe itself, is sustained in volume by outward pressures resulting from the thermal or other internal energy of its parts. If and when this fuel becomes used up, the object will begin to collapse under the mutual gravitational attraction of its parts. As the mutual distances decrease, the attractive and thus contractive forces will become still larger, eventually squeezing the object past the coordinate singularity at $r = 2m$ and on down to the geometric singularity $r = 0$. What happens here is entirely unclear. If the laws of GR remain valid, it appears that the matter will actually be crushed out of existence. However, it may well be that a new kind of physical law will take over at such small distances. Wheeler's analog is that of a water wave, which follows one law while it proceeds toward the shore, but another when it crests in shallower water and breaks up into foam. It may be that space-time can generate its own foam at this point. This may well be the point for GMD to introduce quantum considerations for the proper analysis of this foam, and as the proper way of avoiding the singularity.[14] In any case, it seems clear that a variety of transmutations from matter to curvature or from curvature to matter are possible and compatible with GR. However, the detailed nature of or mechanism for these transmutations is still a mystery and a problem for the theory.

Throughout physics it has been customary to make a sharp distinction between general laws on the one hand, and initial and boundary conditions on the other. The proper task of physical theory was simply to supply these general laws. They would normally contain variables, and one could simply plug in specific observational data as substitution instances for these variables. The mathematics of the laws would then enable one to make equally specific predictions. Awareness of this separation was slow and hard-earned. In pre-Newtonian times the questions, "How will a particle with a certain position and momentum respond to a given force?" and "Why should there be a particle with this initial position and momentum anyway?" would have been considered on a par as questions for mechanics. The first clear suggestion of the value of such a separation appears in Descartes.[15] He suggests that pure reason, working from clear and distinct ideas, should indeed be able to come up with absolutely certain general laws by deduction. However, this was as far as pure reason could go. Observation and even simple experiment were needed to supply the particular data to which these laws could be applied. The idea of separating the two reaches full fruition, of course, in Newton, whose laws indicate very clearly the initial data that has to be supplied. His example has been followed throughout theoretical physics, including quantum mechanics. It has the great advantage of saving physical theories from having to explain everything at once. At some future date we may indeed be able to explain why the initial data have the values that they do; but this will be a separate problem, and may use other laws, even laws quite different from those with which the initial data are concerned.

Lindsay and Margenau have made a careful discussion of the relation of general laws to boundary conditions.[16] They point out that many general boundary conditions will indeed take the form of physical laws which serve as equations of constraint. This situation is illustrated by the initial value equations of GR. Their effect is to place some restrictions on our otherwise complete freedom to specify the initial conditions ad libitum, and they serve to distinguish physically possible from merely mathematically possible situations. However, even within these limitations there will normally be a broad range of magnitudes that we can adjust. Aside from these laws, some of the boundary conditions may restrict our freedom by describing actual boundaries and the phenomena that happen on them, as specific constraints for that particular problem; but constraints of this sort raise no philosophical difficulties.

However, it may well be that in GR we must give up this separability of the two kinds of data and its resultant freedom to specify whatever initial conditions we desire.[17] Our analysis of the initial value problem above has certainly assumed that we could specify any initial geometry and its rate of change that we wished, as long as it was closed and satisfied the initial value equations, but we must look deeper. Initial conditions are usually considered local specifications—in Newton's theory, local to the point of being instantaneous in time, if not in space. However, Newton's theory did assume that the world was infinitely open in both space and time. There was no cosmology, there was no overall history, so that the universe looked the same at all times, and there were no restrictions on the initial conditions. Any combination of particle positions and momenta could count as a possible state of the universe at any given time. GR, on the other hand, does have cosmological implications. It suggests that the universe should be closed, and this assumption is compatible with present astronomical evidence. Furthermore, this universe is not static, but expanding. Extrapolating backward, it appears to have evolved from some initial singular state, and in most GR cosmologies this expansion will eventually be followed by a period of contraction, finally arriving again at a singular state or one in which the ordinary physics must be changed. This cosmology will in turn place restrictions on the possible state of the universe at any given time, and thus the possible initial value data. At most, it could even determine this state uniquely. Once again we find an intimate connection between the local and the global features of the geometry. We cannot just make whatever local assumptions we see fit and later consider their global implications; the global conditions the local as well. If this is so, cosmology is not just a speculative offshoot of astronomy, somewhat outside the mainstream of science; on the contrary, it exerts an influence on everything else. Finally, if theoretical cosmology is a branch of GR, it gives GR a more difficult task but the possibility of a greater reward. It must supply not only the general laws of dynamics for its level but the initial conditions as well, making it a more internally comprehensive theory than any others yet developed. For it may well be that it must answer all the questions in its domain, and not accept anything from outside as a brute fact in no need of an explanation. It is imaginable that if it succeeded, nothing would be left out.

In the Friedmann cosmological solution for a homogeneous universe filled with matter dust but no radiation pressure, one can write down an expression that looks very much like a total energy term in classical physics. It is the sum of terms formally analogous to rest energy, kinetic energy, and potential energy. In classical mechanics one can imagine a similar problem for a collection of particles that bursts apart. In this Newtonian example we can give the corresponding sum any numerical value that we want, and call it the total energy of the system. If this quantity is larger than a certain calculable amount, the particles will fly apart forever, further and further away from each other; if it is smaller, they will eventually be pulled back together by mutual gravitational attractions. Surprisingly enough, this situation does not hold in GR. The 'energy sum', instead of being freely disposable as a constant of the motion, is necessarily zero![18] An expanding Friedmann universe endowed with a given mass necessarily stops its expansion at a characteristic radius, and then contracts.

In fact, it would be rather misleading to speak of this sum as an energy, and thus claim, for example, that the total energy of such a universe must necessarily be zero. For as we have seen above, GR allows a univocal definition of total energy only for a system which is located in an open infinite space which becomes Minkowskian at infinity. For a closed universe like the one we live in it is impossible and meaningless to speak of total energy at all.[19] But, beyond that, we may note that none of the dynamical equations in this cosmology contain any disposable parameters or possible constants of motion. The suspicion arises that perhaps there are no such constants, even for the actual cosmology of the universe, which of course is far more complicated than the simple Friedmann solution. If this were true, then the initial data—at a moment of time-symmetry or any other time—would be cosmologically determined.

How then should we try to characterize the state of the universe, if we cannot do so by constants of motion? Can it be done in quantum-mechanical terms, at least in principle or as an ultimate goal? Such a state is normally characterized by the eigenvalues of such quantities as energy. In addition, all standard versions of quantum mechanics make a sharp dualistic separation between the observer and the system that he is measuring. The system evolves according to the Schrödinger equation until he measures it; then it is mysteriously projected into one of its eigenstates with a probability derivable from its wave or state function. Now there is no intrinsic restriction on how large or small a quantum-mechanical system may be. We can include as much of the observing apparatus as we like, and even the observer, considered qua physical system rather than epistemological agent, can be included in a quantum system measured by an outside, 'higher' observer. It would seem indeed that even the universe as a whole could be considered a quantum system with a wave function of its own—as long as we could find someone outside to observe it! But by hypothesis there is no such observer, since the universe is all-inclusive (neglecting the possibility of a transcendent God).

This fact in turn makes it all the more puzzling how to understand the notion of transitions from one quantum state to another for a closed universe since there is nothing to produce such transitions, or even the meaningfulness of speaking of alternative states here. How could we distinguish one from another?[20] Suppose we accept that the formalism of the Schrödinger equation is appropriate. Then we may want to conclude either (1) that there are different possible eigenstates, but in fact the universe always remains in one of them as it evolves; or (2) that the equation here has one and only one eigensolution. If we choose (1) we are left with the problem of why it should be in this particular state rather than any other, as well as distinguishing the states. If we choose (2) we must explain the extraordinary mathematical fact that the equation has only one solution instead of the expected plurality.

One might of course argue that all this is nonsense, that it is meaningless to talk of 'universal wave functions' or speak of quantum states at all without observers to measure them. But this approach in turn would require setting a rather arbitrary limit on the maximum size of quantum systems—if not the universe as a whole, what is the next largest thing we can use? This is certainly not the place to rehash the Einstein-Bohr controversy, but the insistence on keeping some kind of observer outside the rest of the world does violate the long-standing regulative principle that it is desirable to try to specify the characteristics of the world as a whole, independent of any particular observer. Nevertheless, we as yet do not have any version of quantum mechanics suitable for giving such internal descriptions, which do not introduce the observer as a foreign element, though Everett has made efforts in that direction.[21]

As our final topic in considering the new interpretation of GR that has evolved since 1955, let us turn to the issue of Mach's principle, and whether it should be, can be, or is incorporated into GR. Newton, in defending his notion of absolute space, believed that his bucket-of-water experiment did enable us to single out the class of inertial systems from accelerated systems, and thus true motions from apparent ones, without reference to any other bodies.[22] However, Berkeley pointed out that this conclusion was by no means the only possible one. The experiment could be interpreted, and the inertial system defined, with reference to the system of fixed stars, rather than by introducing any notion of absolute space. Since the stars were directly observable while absolute space was not, Berkeley argued on philosophical grounds that we can and should give up the notion of absolute space.[23] Mach shared many of Berkeley's empiricist-positivist assumptions about the undesirability of giving unobservables such an important role as the determination of inertial motions. Within the Newtonian framework, Newton's first law is 'explained', and the actual motions of bodies not subject to the influence of forces is predicted, by appealing to the Euclidean structure and properties of this absolute space. Inertial forces (centrifugal and Coriolis) are not real or derivable from potentials, but arise as the effects of rotating, noninertial coordinate systems. Mach proposed instead that the inertial behavior of matter, including force-free motion, inertial forces, and inertial as opposed to gravitational mass, should arise and be entirely accounted for by interaction with the other bodies in the universe, even those at very remote distances.[24]

Now this principle can certainly be challenged, even by those with no prior commitments to absolute space. The distant matter of the universe may be observable in principle, but it hardly makes a well-defined totality, and one could imagine it practically impossible to sum up the effect from each individual distant body, even if one could calculate its particular contribution. Thus Bridgman (who was certainly no friend of absolutes!) argued that Mach's principle could never be given sufficient operational significance to be tested. Instead he proposed that we concentrate on finding local criteria for determining inertial systems, without worrying about their deeper origin.[25] In the form stated, the principle is very vague, and Mach never managed to make it more precise. He seems to have regarded it as a very general regulative principle, which might be satisfied in a variety of ways, rather than as a specific law or even a relatively precise 'metanomological' principle. No mention is made of the kind of force or other mechanism whereby this distant matter determines local inertial properties, or whether this must in any case be an instantaneous action at a distance. Certainly it must be a long-range force, and at present only two such are definitely known—gravitation and electromagnetism. The latter is ruled out, since it would seem to depend on the net charge of the universe. This quantity may turn out to be zero, but it is certainly not known on any cosmological scale. And with gravitation, simple use of the inverse-square law will not provide a satisfactory account. Finally, unless it is clear that something is wrong, either theoretically or observationally, with any notion of absolute space, there would seem to be no reason to accept the principle unless one shared Mach's philosophical biases.

In fact, the principle might well have died a natural death except for the fact that Einstein himself took it very seriously. He gave Mach credit for significant influences on his own thinking, especially in connection with the development of SR, and believed that his GR should conform to the principle, so that it would be less than fully satisfactory if it did not. We have seen earlier that his attitude vacillated back and forth between the GMD position that space-time was an all-inclusive medium, with matter only an aspect of it, and the Machian position that space-time was only an intervening variable, clearly essential in making calculations, but in principle eliminable and replaceable by a theory of direct interactions among individual material bodies. He felt an obligation to give the principle a more precise mathematical form, and to derive specific observational predictions from it, which presumably would not follow from a more absolutistic theory. We have argued above that a basic requirement for anyone who defends this position is to provide an integral form of the field equations, comparable in form to Newton's inverse-square law, rather than Poisson's equation, and the character of the theory makes this extremely hard to do.

The language of precisely located bodies exerting forces at great distances is certainly closer to that of Newtonian mechanics than that of GR, with its variable space-time curvatures, so it is not surprising that many alternative theories to GR attempting to incorporate Mach's principle, as well as qualitative accounts of its possible role within GR, have stayed within this framework. In electromagnetic theory, we know that when a charge is accelerated, we have in addition to the Coulomb force qq'/r^2 a new radiative component differing from $qq'a/c^2r$ by only a dimensionless factor close to 1, if a is the magnitude of the acceleration. One might suppose that a similar phenomenon occurs in GR. Because of the c^2 factor in the denominator, this term would be negligibly small compared with the inverse-square term for nearby masses. However, the further away we go the greater will be the proportionate contribution from the $1/r$ term—eventually, it will dominate the expression for total force. Suppose now that we have a body with mass m' acted on by a local force (of whatever origin) to produce an acceleration a relative to the fixed stars. The force will then be $m'a$. Now choose a coordinate system in which that body remains at rest. In this system the net force on it will be zero, but the system of fixed stars will have an acceleration a in the opposite direction. The total radiative force should then be $F = \Sigma_i (Gm_i m'a/c^2 r_i) = m'a(G/c^2)\Sigma_i m_i/r_i$. To balance the forces, we set this equal to the local force $m'a$, getting $\Sigma_i m_i/r_i = c^2/G$, as a mathematical expression of Mach's principle.[26]

Sciama, who believes that GR does not and cannot incorporate Mach's principle, has shown in his 1953 work how this result can be derived if one assumes that the gravitational field derives from a combination of a scalar (as in Newton's theory) and a vector potential, though not the tensor potential of GR. This makes it similar to electromagnetism, which also uses both potentials, combined into a four-vector in covariant representations.[27] However, the law has curious implications. If we approximate the actual sum over all bodies by using the total mass and radius of the universe, we get $M/R = c^2/G$. If we assume that both c and G are universal constants, the law requires that the mass-to-radius ratio remain constant through the history of the universe. Alternatively, we might allow G to vary, and conclude that its value at any given time reflected the state of expansion of the universe and the corresponding mass-to-radius ratio at that time. This approach would allow it to be tied in with theories, such as those developed by Dirac, Jordan, and Dicke, in which the gravitational constant G has a secular variation though remaining constant in space at any given time.[28] If either the M/R ratio or the constancy of G could be determined experimentally with sufficient accuracy, we could then test Mach's principle—or at least this interpretation of it. It would certainly provide an interesting link between the 'constants of nature' and the actual physical structure and content of the universe.

Attempts have been made to follow up this approach and expand the theory, especially by Kaempffer.[29] In addition, there have been efforts to find a still more general theory incorporating the principle while remaining within the Newtonian framework.[30] However, most of these theories do appear to have an ad hoc character, even if this reflects the fact that they have not had time for sufficient development. Aside from the quite specific issue of Mach's principle, which in any case is not an established experimental fact, there seems no reason to prefer them to either Newton's theory or GR. In the experimental sphere, there have indeed been attempts to test the principle or implications of it, like a supposed anisotropy and tensor character of mass. While a negative result here was at first considered evidence against the principle, Dicke has contended that we should expect such anisotropy to affect all bodies equally, and therefore not show differential effects in any experiment.[31]

Einstein himself believed that a theory compatible with Mach's principle should make the following three predictions:[32]

1. The inertia of a body must increase when ponderable masses are piled up in its neighborhood.
2. A body must experience an accelerating force when neighboring masses are accelerated, and, in fact, the force must be in the same direction as that acceleration.
3. A rotating hollow body must generate inside of itself a "Coriolis field," which deflects moving bodies in the sense of the rotation, and a radial centrifugal field as well.

He then claimed that each of these results could be derived from GR. Predictions (2) and (3) seem to be quite well established as legitimate results, though (3) was first proved rigorously by Thirring. Brans has recently shown that (1) does not really follow from GR generally, but is true only as an effect of certain coordinate systems.[33] The exact magnitude of this effect had always been a matter of some dispute.[34]

Nevertheless, Einstein's calculations, or those of anyone else trying to solve these problems within GR, require many additional assumptions. In particular, Einstein assumes that (*A*) the fields are sufficiently weak that they differ only slightly from a background Minkowski space, which for certain purposes (like measuring distances) is actually used as an approximation; and (*B*) the effects of the contributions from different sources can be summed linearly.[35] We know that (*B*) is in general false. The equations are not linear. As a description of the actual effects involved, it may be that the weak-field linear approximation is empirically justified, but it would seem unreasonable to draw philosophical conclusions from something which violates the basic principles of GR. Furthermore, Callaway has emphasized that (*A*) gives the game away right from the start. Even if the fields are in fact weak, why should they differ slightly from the Minkowski solution, rather than some other? If one let them become increasingly weak to the vanishing point, the Minkowski solution would still remain. Space-time would still have a definite structure, and the Minkowski tensor would give this structure.[36]

A strong form of Mach's principle would demand that there should be no limit for the metric tensor in the case of infinitely weak fields. If matter were to disappear, so would space-time itself also be expected to disappear. The question becomes whether matter simply modifies an already existing space-time structure, or whether it is the sole source of that structure. Einstein seems content with the former, as a weaker version of the principle. It at least states that space-time is not absolute in the sense that Newton's was, sitting in aloof splendor and totally unaffected by the matter moving around in it. But unless the strong version is true, space at least retains some character of its own and some degree of independence from matter. On this latter version it would be quite inappropriate to concentrate on vanishingly weak fields; one should see instead how much structure a strong field can give it.

Einstein at least supported the strong version as a hope, or ultimate goal, even if he could only prove the weak one compatible with GR. How much further one can go is still subject to dispute. Davidson claims that at least all of Sciama's ideas can be given a natural home in the richness of GR.[37] However, most others are more skeptical. Solutions like the Taub universe, which are closed and nonflat despite the complete absence of any matter, are quite legitimate within the theory even though they violate the Machian spirit, and Whitrow claims there are many more such solutions.[38] In order to incorporate more fully and naturally the strong Mach principle, while retaining much of the rest of GR, Brans and Dicke have thought it necessary to develop a new theory of gravitation, in which Einstein's tensor field is kept but supplemented by a scalar field corresponding to a gravitational constant that varies in space as well as time. Such a theory is intended to prevent the existence of real solutions in the absence of matter.[39]

Wheeler has pointed out that there are numerous objections that might make it seem undesirable even to try to incorporate Mach's principle into GR at all.[40] We have already mentioned the impropriety of Einstein's assumption of linearity and his use of straightforward integration in adding up contributions from different parts of space-time, when the field equations are essentially nonlinear. The simple $1/r$ term in the integrand is also unjustified—in a generally curved space this would be ambiguous. We might define it as the distance along a geodesic from source to field point, but this would require knowing in advance the structure of space-time and thus the geodesics in it, when in fact our equations are supposed to give us this information. In addition, there may be more than one geodesic between the points, and we can hardly be sure that in the closed universe the effect would always have a $1/r$ character, rather than an expression involving the ratio of r to the radius of the universe R. The problem remains that we simply do not have any integral form for the field equations, nor are we likely to get one with present mathematical techniques. In a similar vein, any attempt to locate the masses responsible for inertia and determine their velocities and accelerations relative to the test particle also seems to presuppose that the geometry is fully known and thus not left to be determined. In the Sciama version, the radiative effect seems to be transmitted instantaneously, when our causality postulate would allow it to be propagated only with the speed of light. Einstein assumed that we must use retarded potentials, as in electromagnetic theory. However, inertia is supposed to be symmetric with respect to time reversal, so there would seem to be no theoretical justification for using retarded rather than advanced, or even some combination of the two. We might get different results in these cases. If we trace the integral back to include everything in the past light-cone, we might very well run into singularities with infinite curvature, which would cause complications. And if space-time is open in these directions, the integral out to infinity might well diverge. Finally, one might ask what really justifies the analogy by which gravitation is assumed to have a $1/r$ radiative component, as well as electromagnetism? There are plenty of other dissimilarities between the two fields.

However, instead of simply dropping Mach's idea as something which stems from an outdated and unwarranted philosophical bias, and which can be properly expressed only in traditional space-particle dualistic terms, Wheeler claims that it still has an important role to play in GR. For we still need some sort of boundary condition if we are to get a unique solution rather than the large class of solutions corresponding to any given value for the matter tensor. We must be able to select those that are physically reasonable as well as merely mathematically compatible with the field equations. Wheeler sees Mach's principle as this boundary condition or selection principle. In its simplest form, all that it requires is that the universe be closed.[41] Einstein had also recognized that Mach's principle corresponded only to a finite universe, bounded in space, and not to an infinite, quasi-Euclidean universe. Insofar as he wished to preserve the principle in some form, he regarded this as a strong argument for a closed universe.[42]

However, Mach's principle appears most clearly in its role as a boundary condition in connection with the initial value problem. Let us recall the main conclusion of Chapter 14: if one specified the geometry and its rate of change on an initial spacelike three-dimensional hypersurface, and the distribution of energy and energy flow on that surface (along with the dynamical equations for this energy), then one could determine *uniquely* the lapse and shift functions, the extrinsic curvature and field momentum, and finally from Einstein's equations the complete four-geometry of space-time. From this one could get the set of all geodesics, and since geodesics correspond to the trajectories of inertial motions, the entire inertial behavior of test particles. The one big proviso was that the initial three-space be properly closed; in an open space one could not guarantee uniqueness without adding further boundary conditions about the behavior at infinity. If this is true, it is certainly a remarkable result. We are not required to say just how the universe is closed, or what size and shape it must take; that it be closed at all seems to be enough. Furthermore, closure is not an arbitrary stipulation imported from outside. As long as we have a sufficient amount of matter or energy-carrying radiation, we can guarantee closure and the satisfaction of the condition. Closure is a universal boundary condition and does not commit us to any particular assumptions. Finally, it seems to be covariant and independent of the coordinate system chosen. The latter will affect its apparent size and shape, but not change its topology. If this is true, GR can refute Hill's claim that while the general laws of GR are covariant, the boundary conditions must always single out particular coordinate systems.[43]

Thus we envisage Mach's principle as taking on this final form: instead of saying that the influence of distant matter should fully determine the inertial properties of test particles, we say that the specification of a sufficiently regular closed three-geometry at an initial instant and its rate of change, along with the density and flow of mass-energy, will fully determine them. In this form Mach's principle is not only compatible with GR, but an essential supplementation of it. One may object to this principle on the ground that it is formulated in very different language from what Mach (or Sciama) had in mind; however, GR certainly can and must be allowed to fit it into its own conceptual framework. One might also object that it allows solutions even in the complete absence of matter (such as the Taub universe) as long as this universe is closed. The statement that nongravitational energy density and flow vanishes everywhere still describes a legitimate distribution. But there is no independent ground for ruling out such a world as physically impossible. What it rules out instead is something like the asymptotically flat Schwarzschild solution for a single isolated mass, except as an approximation. It is both theoretically and empirically better to assume that Schwarzschild solutions are valid only in that local zone where a given mass is the dominant gravitational feature. Unlike the Einstein version, we do not need to presuppose the whole geometry in advance, since to locate our masses we assume only the three-geometry—the four-geometry is yet to be determined. And since our source data refers only to a single initial surface, or 'instant', we do not need to worry about advanced or retarded effects extending far into the past or future. We may not have anything that *looks* directly like Newton's inverse-square law, or like an integral form of the gravitational equations which already incorporates the boundary conditions, but we have something more valuable, a formulation which shows how one very simple and general condition can unlock the vast power of GR for delineating the whole history of the universe from a very sparse amount of prior information.

UNIFICATION 5

18 Epilogue

We have finally come to the end of our study of the general theory of relativity. In the course of it we have started with the first premature attempts to identify matter with space in Plato and Descartes, and the various inadequacies which resulted from their poor and limited conceptions of space. We have then turned to GR itself, examining in detail first the basic principles on which the theory was founded, and then the various implications, many of them novel, exciting, and revolutionary, which were or could have been known in Einstein's lifetime. Finally, we have turned to those new problems and successes which have reopened the theory since 1955 as an area of viable theoretical research, if not of extensive experimental test. It has been my hope to make this a comprehensive account, covering all the developments in GR, both old and new, that have any philosophical significance or interest. It has also been my hope to give an orderly and systematic account of GR and GMD as a history of ideas to someone with sophistication but not full professional status in physics. One cannot study any theory so long or so closely without at least hoping that it will eventually prove correct and capable of overcoming its present difficulties. Nevertheless, my philosophic conscience obligates me to make criticisms where I think they are appropriate. I have done so in many places, and I think that some of these criticisms point to significant weaknesses. The present form of the theory, exactly as it stands, is not wholly adequate as the final version. Many of the philosophical objections which brought down the systems of Plato and Descartes, such as finding a proper geometrical interpretation of dynamical quantities, still retain *some* validity here. Nevertheless GR is not fixed and settled for all time. It has flexibility, it has room for expansion and development, and it has perhaps the most versatile material yet discovered for building theoretical systems—Riemannian space-time. There is research to be done, and I must cheer for its ultimate success.

There are still two areas which I have not touched on except in passing, and which I will simply mention now: the experimental testing of GR and of gravitational phenomena generally, and the prospects for a quantum geometrodynamics. I have been concerned with the structure, content, and overall character of GR, rather than the degree to which it is adequately established experimentally. GR is an excellent example of an area where theory has far outstripped experiment, and probably will continue to do so for the foreseeable future. I have argued above that the three standard noncosmological tests of GR refer only to the Schwarzschild solution, and thus would provide equal confirmation for the host of other theories, classical and relativistic, which could reproduce these results, if only in an ad hoc way or by adjusting convenient parameters. For this reason they could only be a small part of the grounds for acceptance of GR.

With the revival of interest in GR, there have been several recent attempts to improve the experimental situation. One may lump these under four different headings. (1) Attempts to refine the basic experimental foundations of the subject, like the strong and weak principles of equivalence, or the three classical tests.[1] (2) Attempts to measure locally or terrestrially what had previously been considered astronomical effects, like the gravitational red shift, using such phenomena as the Mössbauer effect.[2] (3) Cosmological tests of the predictions of GR about the universe as a whole.[3] (4) And finally, attempts to find new kinds of observational conclusions from GR, whether or not they are really testable as yet.[4] At the very least, none of these experiments provide any strong evidence against GR.

I have avoided reference to the quantum sphere, because it is fair to say that we still do not have a quantum GMD. GMD has been developed as a classical theory, even though its ultimate value for physics will undoubtedly require its being combined or at least harmonized with quantum physics. Wheeler is able to write with buoyant optimism,[5] "General relativity and the quantum theory of the atom, both born in World War I, and surely destined some day to be married in high state. . . ." However, we have yet to see this glorious marriage, much less the offspring of it. The bride and groom are still suspicious, and with some justification, whether they have enough in common to guarantee that they will live together happily ever after. Because of this, each tends to lead his own life and develop at his own pace, and may eventually die celibate. GMD, as we have presented it, is completely deterministic; there is no way of fitting the uncertainty principle into the initial value problem as it presently stands. The geometrical structure of the world is everywhere precise; there are no alternative quantum states of, e.g., the metric tensor, with appropriate probabilities. Geons have deliberately been kept absurdly large just to avoid quantum phenomena; and the charge-to-mass ratios discussed in the Reissner-Nordström solution and the wormhole model have nothing to do with that of any known elementary particle.

One approach to reconciling the two is to express the content of GR in a formalism comparable to that used in quantum mechanics, so that we can speak of canonical conjugate variables, propagators, etc. We have already considered this approach in the P^{ik} of Arnowitt, Deser, and Misner in Chapter 14. Further efforts to set up a formalism and separate out merely coordinate-dependent effects have been made by Anderson, Dewitt, and Misner.[6] The mathematical difficulties here have proved very considerable.

Wheeler has been more concerned with the problem of what happens to the geometry at very small distances, and whether we have the right to suppose that space-time retains a definite and continuous structure, no matter how far down one goes. For him, the crucial distance is $L^* = (hG/c^3)^{1/2} \cong 10^{-33}$ cm. This incredibly small distance, a full 20 orders of magnitude below the Fermi distance which characterizes elementary particles and serves as the present lower limit to our penetration into the very small, is simply written down as the simplest combination of the classical constants of nature c and G and the quantum constant h which will have the dimensions of a length. Wheeler believes that statements about what goes on, on this tiny scale, are still testable in principle, even though one would have to go through a tremendous number of intermediate steps to get even statistical effects from this level up to the point of being observable.[7]

What he proposes is that the continuous structure of space-time does indeed break down here. Instead we have a froth of tiny virtual wormholes, appearing and disappearing in vast quantities with very short lifetimes. At that scale of distances, the energy densities and virtual local curvatures would be far greater than anything ever observed, even in the elementary-particle world. However, they would in general cancel each other out, leaving no more than the usual zero-point fluctuations of the quantum field. Occasionally they would not cancel out completely, but leave a slight unbalanced residue in a region. This would then appear as a new and more stable wormhole mouth, larger than the virtual ones but less sharply curved. Such an object, which on that scale would appear enormous, Wheeler suggests might be an elementary particle. It would not be an irreducible thing, but a statistical effect of what went on at a smaller level. Quantum uncertainties would affect the topological structure of space-time, which would be fully indeterminate like a perfect foam, even though on a larger scale it would appear continuous and well defined.[8] And this foamlike structure is characteristic of all of space-time, not just those regions where there are elementary particles.

All this is extremely speculative, and perhaps nothing more. But the fact remains that there are significant weaknesses in present-day particle theory, and GMD might be as good a source as any of ways of avoiding them. It might, for example, give a more natural account of the 'renormalization' that goes on in getting from the 'undressed' to the experimental electron charge without strange-looking (though effective) mathematical tricks. It might provide an answer to some of Bohm's attempts to look for a sub-quantum level, even though things are certainly not deterministic here.[9] And finally, the character of these fluctuations might shed light on the question of why space-time really does have $3 + 1$ dimensions, rather than any of the other mathematically possible combinations, though it is hardly clear how this might be shown.[10] Čapek has pointed out that there have been a variety of theories which assigned to space-time a discrete or pulsational structure. But the distances at which the structure broke down, and the corresponding minimum lengths and times, were normally considerably larger, near the size of elementary particles, and were not derived from gravitational considerations.[11]

Let us survey just what classical GMD has been able to accomplish. Its basic goal remains the identification of matter with space, which in turn means that all the phenomena traditionally associated with matter and considered conceptually different from space must somehow be incorporated into the natural Riemannian structure of space-time, rather than being imported from outside as 'foreign' entities. This program is often summed up in the slogan "*X* without *X*," requiring that we be able to satisfy all the various roles of *X*, without presupposing the separate existence of *X* from the start. What sorts of things can be substituted for *X* in this slogan? Fletcher gives an impressive list:[12] (1) gravitation, (2) electromagnetism, (3) charge, (4) measuring rods and clocks, (5) physical constants, (6) equations of motion, (7) mass, (8) transmutations, (9) field equations, and (10) boundary conditions. If a quantum GMD can be developed, we may be able to add (11) elementary particles, and (12) observers. As a means of summary and review, let us recall how each of these was done.

(1) Gravitation, as a separate force of nature, was the first to be eliminated. Since it gave the same acceleration to all bodies, it was interpreted as the main source of space-time curvature, and in the absence of other forces motions would proceed inertially along the geodesics of curved space-time. (2) Electromagnetism was incorporated in the equations of "already unified field theory," as developed by Rainich and Misner. It appeared as a special and recognizable mode of curvature, characterized by conditions on the Ricci tensor. (3) Charge was removed by the notion of multiply connected topologies, where electric fields could pass in and out of everywhere-continuous wormhole mouths, without introducing singular sources and sinks for the field. (4) The use of separate rods and clocks, which would either have to have a physical constitution of their own or else be ontologically distinct from ordinary matter, were eliminated by the Marzke theory of measurement, which uses only the trajectories of light and infinitesimal test particles. (5) Physical constants, at least those, such as G and c, with dimensions other than pure lengths, were eliminated

by representing every quantity in terms of units of length. Mass and charge thus appeared as characteristic length parameters, and c and G could be set equal to 1. (6) The geodesic law, as a separate equation of motion, was eliminated by showing that it could be derived from the field equations, whether the moving objects were interpreted as singularities, extended bodies with internal hydrodynamical structure, wormholes, or geons. (7) Mass, in the form of singular point masses, was also removed by the wormhole model. In addition, the geon served as a model for extended massive objects. (8) Transmutations were accounted for in the various geon-field interactions, and the possibility of space-time evolving to cusps or handles, or wormhole mouths moving together and eventually coalescing. (9) The complex set of field equations could themselves be removed and replaced by the much simpler condition that the intrinsic and extrinsic curvature invariants at a given point should differ by the energy density at that point, provided that this was true in every possible coordinate system and spacelike hypersurface passing through that point, according to the covariance principle. (10) Specific boundary conditions were replaced by the general requirement of closure, which in turn could be guaranteed uniquely if the gravitational fields were strong enough. In the much less certain quantum region, (11) particles were interpreted as relatively weak perturbations of the background fluctuation field and foamlike structure of space-time, and (12) observers external to systems could be dispensed with if we could formulate a way of characterizing the state of a closed system, perhaps along the lines of Everett.

By any standard, this is a remarkable series of conceptual accomplishments. But if all these things are in fact eliminated as independent elements in the ontology of GR, what is left to replace them? The answer is, simply, the manifold of space-time itself. In its vast metrical and topological richness, it is presumed capable of fulfilling all these functions by itself, without outside assistance. What we have in fact is a wholly new kind of physical theory, based on what might be called *physical monism.*

The term 'monism' has had various meanings throughout the history of philosophy. As opposed to dualism, it asserts that there is only one *kind* of substance in the world, rather than two (or more) with radically different kinds of essences, like mind and matter or atoms and space. As opposed to pluralism, it asserts that there is only one kind of force, action, or influence, rather than a multitude competing with each other in some uneasy balance. And as opposed to atomism, it asserts that there is only one individual substance with absolute or intrinsic characteristics of its own, rather than a host of logically independent basic entities, each of which has properties of its own, with the properties of larger complexes derivable from those of the atoms. GMD seems to offer us a monism in each of these senses. We no longer have any irreducible matter or other entities different in kind from space-time; this space-time is not a passive arena but the source and medium of all interactions, its parts both acting and being acted upon by each other; and finally, space-time is a unified whole, with global and topological as well as local characteristics. It is not a collection of things, but a single thing—that only thing that is really real. One could call it by such names as pure substance, or being as such.

Nevertheless, GMD differs from other philosophical monisms. Unlike the static, unchanging 'Being' of Parmenides, it is capable of an enormous amount and variety of internal differentiation, while still remaining intelligible. Unlike the 'God-Substance' of Spinoza, it is not capable of supporting radically different attributes which parallel but never interact with each other. All its attributes (in whatever 'modes') are essentially geometrical and capable of interacting with each other. Unlike the 'Absolute' of Hegel and Bradley, it does not contain any mysterious final state which harmonizes all the contradictions of previous stages in a new and essentially unpredictable sort of way. While it includes the entire history as well as spatial extent of the universe, and cannot be represented by the world at any particular instant, it introduces no conceptually new features in its completed totality, and epistemologically its whole structure can be predicted from information at a single instant.

We have reached this monistic conclusion by many routes. The systematic elimination or replacement of various assumed properties of matter, described in the section above, is certainly one of them. In addition, we should recall some of the peculiar features of the theory.

(A) The field equations are nonlinear, a situation unprecedented in either classical or quantum physics. This means that we cannot take the sum of the individual effects from two sources to get a total effect. Conversely, given a total effect we cannot break it down uniquely into the contributions from individual sources. This rules out atomism. The whole cannot be represented as the sum of its parts. Any supposed atoms have no inherent characteristics, like gravitational mass. Their action depends on what other particles are also present, and where they are. Since even the two-body problem has not been given a rigorous solution for bodies of comparable size, we have no reasonable counterpart to Newton's third law, which allows us to interchange source and mass particles at our convenience. Nonlinearity also requires us to give an account of the ultimate constitution of matter, whether discrete or continuous. We cannot start off, like Newton, with single particles and later regard a continuous distribution of mass as a special (limit) case. Each distribution must be solved as a separate case.

(B) We can no longer separate out different kinds of fields and assume that they act independently of each other, so that the resulting acceleration given to test particles just comes from the vector sum of the different kinds of forces. We have shown that electromagnetism has gravitational effects, and gravitation has electromagnetic effects, in terms of the motions they impart to test particles. While a partial separation of the total field into gravitational and electromagnetic components can be made, it is to some extent arbitrary. Neither one is 'reduced' to the other, but both are made aspects of the curvature of space-time. This result rules out pluralism.

(C) The program of eliminating singularities, insofar as it can be fully carried out, removes any restrictions on the completeness of space-time. If there are no regions where its structure breaks down, there is no need for supplementation at these regions.

(D) The use of the T_{ik} tensor leads to a breakdown of the sharp conceptual distinctions between mass, energy, momentum, stress, and other similar properties. These distinctions are especially appropriate to the language of particles moving around in a pre-established spatial frame, and if they cannot be made in an absolute or covariant way there is little justification for keeping the ontology that developed and needed them. In addition, we have seen that there is no necessary correlation between local densities and global conserved quantities, nor any really unambiguous definition of either one.

(E) In emphasizing the dynamical properties of bodies, we must give up most of the perceptual notions of where extended bodies are, along with the related notions of space-filling and impenetrability. GMD embodies the notion that a body is where it acts. Insofar as the curvatures which it generates extend to the whole of space-time, there is a sense in which each body or source is everywhere, and at the same time. The dynamical lengths corresponding to mass and charge are just convenient parameters to characterize the size and shape of the wormhole mouth, and have nothing to do with the perceptual size and shape of the object. Since in this sense every object equally occupies the whole of space, we cannot establish identity or independence by reference to the region of space which that body, and that body alone, occupies. We might still try to maintain a distinction between where matter is and where it is not in terms of whether $T_{ik} \neq 0$ or $T_{ik} = 0$. But this and the corrresponding Einstein tensor are only one possible measure of curvature. Insofar as there is still curvature ($R^i_{jkl} \neq 0$) in the regions of vanishing T_{ik}, there is still dynamical action and not empty space.

(F) Finally, we should mention the possible breakdown of the distinctions between laws and initial conditions in the closure principle. The boundary conditions arise from the content of space-time itself, rather than being freely imposed from outside. We have a single, self-contained process, not one subject to external constraints.

Nevertheless, any monistic theory, if it is to have empirical relevance, must find some way of saving the phenomena of individuality. The at least partial independence of reidentifiable particulars is too well confirmed both by ordinary experience and by sophisticated experiments. The mere fact that two different points in space-time may have different curvatures is not enough to individuate them or allow for coherence of particles through time. GMD has at least some resources for doing this. A wormhole, as a definite, topologically irreducible handle, and a geon can act as extended and coherent units for long periods of time, even though there will ultimately be decay and dissolution. Is this enough? The particles of classical atomism were uncreated and indestructible, and thus had an absolute identity through all time. However, we now know that elementary particle pairs can be created from and annihilated into 'immaterial' radiation. In addition, the Pauli exclusion principle shows that our ordinary notions of identity and individuality are no longer valid in this sphere. It may in fact be a merit rather than a defect in GMD if it shows that there can be no things within space-time which retain their identity forever.

One might still argue for a form of atomism on the following grounds: The various tensors used to characterize different curvatures are all functions of the points in space-time. Thus it is really the individual points, rather than space-time as a whole, that are the substances, and the curvatures are their properties. Space-time is nothing but a general term for the totality, as regions are the names of less inclusive collections of these points, and overall curvatures should be regarded as statistical effects. This might be true if there were no connections between the local and the global structure of space-time, in which case the properties of one point would be logically independent of those of any other point. But we have already pointed out that this is not so. There is an intimate connection between the local and global features, so that the latter influence the former as well as vice versa. Because of this we must consider space-time as our sole substance, albeit not an internally homogeneous one.

One might also try to reject the whole substance-attribute framework in which these distinctions have been made. But this seems hardly justified. Not everything that can truly be predicated of a substance will be an essential property. If, for example, we regarded space-time simply as the carrier of fields which were in no sense geometrical, it would be highly misleading to call them attributes of space-time. We have seen above that this was the motive for introducing ethers as the intermediate substances which could indeed take fields as their essential attributes. But in GMD the fields, both gravitational and electromagnetic, have been fully geometrized. The quantities with which they deal are curvatures and curvatures alone. Since these are indeed the natural attributes of either points or entire manifolds, there is no reason not to use the substance-essential attribute language here. We cannot assign properties at will to space-time; but as long as we remain in the sphere of geometrical properties we need not introduce any other substances.[13]

Grünbaum has argued that far from eliminating absolute space, as Mach and Einstein in his earlier years had certainly hoped to do, GR preserves it in a strong form.[14] Others have recognized this fact, and often accepted it cheerfully.[15] It is certainly true that space is hardly eliminated if indeed it is the only surviving substance; on the contrary, it eliminates everything else. But at this stage the absolute-relative dichotomy for space is less important or enlightening than the arena-everything dichotomy. Space-time must be absolute rather trivially, for there is nothing outside of it to be relative to. What is essential is that Einstein's space-time has indeed turned out to be such an all-encompassing medium, far more than even he had expected. The work is still largely programmatical and not yet complete, but it has already achieved extraordinary success. In any case, it is hardly absolute in the sense of being isolated and unaffected by anything that happened inside or outside of it, as was Newton's. Both its overall structure and the properties of its individual parts are constantly changing in a complex series of mutual interactions.

It would be hard to underestimate the revolutionary and unprecedented character of a physical theory based on monism. Since the dawn of modern science (and indeed long before) any attempts to understand the world had presupposed a plurality of independent things with definite properties of their own, and the possibility of understanding wholes through analysis into their parts. This presumption certainly is borne out by ordinary experience, and it has worked so well in previous science that it has become a well-established regulative principle. The principle has indeed affected the structure of our very language so completely that it is hard to find adequate expression for the new ideas of GR. Because of the difficulty of breaking outside of this framework, and the fact that any experimental tests and comparisons with classical theories have required a translation back to more familiar concepts, the true character of the theory has not been understood until recently, and then slowly and with difficulty. It is not clear that we could ever develop a language, mathematical system, or way of thinking in which ideas of monism and interdependence would seem more natural, though perhaps the theory will always appear puzzling and paradoxical until then. But I believe that it is not impossible. The human mind has shown a remarkable flexibility in eventually grasping ideas from new physical theories. It may well be that Einstein himself was not fully aware of the conceptual implications of his own theory. But this is hardly a criticism of him. At least it was his genius that gave us this basis. If he could not see himself, he gave others the means to see. To do it all himself would have required a super-human genius.

One should still be careful not to push the implications too far. Certainly GR has not explained everything yet, nor is there any reason to believe it could ever do so. "Everything" here is a very large and vague totality. At most we may say that GR is a description and account of a level of being. Through its monism it gives to that level a far greater degree of unity and internal coherence than any other level postulated so far. The level does attempt to encompass the whole universe and leave nothing out. However, we must proceed on a step-by-step basis to see just how many phenomena described at other levels can be grounded in it. It may provide some basis for reproducing the distinctions of other levels, but may do so in a way that sets up a very different system of conceptual classifications. In accepting GR we are not committed to saying that the world is really monistic in any absolute or complete sense, or that everything in it is dependent on everything else. All we have shown is that at least one level, and indeed a very basic one, has a monistic character. If we could fully ground all the phenomena of some other level in it, we could reduce and eliminate that other level. But until this has in fact been done, there is no reason to assume in advance that it will be. In addition, the level of GMD might yet be grounded in some other. There are certainly puzzles remaining in the theory. Some of them may be specious, reflecting the peculiarities of our language and our imperfect models rather than the world. Some may eventually become anomalies which will lead to the downfall of the theory. As yet, however, there seems to be enough flexibility in the theory to move in several different directions to incorporate new ideas. In any case, we are not entitled to assume that in GMD we have found the ultimate physical theory, the one that will answer all questions both now and in the future. But in the meantime it must stand as one of the greatest intellectual creations of man.

Long ago a little stream began to gurgle in the mountain wilderness. Slowly but steadily it grew, and it was joined by many others. Together they formed a mighty river, majestically and inexorably following its course to the sea. It flowed through fields and forests, through hills and valleys, through crowded cities and untouched lands. It belonged to all and to none of them. We have followed that river throughout its course. We have seen it grow; we have enjoyed its surprises. And we have gone slowly and carefully to admire the scenery along its route. Much of it has been very beautiful, both by itself and in its reflection in the water. We have tried to see it all, and to sketch it to preserve it in memory. Yet we have never been willing to remain forever at any one place, no matter what its charms. The river beckons us on, and there is always more to see. And now at last we have reached the sea. The land is behind us; we have taken our last view and tried to recollect our journey as a whole. The voyage has been long. In many places the currents have been rough and turbulent, and the passage has been fatiguing for both the pilot and his passengers. But at last we have reached our destination; our journey is over. The pilot can only hope that his passengers have enjoyed the glories of the scenery, to compensate for the hardships. Now it is time for others to take them across the sea, or lead them back again to the land.

Chapter 2

1. J. J. C. Smart, *Philosophy and Scientific Realism,* especially ch. 2, Humanities Press, New York, 1963.

2. Thomas S. Kuhn, *The Structure of Scientific Revolutions,* esp. ch. 10, 11, 13, University of Chicago Press, Chicago, 1962.

3. Immanuel Kant, *Critique of Pure Reason* (Norman Kemp Smith translation), A462–A567, B490–B595, Macmillan, London, 1933.

4. Stephen Körner, "On Philosophical Arguments in Physics," in Stephen Körner, ed., *Observation and Interpretation in the Philosophy of Physics* (Colston Papers, no. 9), pp. 97–102, Dover, New York, 1962. Cf. Paul K. Feyerabend, "Realism and Instrumentalism: Comments on the Logic of Factual Support," in Mario Bunge, ed., *The Critical Approach to Science and Philosophy,* pp. 280–308, Free Press, Glencoe, Ill., 1964.

5. Leonard K. Nash, *The Nature of the Natural Sciences,* pp. 84–113, Little, Brown, Boston, 1963.

6. Stephen Toulmin, *Foresight and Understanding,* ch. 3, 4, Harper Torchbooks, Harper & Row, New York, 1963.

7. Arthur Pap, "Does Science Have Metaphysical Presuppositions?" in Herbert Feigl and May Brodbeck, eds., *Readings in the Philosophy of Science,* pp. 21–34, Appleton-Century-Crofts, New York, 1953.

8. See, e.g., Bertrand Russell, *Mysticism and Logic,* p. 155, Anchor Books, Doubleday, Garden City, N.Y., 1957.

9. Rudolf Carnap, "Testability and Meaning," in Feigl and Brodbeck, pp. 78–81.

10. See the discussion in Nelson Goodman, *The Structure of Appearance,* ch. 5, pp. 114–146, Harvard University Press, Cambridge, Mass., 1951; Victor Kraft, *The Vienna Circle* (Arthur Pap, trans.), pp. 86–95, Philosophical Library, New York, 1953.

11. Wilfred Sellars, *Science, Perception, and Reality* (hereafter abbreviated *SPR*), Humanities Press, New York, 1963.

12. Sellars, "Phenomenalism," *SPR,* p. 77.

13. Sellars, *SPR,* pp. 82f.

14. *Ibid.,* p. 91.

15. Sellars, throughout "Phenomenalism" and elsewhere in *SPR,* esp. "Empiricism and the Philosophy of Mind," pp. 127–197; also published in *Minnesota Studies in the Philosophy of Science,* vol. 1, University of Minnesota Press, Minneapolis, 1956.

16. Smart, *Philosophy and Scientific Realism,* pp. 16–27.

17. Sellars, *SPR,* p. 96.

18. See Percy W. Bridgman, *The Way Things Are,* esp. ch. 6, 7, Viking Press, New York, 1961; *Reflections of a Physicist,* Philosophical Library, New York, 1955.

19. Percy W. Bridgman, *The Logic of Modern Physics,* esp. ch. 1, Macmillan, New York, 1927.

20. Toulmin, *Foresight and Understanding,* ch. 2, pp. 18–43; cf. W. H. Watson, *Understanding Physics Today,* ch. 1, Cambridge University Press, London, 1963.

21. Carl G. Hempel and Paul Oppenheim, "Studies in the Logic of Explanation," in Feigl and Brodbeck, pp. 319–353; cf. Hempel, "Deductive-Nomological vs. Sta-

tistical Explanation," *Minnesota Studies in the Philosophy of Science,* vol. 3, pp. 98–169, University of Minnesota Press, Minneapolis, 1962.

22. Herbert Feigl, "Some Remarks on the Nature of Scientific Explanation," in Feigl and Sellars, *Readings in Philosophical Analysis,* pp. 510–514, Appleton-Century-Crofts, New York, 1949.

23. Michael Scriven, "The Limits of Physical Explanation," in Bernard Baumrin, ed., *Philosophy of Science—The Delaware Seminar,* vol. 2, pp. 107–138, esp. pp. 108–116, Interscience, New York, 1963; Scriven, "The Key Property of Physical Laws: Inaccuracy," in Herbert Feigl and Grover Maxwell, eds., *Current Issues in the Philosophy of Science,* pp. 91–102, Holt, Rinehart, and Winston, New York, 1961.

24. Wilfred Sellars, "The Language of Theories," in Feigl and Maxwell, *Current Issues,* p. 71; also in *SPR,* p. 121; see also Pierre Duhem, *The Aim and Structure of Physical Theory* (Philip Wiener, trans.), part 2, ch. 3–5, Princeton University Press, Princeton, 1954.

25. Sellars, *SPR,* p. 97.

26. Sellars, "The Language of Theories," in *SPR,* p. 126; in Feigl and Maxwell, p. 76.

27. David Bohm, in Körner, *Observation and Interpretation,* pp. 33–41; cf. Bohm, *Phys. Rev.,* vol. 85 (1952), pp. 166–193.

28. Mario Bunge, *Metascientific Queries,* ch. 5, pp. 108–123, Charles C Thomas, Springfield, Ill., 1959; *The Myth of Simplicity,* pp. 36–50, Prentice-Hall, New York, 1963.

29. Rudolf Carnap, "Logical Foundations of the Unity of Science," in Feigl and Sellars, pp. 408–424; also *International Encyclopedia of Unified Science,* vol. 1, no. 1, pp. 42–63, University of Chicago Press, Chicago, 1955; cf. Ernest Nagel, *The Structure of Science,* ch. 13, Harcourt, Brace, & World, New York, 1961; Carl G. Hempel, "Operationism, Observation, and Scientific Terms," in Arthur Danto and Sidney Morgenbesser, eds., *Philosophy of Science,* pp. 101–120, Meridian Books, New York, 1960.

30. Wilfred Sellars, "Philosophy and the Scientific Image of Man," in *SPR,* pp. 1–40; also in Robert G. Colodny, ed., *Frontiers of Science and Philosophy,* pp. 35–78, University of Pittsburgh Press, Pittsburgh, 1962.

31. Arthur S. Eddington, *The Nature of the Physical World,* pp. xi–xiii, Cambridge University Press, London, 1928; Sellars, *SPR,* p. 118; Feigl and Maxwell, p. 69.

32. Bunge, *Metascientific Queries,* pp. 114–119.

33. Smart, *Philosophy and Scientific Realism,* pp. 13–15 et passim.

34. Bunge, *Metascientific Queries,* pp. 119–122.

35. Paul Feyerabend, "How to Be a Good Empiricist," in Baumrin, pp. 3–41; "Explanation, Reduction, and Empiricism," in *Minnesota Studies.* vol. 3, pp. 28–97.

36. Kuhn, ch. 10, pp. 110–135.

37. Hilary Putnam, "What Theories Are Not," in Ernest Nagel, Patrick Suppes, and Alfred Tarski, eds., *Logic, Methodology, and Philosophy of Science,* pp. 240–251, esp. p. 240, Stanford University Press, Stanford, 1962.

38. Wilfred Sellars, "Theoretical Explanation," in Baumrin, pp. 61–79, esp. p. 63.

39. Putnam, "What Theories Are Not," p. 241.

40. *Ibid.,* p. 243.

41. Grover Maxwell, "The Ontological Status of Theoretical Entities," *Minnesota Studies,* vol. 3, pp. 3–27, esp. p. 7.

42. Norwood Russell Hanson, *Patterns of Discovery,* ch. 1, pp. 4–30, Cambridge University Press, London, 1958.

43. Kuhn, pp. 111–115.

44. Karl R. Popper, "Philosophy of Science: a Personal Report," in C. A. Mace, ed., *British Philosophy in the Mid-Century,* pp. 155–194, George Allen & Unwin, London, 1957; public lecture at Princeton University, March, 1963.

45. Hilary Putnam, "The Analytic and the Synthetic," *Minnesota Studies,* vol. 3, pp. 358–397, esp. pp. 376–381; cf. Nagel, Suppes, and Tarski, p. 247.

46. Putnam, in Nagel, Suppes, and Tarski, p. 241.

47. Sellars, in Baumrin, pp. 61, 71–77.

Chapter 3

1. Henry Veatch, *Intentional Logic,* p. 14, Yale University Press, New Haven, 1952.

2. I have discussed these issues in more detail in an unpublished paper on "Aristotelian and Modern Logic," January, 1962.

3. J. O. Urmson, *Philosophical Analysis,* pp. 130–146, Oxford University Press, London, 1958.

4. Richard B. Braithwaite, "Models in the Empirical Sciences," in Nagel, Suppes, and Tarski, pp. 224–232; cf. Braithwaite, *Scientific Explanation,* pp. 50–115, Cambridge University Press, London, 1955.

5. Cf. Robert B. Lindsay and Henry Margenau, *Foundations of Physics,* pp. 79–159, Dover, New York, 1957.

6. Willard Van Orman Quine, "Two Dogmas of Empiricism," *From a Logical Point of View,* pp. 20–46, Harvard University Press, Cambridge, Mass., 1953.

7. Putnam, "The Analytic and the Synthetic," *Minnesota Studies,* vol. 3, pp. 358–397.

8. Maxwell, "The Ontological Status of Theoretical Entities," *Minnesota Studies,* vol. 3, pp. 3–27, esp. pp. 22–24; "Meaning Postulates in Scientific Theories," in Feigl and Maxwell, pp. 169–183.

9. Willard Van Orman Quine, "Logical Truth," in Sidney Hook, ed., *American Philosophers at Work,* pp. 121–134, esp. p. 129, Criterion Books, New York, 1956.

10. Putnam, in *Minnesota Studies,* vol. 3, pp. 376–381, 392f.

11. Dudley Shapere, "Space, Time, and Language," in Baumrin, pp. 137–170, esp. pp. 158f.

12. Alfred J. Ayer, "Basic Propositions," in Max Black, ed., *Philosophical Analysis,* pp. 57–71, Prentice-Hall, New York, 1950.

13. Feyerabend, "How to Be a Good Empiricist," in Baumrin, pp. 3–40.

14. Cf. R. Taton, *Reason and Chance in Scientific Discovery,* Spectrum Books, Prentice-Hall, Englewood Cliffs, N.J., 1964.

15. Popper, *The Logic of Scientific Discovery,* pp. 27–33, Basic Books, New York, 1959.

16. Kuhn, *Structure of Scientific Revolutions,* ch. 6, pp. 52–65.

17. *Ibid.,* ch. 12, pp. 143–158.

18. Bridgman, *The Logic of Modern Physics,* p. 2.

Chapter 4

1. Ernest Nagel, *The Structure of Science,* pp. 90–97, 106–117, Harcourt, Brace, & World, New York, 1961.

2. Sellars, "The Language of Theories," *SPR*, pp. 123–126; also in Feigl and Maxwell, pp. 74–77.

3. Cf. Nagel, pp. 114, 173.

4. Pierre Duhem, *The Aim and Structure of Physical Theory,* pp. 55–106, esp. p. 70, Princeton University Press, Princeton, 1954.

5. Nagel, ch. 7, pp. 153–202.

6. Bunge, *Metascientific Queries,* pp. 153–172.

7. E.g., Bridgman, *The Logic of Modern Physics.*

8. Mary Hesse, *Models and Analogies in Science,* ch. 1, Sheed & Ward, London, 1963.

9. Max Black, *Models and Metaphors,* pp. 236–237, Cornell University Press, Ithaca, 1962.

10. *Ibid.,* p. 239.

11. Romano Harré, *An Introduction to the Logic of the Sciences,* pp. 82–110, Macmillan, London, 1960.

12. *Ibid.,* p. 99.

13. Norman R. Campbell, *Foundations of Physics* (formerly entitled *Physics: The Elements*) p. 129, Dover, New York, 1957.

14. Nagel, *Structure of Science,* pp. 115–117.

15. Kuhn, *Structure of Scientific Revolutions,* ch. 2–5, esp. ch. 5.

16. *Ibid.,* pp. 26f.

17. John A. Wheeler, in Hong-Yee Chiu and William F. Hoffman, eds., *Gravitation and Relativity,* p. 315, Benjamin, New York, 1964.

18. Sellars, in *SPR,* pp. 118f.; in Feigl and Maxwell, pp. 68f.

19. Nagel, *Structure of Science,* ch. 7, pp. 153–174.

20. Bunge, *Metascientific Queries,* pp. 153–172.

21. *Ibid.*, pp. 162f., 171.

22. Feyerabend, "How to Be a Good Empiricist," in Baumrin, pp. 26–30.

23. Sellars, "Philosophy and the Scientific Image of Man," in *SPR,* p. 37; also in Colodny, *Frontiers of Science and Philosophy,* pp. 73–75.

24. Immanuel Kant, *Critique of Pure Reason,* A435–A439, B463–B467.

Chapter 5

1. Stephen Toulmin, *Philosophy of Science,* ch. 4, pp. 105–140, Harper Torchbooks, Harper & Row, New York, 1960.

2. Kuhn, *Structure of Scientific Revolutions,* p. 141.

3. Dudley Shapere, review of Kuhn, in *Philosophical Review,* vol. 73 (1964), pp. 383–394.

4. Kuhn, p. 141.

5. Putnam, in *Minnesota Studies,* vol. 3, p. 379.

6. Solomon Bochner, "Aristotle's Physics and Today's Physics," *International Philosophical Quarterly,* vol. 4 (May, 1963), pp. 217–244, esp. pp. 221f.; also in Bochner,

The Role of Mathematics in the Rise of Science, ch. 4, pp. 143–178, Princeton University Press, Princeton, 1966.

7. Plato, *Phaedo,* 97c–99c; R. Hackforth, *Plato's Phaedo,* pp. 124–127, Cambridge University Press, London, 1952.

8. Sellars, Hanson, et al., in Ernan McMullin, *The Concept of Matter,* p. 77, University of Notre Dame Press, Notre Dame, 1963.

9. Leonard J. Eslick, "The Material Substrate in Plato," in McMullin, pp. 59–74.

10. Albert Einstein, in Max Jammer, *Concepts of Space,* p. xiv, Harper & Row, New York, 1954.

11. Cf. Désiré Nys, "The Aristotelian-Thomistic Theory of Space and Time," in Henry J. Koren, ed., *Readings in the Philosophy of Nature,* pp. 244–261, Newman Press, Westminster, Md., 1961.

12. Jammer, *Concepts of Space,* pp. 9f.

13. Alfred North Whitehead, *Adventures of Ideas,* p. 138, and elsewhere, Macmillan, New York, 1933.

14. Eslick, in McMullin, pp. 39–74.

15. Paul Shorey, *What Plato Said,* pp. 337–339, University of Chicago Press, Chicago, 1933.

16. David Ross, *Plato's Theory of Ideas,* pp. 123–128, Clarendon Press, Oxford, 1951.

17. Jammer, *Concepts of Space,* pp. 12–14.

18. Alfred E. Taylor, *A Commentary on Plato's Timaeus,* esp. p. 347, Clarendon Press, Oxford, 1928; Taylor, *Plato: The Man and His Work,* Meridian Books, New York, 1956.

19. Eduard Zeller, *Outlines of the History of Greek Philosophy,* pp. 162–167, Meridian Books, New York, 1955.

20. Edmund Whittaker, *From Euclid to Eddington,* pp. 5f., Dover, New York, 1958.

21. George S. Claghorn, *Aristotle's Criticism of Plato's "Timaeus,"* pp. 5–20, Martinus Nijhoff, The Hague, 1954.

22. Robin G. Collingwood, *The Idea of Nature,* pp. 72–79, Oxford University Press, London, 1945.

23. Francis M. Cornford, *Plato's Cosmology,* pp. 181, 229, Liberal Arts Press, New York, 1957.

24. I have discussed this further in an unpublished paper on "Plato's Concept of Matter," esp. pp. 20–26, May, 1961.

25. Cornford, *Plato's Cosmology,* pp. 171f.

26. Paul Friedländer, *Structure and Destruction of the Atom According to Plato's Timaeus,* University of California Press, Berkeley, 1949.

27. S. Sambursky, *The Physical World of the Greeks,* pp. 42–44, Macmillan, New York, 1956.

Chapter 6

1. Ralph M. Blake, "The Role of Experience in Descartes' Theory of Method," in Blake, Curt J. Ducasse, and Edward H. Madden, *Theories of Scientific Method,* pp. 75–103, esp. p. 87, University of Washington Press, Seattle, 1960.

2. See e.g., Fernand Renoirte, "Critique of Dynamism," in Koren, *Readings in the Philosophy of Nature,* pp. 108–110, esp. p. 108.

3. Richard Rorty has made this point in "Realism, Categories, and the Linguistic Turn," *International Philosophical Quarterly,* vol. 2 (1962), pp. 307–322.

4. John E. Boodin, "The Discovery of Form," in Philip P. Wiener and Aaron Noland, eds., *Roots of Scientific Thought,* pp. 57–72, Basic Books, New York, 1957.

5. Rorty, *"The Irreducibility of Substance,"* p. 13–14 (unpublished); cf. Aristotle, *Metaphysics,* 7:16.

6. Rene Descartes, *Principles of Philosophy,* I:51; in Elizabeth S. Haldane and G. R. T. Ross, *Philosophical Works of Descartes,* vol. 1, p. 239, Dover, 1931.

7. Descartes, *Principles,* I:60; Haldane and Ross, p. 243.

8. *Principles,* I:60–62; Haldane and Ross, pp. 243–245.

9. *Principles,* I:53; Haldane and Ross, p. 240.

10. *Principles,* II:10–15; Haldane and Ross, pp. 259–262; Descartes, *Rules for the Direction of the Mind,* in Haldane and Ross, pp. 55–60.

11. *Principles,* II:4; Haldane and Ross, pp. 255f.

12. Edwin A. Burtt, *The Metaphysical Foundations of Modern Science,* p. 106, Doubleday, New York, 1955.

13. *Ibid.,* p. 117.

14. Charles C. Gillispie, *The Edge of Objectivity,* pp. 87f.; Princeton University Press, Princeton, 1960.

15. Blake, in Blake, Ducasse, and Madden, p. 83.

16. See J. F. Scott, *The Scientific Work of Descartes,* p. 162, Taylor and Francis, London, 1952; also Gillispie, p. 93.

17. Descartes, *Principles,* II:64; Haldane and Ross, p. 269.

18. Blake, in Blake, Ducasse, and Madden, p. 95.

19. E. J. Dijksterhuis, *The Mechanization of the World-Picture,* p. 409, Clarendon Press, Oxford, 1961.

20. Descartes, *Principles,* II:16; Haldane and Ross, p. 262.

21. Edmund Whittaker, *From Euclid to Eddington,* p. 11.

22. Milič Čapek, *The Philosophical Impact of Contemporary Physics,* pp. 111–116, Van Nostrand, Princeton, 1961.

23. Descartes, *Principles,* II:33.

24. *Ibid.,* III:63.

25. Čapek, *Philosophical Impact,* p. 106.

26. Descartes, *Principles,* II:32, 33.

27. *Ibid.,* II:36.

28. *Ibid.,* II:22, 23.

29. Whittaker, *From Euclid to Eddington,* p. 11 footnote.

30. Descartes, *Principles,* II:20.

31. Alexandre Koyré, *From the Closed World to the Infinite Universe,* pp. 99–124, Harper Torchbooks, Harper & Row, New York, 1958.

32. Descartes, *Principles,* I:26–27; Haldane and Ross, pp. 229–230; also *Principles,* II:21.

33. Gillispie, *Edge of Objectivity,* pp. 83f.

34. Descartes, *Principles,* II:37, 39.

35. *Ibid.,* II:36.

36. Scott, *Scientific Work of Descartes,* p. 163; Descartes, *Principles,* II:40–63.

37. P. Henry Van Laer, "Extension as the Criterion of Matter," in Koren, *Readings in the Philosophy of Nature,* pp. 216–223.

38. Koyré, *From the Closed World to the Infinite Universe,* pp. 110–154.

39. *Ibid.,* p. 132.

40. Descartes, *Principles,* I:61, 65; Haldane and Ross, pp. 244–246; Norman Kemp Smith, *Studies in the Cartesian Philosophy,* p. 75, Macmillan, London, 1902.

41. *Principles,* II:23.

42. Marie Boas Hall, "Matter in Seventeenth Century Science," in McMullin, *The Concept of Matter,* pp. 344–367, esp. p. 351.

43. Kemp Smith, *Studies,* pp. 69–70, 75.

44. Čapek, *Philosophical Impact,* ch. 5, pp. 67–78.

45. Descartes, *Principles,* II:25; Haldane and Ross, p. 266.

46. *Principles,* I:21; Haldane and Ross, pp. 227–228.

47. *Principles,* I:57; Haldane and Ross, p. 242.

48. F. Renoirte, "Critique of Mechanism," in Koren, pp. 115–120, esp. p. 116.

49. Descartes, *Principles,* III:48–53; cf. Scott, pp. 172f.

50. Čapek, *Philosophical Impact,* p. 114.

51. Dijksterhuis, *Mechanization,* p. 414.

52. Descartes, *Principles,* II:5–7, 19.

53. Max Jammer, *Concepts of Force,* pp. 103–108, Harper Torchbooks, Harper & Row, New York, 1962.

54. Stephen Toulmin and June Goodfield, *The Architecture of Matter,* p. 159, Hutchinson, London, 1962.

55. Koyré, p. 290, note 7.

56. Descartes, in Haldane and Ross, pp. 61f.; Burtt, *Metaphysical Foundations,* pp. 109f.

Chapter 7

1. Max Jammer, *Concepts of Space,* p. 97.

2. Hakan Törnebohm, *A Logical Analysis of the Theory of Relativity,* pp. 54–55, Almqvist and Wiksell, Stockholm, 1952.

3. Plato, *Timaeus,* 56e–57a; Cornford, *Plato's Cosmology,* pp. 224–230.

4. Marie Boas Hall, in McMullin, *The Concept of Matter,* pp. 363f.

5. Mary Hesse, *Models and Analogies in Science,* p. 12, Sheed & Ward, London, 1963.

6. Jammer, *Concepts of Mass,* pp. 64f., Harvard University Press, Cambridge, Mass., 1961; Burtt, *Metaphysical Foundations,* p. 241.

7. Čapek, *Philosophical Impact,* pp. 59f.

8. Kuhn, *Structure of Scientific Revolutions,* p. 72; cf. Jammer, *Concepts of Space,* pp. 114–124.

9. Čapek, pp. 83–89.

10. Roger J. Boscovich, selection from "Theory of the Philosophy of Nature," in Koren, *Readings in the Philosophy of Nature,* pp. 102–104; also see Jammer, *Concepts of Force,* pp. 170–178; Mary Hesse, *Forces and Fields,* pp. 163–166; Thomas Nelson & Sons, New York, 1961.

11. Hesse, *Forces and Fields,* p. 170; cf. Adolf Grünbaum, *Philosophical Problems of Space and Time,* pp. 330–337, Knopf, New York, 1963.

12. Immanuel Kant, selections from "Physical Monadology" and "Metaphysical Foundations of Physical Science," in Koren, pp. 105–108; Hesse, *Forces and Fields,* pp. 170–180; Jammer, *Concepts of Force,* pp. 179–182.

13. J. P. Vigier, "Determinism and Indeterminism in a New 'Level' Conception of Matter," in Nagel, Suppes, and Tarski, pp. 262–264, esp. p. 262.

14. Čapek, *Philosophical Impact,* p. 79.

Chapter 8

1. Albert Einstein, "Autobiographical Notes," in Paul A. Schilpp, ed., *Albert Einstein: Philosopher-Scientist,* p. 65, Tudor, New York, 1951.

2. *Ibid.,* p. 73.

3. See, e.g., Carnap, "Testability and Meaning," in Feigl and Brodbeck, pp. 47–92.

4. Hesse, *Forces and Fields,* pp. 216–219.

5. Albert Einstein and Leopold Infeld, *The Evolution of Physics,* pp. 125–153, Simon and Schuster, New York, 1938.

6. *Ibid.,* pp. 143, 146.

7. Hesse, *Forces and Fields,* pp. 203–206; also see her article "Action at a Distance," in McMullin, *The Concept of Matter,* pp. 372–390, esp. pp. 379f.

8. Albert Einstein, *The Meaning of Relativity,* 5th edition, p. 140, Princeton University Press, Princeton, 1955.

9. Charles W. Misner, "Mass as a Form of Vacuum," in McMullin, pp. 596–608, esp. p. 596.

10. See, e.g., Jammer, *Concepts of Mass,* pp. 144–153.

11. Törnebohm, *A Logical Analysis of the Theory of Relativity,* pp. 56f.

Chapter 9

1. Jammer, *Concepts of Mass,* p. 125; also see Cecil B. Mast, "A Note on Three Concepts of Mass," in McMullin, *The Concept of Matter,* p. 574, and McMullin's introduction, pp. 23–25.

2. For references to the experimental literature available in Einstein's time, see Jammer, *Concepts of Mass,* p. 125.

3. Einstein, *The Meaning of Relativity,* p. 140.

4. Kuhn, *Structure of Scientific Revolutions,* pp. 143–158.

5. Robert H. Dicke, "Mach's Principle and Equivalence," in C. Møller, ed., *Evidence for Gravitational Theories,* pp. 1–50, esp. pp. 15–31, Academic Press, New York, 1962; also see Hong-Yee Chiu and William F. Hoffman, eds., *Gravitation and Relativity,* pp. 12–14, Benjamin, New York, 1964.

6. P. G. Roll, R. Krotkov, and R. H. Dicke, "The Equivalence of Inertial and Passive Gravitational Mass," *Ann. Phys.* (N.Y.) vol. 26 (1964), pp. 442–517; Dicke, "The Eötvös Experiment," *Scientific American,* December 1961, pp. 84–94.

7. Dicke, "Remarks on the Observational Basis of General Relativity," in Chiu and Hoffman, pp. 1–16, esp. p. 14.

8. Also cf. Dicke, in Møller, *Evidence for Gravitational Theories* p. 17; A. d'Abro, *The Evolution of Scientific Thought,* 2nd edition, p. 241, Dover, New York, 1950.

9. Joseph J. Weber, *General Relativity and Gravitational Waves,* p. 10; Interscience, New York, 1961.

10. Herman Bondi, "Negative Mass in General Relativity," *Rev. Mod. Phys.,* vol. 29 (1957), pp. 423–428.

11. L. I. Schiff, "Gravitational Properties of Antimatter," *Proc. Natl. Acad. Sci. U.S.,* vol. 45 (1959), pp. 69–80.

12. A. Schild, "The Principle of Equivalence," *The Monist,* vol. 47, no. 1 (1962), pp. 20–39.

13. *Ibid.,* p. 35. The argument is given in full mathematical detail in Schild, "Gravitational Theories of the Whitehead Type and the Principle of Equivalence," in Møller, *Evidence for Gravitational Theories,* pp. 69–115.

14. Hans Reichenbach, *The Philosophy of Space and Time,* p. 227, Dover, New York, 1957.

15. Albert Einstein, *The Principle of Relativity,* p. 117, Dover, New York, 1923.

16. Weber, *General Relativity and Gravitational Waves,* p. 15.

17. Richard C. Tolman, *Relativity, Thermodynamics, and Cosmology,* p. 166, Clarendon Press, Oxford, 1934.

18. Hermann Minkowski, "Space and Time," in Einstein, *The Principle of Relativity,* pp. 73–91.

19. Peter G. Bergmann, *Introduction to the Theory of Relativity,* p. 159, Prentice-Hall, New York, 1942.

20. *Ibid.,* p. 158.

21. Cf. the discussion in d'Abro, *The Evolution of Scientific Thought,* 2nd edition, pp. 262–266, Dover, New York, 1950.

22. Erich Kretschmann, "Ober den Physikalischen Sinn der Relativitätspostulate, A. Einstein's neue und seine Ursprüngliche Relativitätstheorie," *Ann. Physik,* vol. 53 (1917), pp. 575–614.

23. Tolman, *Relativity, Thermodynamics, and Cosmology,* p. 167.

24. *Ibid.,* p. 168.

25. d'Abro, *The Evolution of Physics,* p. 462; Tolman, p. 168.

26. Mario Bunge, *Foundations of Physics,* p. 217; Springer-Verlag, New York, 1967; E. Cartan, *Annales École Normale Supérieure,* vol. 40 (1923), p. 325; vol. 41 (1924), p. 1; Peter Havas, *Rev. Mod. Phys.,* vol. 36 (1964), p. 938.

27. Percy W. Bridgman, *The Nature of Physical Theory,* ch. 7, pp. 72–92, Dover, New York, 1936; also Bridgman, "Einstein's Theories and the Operational Point of View," in Schilpp, *Albert Einstein: Philosopher-Scientist,* pp. 333–354.

28. Bridgman, *The Logic of Modern Physics,* p. 2.

29. Bridgman, in Schilpp, pp. 333–354.

30. Bridgman, *The Nature of Physical Theory,* pp. 82f.

31. Einstein, "Autobiography," in Schilpp, pp. 67–69.

32. This is the central thesis of Bridgman's last book, *The Way Things Are,* Viking Press, New York, 1961.

33. V. Fock, "Three Lectures on Relativity Theory," *Rev. Mod. Phys.,* vol. 29 (1957), pp. 325f.; also see Fock, *The Theory of Space, Time, and Gravitation,* esp. ch. 6, Pergamon Press, New York, 1959.

34. J. L. Anderson, "Relativity Principles and the Role of Coordinates in Physics," in Chiu and Hoffman, *Gravitation and Relativity,* pp. 175–194, esp. pp. 188f.

35. cf. Robert M. Palter, "Copernicanism, Old and New," *The Monist,* vol. 48 (1964), pp. 143–184, esp. p. 178.

36. Fock, *Rev. Mod. Phys.,* vol. 29 (1957), p. 326.

37. Bridgman, *The Nature of Physical Theory,* pp. 73–80.

38. Hakan Törnebohm, *A Logical Analysis of the Theory of Relativity,* pp. 68f., Almqvist and Wiksell, Stockholm, 1952.

39. Eugene P. Wigner, "Events, Laws of Nature, and Invariance Principles," *Science,* vol. 145 (1964), pp. 995–999 (Nobel Prize lecture).

40. Törnebohm, p. 218–223.

41. Mario Bunge, *The Myth of Simplicity,* ch. 12, pp. 204–230, Prentice-Hall, New York, 1963.

42. Törnebohm, p. 229.

43. Bridgman, *The Nature of Physical Theory,* p. 81.

44. René Dugas, *A History of Mechanics* (J. R. Maddox, trans.), p. 523, Neuchâtel, Switzerland, 1955.

45. William Kingdon Clifford, "On the Space Theory of Matter," *Proc. Cambridge Phil. Soc.,* vol. 2 (1876), pp. 157f.

46. Clifford, *Mathematical Papers of William Kingdon Clifford,* pp. 21f., Macmillan, London, 1882; Clifford, *The Common Sense of the Exact Sciences,* pp. 193–204, Knopf, New York, 1946.

47. Clifford, *The Common Sense of the Exact Sciences,* p. 202.

48. Reichenbach, *The Philosophy of Space and Time,* p. 250.

49. Erwin Schrödinger, *Space-Time Structure,* Cambridge University Press, London, 1950; Einstein, *The Meaning of Relativity,* pp. 133–163.

50. See, e.g., d'Abro, *The Evolution of Physics,* pp. 330–340.

51. Robert Marzke and John A. Wheeler, "Gravitation as Geometry—I," in Chiu and Hoffman, *Gravitation and Relativity,* pp. 58–62.

52. John A. Wheeler, "Gravitation as Geometry—II," in Chiu and Hoffman, pp. 68, 81, 88 note 5.

53. Reichenbach, *The Philosophy of Space and Time;* Reichenbach, "The Philosophical Significance of the Theory of Relativity," in Schilpp, p. 287–312; Adolf Grünbaum, *Philosophical Problems of Space and Time,* esp. pp. 81–152, Knopf, New York, 1963; Grünbaum, "Geometry, Chronometry, and Empiricism," pp. 405–526 in *Minnesota Studies,* Vol. 3.

54. Reichenbach, *The Philosophy of Space and Time,* p. 256.

55. Bridgman, *The Nature of Physical Theory,* p. 73; and in Schilpp, p. 338.

56. Anderson, in Chiu and Hoffman, *Gravitation and Relativity,* pp. 185–194.

57. Fock, *The Theory of Space, Time, and Gravitation,* Introduction and ch. 5–7.

58. Alfred North Whitehead, *The Principle of Relativity,* Cambridge University Press, London, 1922; Whitehead, *The Interpretation of Science,* Liberal Arts Press, 1961.

59. Anderson, in Chiu and Hoffman, p. 192.

60. Eugene Wigner, "Invariance in Physical Theory," *Proc. Am. Phil. Soc.,* vol. 93 (1949), pp. 521–526; "Symmetry and Conservation Laws," *Proc. Natl. Acad. Sci. U.S.,* vol. 51 (1964), pp. 956–965.

Chapter 10

1. Niels Bohr and L. Rosenfeld, "Field and Charge Measurements in Quantum Electrodynamics," *Phys. Rev.,* vol. 78 (1950), pp. 794–798.

2. Einstein, "Autobiography," in Schilpp, *Albert Einstein: Philosopher-Scientist,* p. 59.

3. Einstein, "Replies to Objections," in Schilpp, pp. 685f.

4. Robert F. Marzke, *The Theory of Measurement in General Relativity,* A.B. senior thesis, Princeton University, 1959.

5. J. L. Synge, *Relativity: The General Theory,* p. ix, North Holland Publishing Co., Amsterdam, 1960.

6. For good discussions and mathematical derivations, see Lev Landau and E. Lifschitz, *The Classical Theory of Fields* (Morton Hammermesh, trans.), pp. 256–260, Addison-Wesley, Reading, Mass., 1951; and Weber, *General Relativity and Gravitational Waves,* pp. 39–41.

7. Landau and Lifschitz, p. 259.

8. Marzke and Wheeler, in Chiu and Hoffman, *Gravitation and Relativity,* pp. 46–48.

9. Saul A. Basri, "Operational Foundations of Einstein's General Theory of Relativity." *Rev. Mod. Phys.,* vol. 37 (1965), pp. 288–315, esp. pp. 291–294.

10. H. Salecker and Eugene Wigner, "Quantum Limitations on the Measurement of Space-Time Distances," *Phys. Rev.,* vol. 109 (1958), pp. 571–577; Wigner, "Relativistic Invariance and Quantum Phenomena," *Rev. Mod. Phys.,* vol. 29 (1957), pp. 255–268.

11. J. L. Synge, "Relativity Based on Chronometry," pp. 441–448 in *Recent Developments in General Relativity,* Macmillan, New York, 1962; also Synge, *Relativity: The General Theory.*

12. E. Newman and J. N. Goldberg, "The Measurement of Distance," in *Colloquie sur la Théorie de la Relativité,* pp. 37–41, Centre Belge de Recherches Mathematiques, Louvain, 1960.

13. John G. Fletcher, "Geometrodynamics," in Louis Witten, *Gravitation: An Introduction to Current Research,* pp. 412–437, esp. p. 418, John Wiley & Sons, New York, 1962.

14. Marzke and Wheeler, in Chiu and Hoffman, *Gravitation and Relativity,* p. 48; Wheeler, in Bryce Dewitt and C. Dewitt, *Relativity, Groups, and Topology,* p. 342, Gordon and Breach, New York, 1964.

15. For more detailed discussions and the calculations required see Marzke and Wheeler, in Chiu and Hoffman, pp. 48–58; Fletcher, in Witten, pp. 419–423; Wheeler, *Geometrodynamics,* pp. 9–13, Academic Press, New York, 1962.

16. See, e.g., Wheeler, "Problems on the Frontiers between General Relativity and Differential Geometry," *Rev. Mod. Phys.,* vol. 34 (1962), pp. 877f.; Wheeler, in Chiu and Hoffman, pp. 74–76; Wheeler, in Dewitt and Dewitt, pp. 342f.; Synge, *Relativity: The General Theory,* pp. 19–25.

17. Wheeler, *Rev. Mod. Phys.,* vol. 34 (1962), p. 878.

18. Wigner, *Rev. Mod. Phys.,* vol. 29 (1957), pp. 255–269; *Phys. Rev.,* vol. 120 (1960), p. 643.

19. Bruno Bertotti, in Møller, *Evidence for Gravitational Theories,* pp. 195–199.

20. Synge, *Relativity: The General Theory,* pp. 156–158; Synge, in *Recent Developments in General Relativity* (1962), pp. 441–448.

21. Saul A. Basri, *Rev. Mod. Phys.,* vol. 37 (1965), pp. 304–315.

Chapter 11

1. Einstein, "Autobiography," in Schilpp, *Albert Einstein: Philosopher-Scientist,* p. 31.

2. Ronald Adler, Maurice Bazin, and Menahem Schiffer, *Introduction to General Relativity,* ch. 2, pp. 39–62, McGraw-Hill, New York, 1965; Erwin Schrödinger, *Expanding Universes,* pp. 41–45, Cambridge University Press, London, 1956; J. L. Anderson, in Chiu and Hoffman, *Gravitation and Relativity,* pp. 25–31.

3. Reichenbach, *The Philosophy of Space and Time,* pp. 1-28; Reichenbach, in Schilpp, pp. 287–311; Grünbaum, *Philosophical Problems of Space and Time,* pp. 106–151.

4. R. L. Dicke, in Chiu and Hoffman, pp. 8–11.

5. Whitehead, *The Principle of Relativity;* Robert M. Palter, *Whitehead's Philosophy of Science,* pp. 188–213; University of Chicago Press, Chicago, 1960.

6. For more mathematical details, see Wheeler, in Chiu and Hoffman, pp. 68–71; Dicke, in Chiu and Hoffman, pp. 8f.; Adler, Bazin, and Schiffer, pp. 51–59; and other standard texts.

7. Anderson, in Chiu and Hoffman, pp. 25–31; Adler, Bazin, and Schiffer, pp. 39–53, esp. p. 45.

8. Synge, *Relativity: The General Theory,* pp. 109–111.

9. Tolman, *Relativity, Thermodynamics, and Cosmology,* p. 260.

10. Weber, *General Relativity and Gravitational Waves,* pp. 54f.

11. D. M. Chase, *Phys. Rev.,* vol. 95 (1954), pp. 243–246.

12. Wheeler, in Dewitt and Dewitt, pp. 328–337; Wheeler, in Chiu and Hoffman, pp. 69f.; *Rev. Mod. Phys.,* vol. 34 (1962), p. 876.

13. See, e.g., Törnebohm, *Logical Analysis of the Theory of Relativity,* pp. 82–86, 93–98; Adler, Bazin, and Schiffer, pp. 261–275; Lindsay and Margenau, *Foundations of Physics,* pp. 377–384.

14. Dicke, "Mach's Principle and Equivalence," pp. 1–50 in Møller, *Evidence for Gravitational Theories;* and elsewhere.

15. Einstein, "Autobiography," in Schilpp, p. 23.

16. *Ibid.,* p. 89.

17. Einstein, *The Meaning of Relativity,* pp. 83f.

18. Wolfgang Pauli, *Theory of Relativity* (G. Field, trans.), p. 48, Pergamon Press, New York, 1958.

19. *Ibid.,* pp. 159–161; and other standard texts.

20. Einstein, *The Meaning of Relativity,* appendix 1, pp. 109–132.

21. Arthur S. Eddington, *The Mathematical Theory of Relativity,* pp. 141–144, Cambridge University Press, London, 1960.

22. Wheeler, in Chiu and Hoffman, p. 81.

23. See, e.g., Herbert Goldstein, *Classical Mechanics,* pp. 47–55, Addison-Wesley, Reading, Mass., 1950.

24. Erwin Schrödinger, *Space-Time Structure,* p. 99, Cambridge University Press, London, 1950.

25. Schrödinger, "The General Theory of Relativity and Wave Mechanics," pp. 65–74, in *Scientific Papers Presented to Max Born,* Hafner, New York, 1953.

26. Arthur S. Eddington, *The Nature of the Physical World,* p. 156, Cambridge University Press, London, 1953; Čapek, *Philosophical Impact,* pp. 178–180.

27. Čapek, *Philosophical Impact,* pp. 180f.

28. Cecil Mast, "Matter and Energy in Scientific Theory," in McMullin, *The Concept of Matter,* pp. 592–595.

29. d'Abro, *The Evolution of Scientific Thought,* pp. 327–329.

30. Leopold Infeld, "On Equations of Motion in General Relativity Theory," pp. 206–209 in *Bern Jubilee of Relativity Theory, Helvetica Physica Acta,* suppl. vol. 4, 1956; Infeld, *Rev. Mod. Phys.,* vol. 29 (1957), p. 398; Jammer, *Concepts of Mass,* p. 211.

31. Einstein, "Autobiography," in Schilpp, p. 75.

32. Einstein, *The Meaning of Relativity,* pp. 82, 106, 165.

33. Einstein, "Replies to Criticisms," in Schilpp, p. 684.

34. Pauli, *Theory of Relativity,* pp. 45–47; Anderson, in Chiu and Hoffman, pp. 33f.

35. See Wheeler, in Chiu and Hoffman, pp. 76, 79, 81, for a slightly different way of showing the geometrical significance of G_{ik}, derived from Cartan.

36. J. Géhéniau and R. Debever, "Les Quatorze Invariantes de Courbure de l'Espace Riemannien à Quatre Dimensions," in *Bern Jubilee of Relativity,* pp. 101–105.

37. Whittaker, *From Euclid to Eddington,* pp. 39f.

38. Eddington, *The Mathematical Theory of Relativity,* p. 149.

39. H. P. Robertson, "Geometry as a Branch of Physics," in Schilpp, pp. 313–332.

40. Wheeler, *Geometrodynamics,* p. 12.

41. Cf. the discussion in Reichenbach, *The Philosophy of Space and Time,* pp. 37–58.

42. I. J. Good, "Winding Space," in Good, *The Scientist Speculates,* pp. 330–337, Capricorn Books, New York, 1965.

43. Dugas, *History of Mechanics,* p. 533.

44. John C. Graves, *Singularities in Solutions of the Field Equations of General Relativity,* ch. 8, pp. 56–67; A.B. senior thesis, Princeton University, 1960 (unpublished); cf. Edward Kasner, "Finite Representation of the Solar Gravitational Field in a Flat Space of Six Dimensions," *Am. J. Math.,* vol. 43 (1921), pp. 130–133; Tolman, *Relativity, Thermodynamics, and Cosmology,* p. 346.

Chapter 12

1. Wheeler, *Geometrodynamics,* p. 13.

2. Lindsay and Margenau, *Foundations of Physics,* pp. 372f.

3. Kuhn, *Structure of Scientific Revolutions,* pp. 100f.

4. Cf. the discussion in d'Abro, *The Evolution of Scientific Thought,* pp. 285f.

5. *Ibid.*, pp. 279f.

6. John A. Wheeler, unpublished lecture notes on relativity, 1960, section 3.

7. Immanuel Kant, *Prolegomena to any Future Metaphysics,* p. 68, Liberal Arts Press, New York, 1950; cf. Toulmin, *Foresight and Understanding,* pp. 83–89; Jammer, *Concepts of Mass,* pp. 218f.

8. G. T. Whitrow, "Why Physical Space Has Three Dimensions," *Brit. J. Phil. Sci.,* vol. 6 (1955), pp. 13–32; B. Abramenko, "On Dimensionality and Continuity of Physical Space and Time," *Brit. J. Phil. Sci.,* vol. 9 (1958), pp. 89–110.

9. Hermann Weyl, *Philosophy of Mathematics and Natural Science,* p. 136, Princeton University Press, Princeton, 1949; Dugas, *History of Mechanics,* pp. 519–521.

10. Weyl, *Philosophy of Mathematics and Natural Science,* p. 136.

11. Victor Lenzen, *Causality in Natural Science,* p. 80; Charles C Thomas, Springfield, Ill., 1954; P. Jordan, "Kausalität und Statistik in der Modernen Physik," *Die Naturwissenschaften,* vol. 15 (1927), p. 105.

12. Törnebohm, *Logical Analysis of the Theory of Relativity,* p. 87.

13. Norman R. Campbell, *Foundations of Science,* ch. 14, pp. 361–402, esp. pp. 386, 389, Dover, New York, 1957.

14. *Ibid.,* p. 391.

15. Törnebohm, pp. 86–90, 102–104.

16. *Ibid.,* p. 104.

17. John G. Fletcher, "Geometrodynamics," in Witten, pp. 412–437; quote, p. 423.

18. Burtt, *Metaphysical Foundations of Modern Science,* pp. 109f.

19. Synge, *Relativity: The General Theory,* pp. 265–269; cf. Tolman, *Relativity, Thermodynamics, and Cosmology,* pp. 239–241.

20. Saul A. Basri, *Rev. Mod. Phys.,* vol. 37 (1965), pp. 288–315, esp. pp. 313–315.

21. Synge, p. 268.

22. H. P. Robertson, in Schilpp, *Albert Einstein: Philosopher-Scientist,* pp. 313–332.

23. For more mathematical details, see Graves, *Singularities,* ch. 10, pp. 79–94; John C. Graves and Dieter R. Brill, "Oscillatory Character of the Reissner–Nordström Metric for an Ideal Charged Wormhole," *Phys. Rev.,* vol. 120 (1960), pp. 1507–1513. A similar but somewhat more limited analysis appears in Adler, Bazin, and Schiffer, *Introduction to General Relativity,* p. 184.

24. See any standard text, e.g., Adler, Bazin, and Schiffer, pp. 177–187.

25. A. Schild, in Møller, *Evidence for Gravitational Theories,* pp. 69–115; *The Monist,* vol. 47 (Fall, 1962), pp. 20–39.

26. Adler, Bazin, and Schiffer, pp. 202–206.

Chapter 13

1. Peter Bergmann, "Non-linear Field Theories," *Phys. Rev.,* vol. 75 (1949), pp. 680–686; quote p. 681.

2. Einstein, "Autobiography," in Schilpp, *Albert Einstein: Philosopher-Scientist,* pp. 77–79.

3. Wheeler, *Geometrodynamics,* pp. 300f.; Richard Lindquist and Wheeler, *Rev. Mod. Phys.,* vol. 29 (1957), p. 440; Wheeler, *Rev. Mod. Phys.,* vol. 33 (1961), p. 65.

4. Cf. the discussion in Reichenbach, *The Theory of Relativity and A Priori Knowledge,* ch. 3, pp. 22–33; University of California Press, Berkeley, 1965.

5. Wheeler, *Geometrodynamics,* p. 80, footnote 60.

6. Peter G. Bergmann, *Phys. Rev.,* vol. 75 (1949), p. 680; quote from his abstract.

7. Leopold Infeld, "On Equations of Motion in General Relativity Theory," pp. 206–209 in *Bern Jubilee of Relativity Theory,* 1956.

8. For a good bibliography, which carefully distinguishes the contributions at each step, see John A. Wheeler, *Rev. Mod. Phys.,* vol. 33 (1961), p. 63; and *Geometrodynamics,* p. 24.

9. D. M. Chase, "The Equations of Motion of Charged Test Particles in General Relativity," *Phys. Rev.,* vol. 95 (1954), pp. 243–246.

10. For an unusually clear, if nonrigorous, discussion, see Adler, Bazin, and Schiffer, pp. 296–302.

11. Einstein, "Autobiography," in Schilpp, pp. 79–81; Jammer, *Concepts of Force,* pp. 262f.; Jammer, *Concepts of Mass,* pp. 215–217.

12. Leopold Infeld and Jerzy Plebanski, *Motion and Relativity,* esp. ch. 1, pp. 14–23; Pergamon, Warsaw, 1960; Joshua N. Goldberg, "The Equations of Motion," in Witten, *Gravitation: An Introduction to Current Research,* pp. 102–129; Infeld, "Equations of Motion in General Relativity Theory and the Action Principle," *Rev. Mod. Phys.,* vol. 29 (1957), pp. 398–412.

13. V. Fock, *The Theory of Space, Time, and Gravitation,* ch. 6; Fock, *Rev. Mod. Phys.,* vol. 29 (1957), pp. 327–330; Weber, *General Relativity and Gravitational Waves,* pp. 153–157; Weber, "Gravitational Radiation Experiments," in Dewitt and Dewitt, *Relativity, Groups, and Topology,* pp. 875–880.

14. Wheeler, *Geometrodynamics,* pp. 49, 301–302.

15. Richard Lindquist and John A. Wheeler, "Dynamics of a Lattice Universe by the Schwarzschild Cell Method," *Rev. Mod. Phys.,* vol. 29 (1957), pp. 432–443; Wheeler, in Dewitt and Dewitt, pp. 370–379; Wheeler, in Chiu and Hoffman, *Gravitation and Relativity,* pp. 308–313, 344.

16. F. K. Manasse, *J. Math. Phys.,* vol. 4 (1963), pp. 735–761; Tullio Regge and John A. Wheeler, *Phys. Rev.,* vol. 108 (1957), pp. 1063–1069.

17. Eugene Wigner, "Symmetry and Conservation Laws," *Proc. Natl. Acad. Sci.,* vol. 51 (1964), pp. 956–965; and in many of his other writings; cf. John G. Fletcher, "Local Conservation Laws in Generally Covariant Theories," *Rev. Mod. Phys.,* vol. 32 (1960), pp. 65–87; Andrzej Trautman, in Witten, p. 197; Goldstein, *Classical Mechanics,* pp. 47–55.

18. Synge, *Relativity: The General Theory,* pp. 41–46, 229–232.

19. Cf. Adler, Bazin, and Schiffer, *Introduction to General Relativity,* p. 234.

20. See, e.g. Weber, *General Relativity and Gravitational Waves,* pp. 70–86; Landau and Lifschitz, *The Classical Theory of Fields,* pp. 316–323.

21. Landau and Lifschitz, p. 321.

22. Andrzej Trautman, "Conservation Laws in General Relativity," in Witten, pp. 169–198.

23. John C. Graves and James E. Roper, "Measuring Measuring Rods," *Philosophy of Science,* vol. 32 (1965), pp. 39–56; cf. Hilary Putnam, "The Analytic and the Synthetic," pp. 358–397 in *Minnesota Studies,* vol. 3.

24. Christian Møller, pp. 15–21 in *Les Théories Relativistes de la Gravitation,*

Royaumont Conference, 1959, Centre National de la Recherche Scientifique, Paris, 1962; Møller, in *Evidence for Gravitational Theories,* p. 253.

25. Christian Møller, "Tetrad Fields and Conservation Laws in General Relativity," in *Evidence,* pp. 252–263, esp. p. 254.

26. Wheeler, in Dewitt and Dewitt, *Relativity, Groups, and Topology,* pp. 435f.; Landau and Lifschitz, *The Classical Theory of Fields,* pp. 319f.

27. Kip S. Thorne, *Phys. Rev.,* vol. 138 (1965), pp. B251–B266.

28. Jammer, *Concepts of Mass,* pp. 204, 214.

29. McMullin, in *The Concept of Matter,* pp. 26–27; Mast, in *The Concept of Matter,* pp. 574–577.

30. Einstein, "Autobiography," in Schilpp, *Albert Einstein: Philosopher-Scientist,* p. 81.

31. A. H. Taub, *Ann. Math.,* vol. 53 (1951), p. 473.

32. John A. Wheeler, *Rev. Mod. Phys.,* vol. 33 (1961), p. 64; *Geometrodynamics,* pp. 63f.

33. Martin Kruskal, *Phys. Rev.,* vol. 119 (1960), p. 1743.

34. Graves, *Singularities,* ch. 6, 7; John C. Graves and Dieter R. Brill, *Phys. Rev.,* vol. 120 (1960), pp. 1507–1513.

35. Albert Einstein and N. Rosen, "The Particle Problem in the General Theory of Relativity," *Phys. Rev.,* vol. 48 (1935), pp. 73–77.

36. Dugas, *History of Mechanics,* p. 526; Hermann Weyl, *Space-Time-Matter,* pp. 259f., Dover, New York, 1922.

37. Einstein and Rosen, "The Particle Problem in the General Theory of Relativity."

38. Graves and Brill, *Phys. Rev.,* vol. 120 (1960), pp. 1507–1513. Graves, *Singularities,* ch. 9; Wheeler, *Rev. Mod. Phys.,* vol. 33 (1961), pp. 68f.

39. Arthur Komar, *Phys. Rev.,* vol. 104 (1956), pp. 544–546; Wheeler, *Geometrodynamics,* p. 53.

40. Dieter R. Brill, *Ann. Phys.,* vol. 7 (1959), pp. 466–483; Brill, "Time-Symmetric Gravitational Waves," in *Les Théories Relativistes de la Gravitation,* pp. 147–153; Wheeler, *Geometrodynamics,* pp. 54–58.

41. A. H. Taub, *Ann. Math.,* vol. 53 (1951), p. 484; A. Papapetrou, in *Les Théories Relativistes de la Gravitation,* pp. 193–198.

42. Wheeler, *Geometrodynamics,* pp. 59–61; L. C. Shepley, *Proc. Natl. Acad. Sci.,* vol. 52 (1964), pp. 1403–1409.

43. Wheeler, *Geometrodynamics,* p. 63.

44. John A. Wheeler, "The Universe in the Light of General Relativity," pp. 40–76, esp. pp. 59–62, in *The Monist,* vol. 47 (1962); *Geometrodynamics,* p. 66.

Chapter 14

1. Cf. P. A. M. Dirac, "The Evolution of the Physicist's Picture of Nature, "*Scientific American,* May, 1963.

2. Olivier Costa de Beauregard, "Time in Relativity Theory: Arguments for a Philosophy of Being," in J. T. Fraser, *The Voices of Time,* pp. 417–433, Braziller, New York, 1966.

3. Emile Meyerson, *Identity and Reality* (Kate Loewenberg, trans.), ch. 6, pp. 215–233, George Allen and Unwin, London, 1930.

4. J. J. C. Smart, "Spatialising Time," pp. 163–168, in Richard M. Gale, *The Philosophy of Time,* Anchor Books, Doubleday, Garden City, N.Y., 1967.

5. Willard Van Orman Quine, *Word and Object,* pp. 170–176, The M.I.T. Press, Cambridge, Mass., 1960.

6. Čapek, *Philosophical Impact,* ch. 11–13, pp. 158–243; Čapek, "Time in Relativity Theory: Arguments for a Philosophy of Becoming," in Fraser, *The Voices of Time,* pp. 434–454.

7. Cf. the tabulation given in Wheeler, *Rev. Mod. Phys.,* vol. 34 (1962), p. 881; and those given in all standard treatises.

8. B. Kent Harrison, *Exact 3-Valued Solutions of the Field Equations of General Relativity,* Ph.D. dissertation, Princeton University (unpublished); and elsewhere.

9. Jürgen Ehlers and Wolfgang Kundt, "Exact Solutions of the Gravitational Field Equations," in Witten, *Gravitation: An Introduction to Current Research,* pp. 49–101.

10. E. A. Power and John A. Wheeler, "Thermal Geons," in Wheeler, *Geometrodynamics,* pp. 187–189; also in *Rev. Mod. Phys.,* vol. 29 (1957), p. 481.

11. Cf. Andre Lichnerowicz, *Théories Relativistes de la Gravitation et de l'Electromagnetisme,* Masson et Cie., Paris, 1955; R. Arnowitt, S. Deser, and C. W. Misner, "The Dynamics of General Relativity," in Witten, pp. 227–265.

12. E. L. Hill, "Quantum Physics and the Relativity Theory," in Feigl and Maxwell, *Current Issues in the Philosophy of Science,* p. 444.

13. Hill, in Feigl and Maxwell, p. 433.

14. Wheeler, in Dewitt and Dewitt, *Relativity, Groups, and Topology,* pp. 338–345.

15. R. K. Sachs, "Gravitational Radiation," in Dewitt and Dewitt, p. 527.

16. R. Arnowitt, S. Deser, and C. W. Misner, "The Dynamics of General Relativity," in Witten, pp. 227–265; also in *Recent Developments in General Relativity,* pp. 127–136; *Phys. Rev.,* vol. 113 (1959), p. 745; vol. 116 (1959), pp. 1322–1330; vol. 117 (1960), pp. 1595–1602; vol. 118 (1960), pp. 1100–1104; *Nuovo Cimento,* vol. 19 (1961), pp. 668–681; *J. Math. Phys.,* vol. 1 (1960), pp. 434–439.

17. See Wheeler, in Dewitt and Dewitt, pp. 345–349 and in Wheeler's many other writings on the subject, e.g., in Chiu and Hoffman, *Gravitation and Relativity,* pp. 315–318.

18. E. Cartan, *Bulletin de la Société Mathématique de France,* vol. 59 (1931), p. 88; Andre Lichnerowicz, *Problèmes Globaux en Méchanique Relativiste,* Paris, 1939; Yvonne Bruhat, "The Cauchy Problem," in Witten, pp. 130–168.

19. Arnowitt, Deser, and Misner, variously discussed throughout the references under footnote 16 above.

20. Wheeler, *Rev. Mod. Phys.,* vol. 33 (1961), p. 66.

21. Wheeler, in Dewitt and Dewitt, p. 350.

22. Wheeler, in Chiu and Hoffman, p. 326.

23. Wheeler, in Dewitt and Dewitt, pp. 350–354.

24. David Howland Sharp, A.B. senior thesis, Princeton Univ., 1960; Ralph Baierlein, Sharp, and Wheeler, *Phys. Rev.,* vol. 126 (1962), p. 1864; Wheeler, *Geometrodynamics,* p. 43; Wheeler, *Rev. Mod. Phys.,* vol. 34 (1962), p. 890.

25. Baierlein, Sharp, and Wheeler, *Phys. Rev.,* vol. 126 (1962), p. 1864; Wheeler, in Dewitt and Dewitt, pp. 358–361; Wheeler, in Chiu and Hoffman, pp. 329–334; Wheeler, in Thomas Gold, *The Nature of Time,* pp. 96f.; Cornell University Press, Ithaca, 1967.

26. John A. Wheeler and P. Bergmann, discussion in Gold, *The Nature of Time,* pp. 99–102.

27. Wheeler, in Chiu and Hoffman, pp. 320–322; in Dewitt and Dewitt, pp. 364–367.

28. Wheeler, in Dewitt and Dewitt, pp. 362–364; in Chiu and Hoffman, pp. 310f., 334.

29. Bergmann, in Gold, p. 99.

30. Wheeler, *Geometrodynamics,* pp. 40f.; in Chiu and Hoffman, p. 83; in Dewitt and Dewitt, p. 437.

31. Brill, *Ann. of Phys.,* vol. 7 (1959), pp. 466–483; Huzihiro Araki, *Ann. Phys.,* vol. 7 (1959), pp. 456–465; Wheeler, in *Geometrodynamics,* pp. 54–58; Brill, in *Les Théories Relativistes de la Gravitation,* pp. 147–153.

32. Charles W. Misner, "The Method of Images in Geometrostatics," *Ann. Phys.,* vol. 24 (1963), pp. 102–117.

33. Wheeler, in Dewitt and Dewitt, pp. 452–463.

Chapter 15

1. Einstein, "Autobiography," in Schilpp, *Albert Einstein: Philosopher-Scientist,* p. 81.

2. For good summaries, see Pauli, *Theory of Relativity,* pp. 227–232; Valentine Bargmann, *Rev. Mod. Phys.,* vol. 29 (1957), pp. 169–171.

3. Bargmann, *Rev. Mod. Phys.,* vol. 29 (1957), p. 171.

4. See, e.g., Dicke, in Dewitt and Dewitt, pp. 195–316; Brill, "Review of Jordan's Extended Theory of Gravitation," in Møller, *Evidence for Gravitational Theories,* pp. 51–68; P. Jordan, *Schwerkraft und Weltall,* F. Vieweg, Brunswick, 1955.

5. Pauli, *Theory of Relativity,* p. 230.

6. *Ibid.*

7. See Edmund A. Whittaker, *A History of the Theories of Aether and Electricity,* vol. 2, pp. 188–192, Harper Torchbooks, Harper & Row, 1960; Weyl, *Space-Time-Matter,* ch. 4, pp. 282–295; Eddington, *The Mathematical Theory of Relativity,* ch. 7, pp. 196–241; Schrödinger, *Space-Time Structure,* Cambridge University Press, London, 1950.

8. Einstein, *The Meaning of Relativity,* 5th Edition, Appendix 2, pp. 133–166; Pauli, *Theory of Relativity,* pp. 226f.; Bargmann, *Rev. Mod. Phys.,* vol. 29 (1957) p. 172.

9. Einstein, "Autobiography," in Schilpp, pp. 91–95.

10. M. A. Tonnelat, *Les Principes de la Théorie Électromagnétique et de la Relativite,* Masson et Cie., Paris, 1959; V. Hlavaty, *Geometry of Einstein's Unified Field Theory,* Gröningen, Germany, 1957.

11. Dominic G. B. Edelen, *The Structure of Field Space,* University of California Press, Berkeley, 1962.

12. J. Callaway, *Phys. Rev.,* vol. 92 (1953), p. 1567.

13. C. P. Johnson, *Phys. Rev.,* vol. 89 (1953), p. 320.

14. P. Bergmann, in *Les Théories Relativistes de la Gravitation,* p. 465.

15. Pauli, *Theory of Relativity,* p. 227.

16. Bargmann, *Rev. Mod. Phys.,* vol. 29 (1957), p. 173.

17. Cornelius Lanczos, *Rev. Mod. Phys.,* vol. 29 (1957), pp. 337–350.

18. The discussion here follows the mathematical development in the standard treatises, e.g., Wheeler, *Geometrodynamics,* pp. 13–24; Misner and Wheeler, *Ann. Phys.,* vol. 2 (1957), pp. 525–603 (reprinted in *Geometrodynamics,* pp. 225–253); Witten, in Witten, *Gravitation: An Introduction to Current Research,* pp. 382–398; Adler, Bazin, and Schiffer, *Introduction to General Relativity,* pp. 418–441; Weber, *General Relativity and Gravitational Waves,* pp. 146–153. Since the accounts are similar, I have not given credit for individual points.

19. Wheeler, *Geometrodynamics,* pp. 244–249; Witten, "A Geometric Theory of the Electromagnetic and Gravitational Fields," in Witten, pp. 382–398, esp. pp. 388f.

20. Wolfgang Kundt, work discussed in lecture notes by Wheeler, 1960.

21. Asher Peres, "On Geometrodynamics and Null Fields," *Ann. Phys.,* vol. 14 (1961), pp. 419–439; Witten, in Witten, pp. 394–396.

22. Gerald Rosen, "Geometrical Significance of the Einstein-Maxwell Equations," *Phys. Rev.,* vol. 114 (1959), pp. 1179–1181; Wheeler, *Geometrodynamics,* p. 249.

23. C. Lanczos, *Phys. Rev.,* vol. 39 (1932), pp. 716–736; Witten, *Phys. Rev.,* vol. 115 (1959), pp. 206–215.

24. David Sharp, "Variational Principle for Geometrodynamics," *Phys. Rev. Letters,* vol. 3 (1959), pp. 108–110; Wheeler, *Geometrodynamics,* p. 22.

25. Sharp, "Variational Principle"; Wheeler, *Geometrodynamics,* p. 306.

26. Fletcher, in Witten, p. 415.

27. Roger Penrose, unpublished; Wheeler, *Geometrodynamics,* pp. 22f.; Witten, *Phys. Rev.,* vol. 120 (1960), p. 635.

28. David Sharp, "Remarks on the Problem of Penrose," unpublished; Wheeler, *Geometrodynamics,* p. 24.

Chapter 16

1. G. Y. Rainich, "Electrodynamics in the General Relativity Theory," *Proc. Natl. Acad. Sci. U.S.,* vol. 10 (1924), pp. 124–127, 294–298; *Trans. Am. Math. Soc.,* vol. 27 (1925), pp. 106–136.

2. C. W. Misner and John A. Wheeler, "Classical Physics as Geometry," *Ann. Phys.,* vol. 2 (1957), pp. 525–603; reprinted in Wheeler, *Geometrodynamics,* pp. 225–307.

3. G. Y. Rainich, *Mathematics of Relativity,* John Wiley & Sons, New York, 1950.

4. For details, see Tolman, *Relativity, Thermodynamics, and Cosmology,* pp. 265–267; Adler, Bazin, and Schiffer, pp. 396–401; or any other standard texts.

5. For derivations, see Graves, *Singularities,* ch. 10; Graves and Brill, *Phys. Rev.,* vol. 120 (1960), pp. 1512f.

6. Graves, *Singularities,* ch. 4–7; Graves and Brill, *Phys. Rev.,* vol. 120 (1960) pp. 1508–1510.

7. Graves, *Singularities,* ch. 2–3.

8. *Ibid.,* ch. 6, pp. 44–46.

9. Einstein and Rosen, *Phys. Rev.,* vol. 48 (1935), pp. 73–77.

10. Wheeler, *Geometrodynamics,* pp. 47f.

11. Weyl, *Philosophy of Mathematics and Natural Science,* p. 91, Princeton University Press, Princeton, 1949.

12. R. W. Bass and L. Witten, "Remarks on Cosmological Models," *Rev. Mod. Phys.,* vol. 29 (1957), pp. 452f.

13. Wheeler, *Geometrodynamics,* p. 254.

14. Graves, *Singularities,* ch. 9; Wheeler, *Geometrodynamics,* pp. 298–300.

15. Wheeler, *Geometrodynamics,* pp. 49f., 222–232; Graves and Brill, *Phys. Rev.,* vol. 120 (1960), p. 1512.

16. Misner, *Ann. Phys.,* vol. 24 (1963), p. 114.

17. R. W. Fuller and John A. Wheeler, "Causality and Multiply Connected Space-Time," *Phys. Rev.,* vol. 128 (1962), pp. 919–929.

18. Graves and Brill, *Phys. Rev.,* vol. 120 (1960), p. 1512.

19. Fuller and Wheeler, "Causality and Multiply Connected Space-Time."

20. Wheeler, *Geometrodynamics,* pp. 64–66; Fuller and Wheeler, "Causality and Multiply Connected Space-Time."

21. Graves, *Singularities,* ch. 9; Graves and Brill, *Phys. Rev.,* vol. 120 (1960), p. 1512.

22. Wheeler, in Nagel, Suppes, and Tarski, *Logic, Methodology and Philosophy of Science,* pp. 371f.

23. Morton R. Dubman, *Geometrical Representation of Ideal Mass Particles in the General Theory of Relativity,* A.B. senior thesis, Princeton University, 1957 (unpublished).

24. Misner, "Wormhole Initial Conditions," *Phys. Rev.,* vol. 118 (1960), p. 1110; Wheeler, *Rev. Mod. Phys.,* vol. 33 (1961), pp. 70f.; Wheeler, *Geometrodynamics,* pp. 49, 301–302.

25. Jammer, *Concepts of Force,* p. 191.

26. Kenneth W. Ford, "Magnetic Monopoles," *Scientific American,* December, 1963, pp. 122–131.

27. Wheeler, *Geometrodynamics,* pp. 303–305; Fletcher, in Witten, pp. 417f.

28. Wheeler, *Geometrodynamics,* pp. 274–278.

29. Witten, in Witten, p. 402.

30. Wheeler, "Absence of a Gravitational Analog to Electric Charge," pp. 1–20 in Robert Wasserman and Charles P. Wells, eds., *Relativistic Fluid Mechanics and Magnetohydrodynamics,* Academic Press, New York, 1963; Wheeler, in Chiu and Hoffman, p. 347.

31. Wheeler, in Wasserman and Wells, pp. 17–20; *Rev. Mod. Phys.,* vol. 34 (1962), p. 891.

32. David Finkelstein and C. W. Misner, "Some New Conservation Laws," *Ann. Phys.,* vol. 6 (1959), pp. 230–243; Bergmann, in *Les Théories Relativistes de la Gravitation,* summary, p. 464.

Chapter 17

1. A. H. Taub, *Ann. Math.,* vol. 53 (1951), pp. 472–490.

2. Joseph J. Weber and John A. Wheeler, *Rev. Mod. Phys.,* vol. 29 (1957), pp. 509–516.

3. Dieter Brill, *Ann. Phys.,* vol. 7 (1959), pp. 466–483; Brill, "Time-Symmetric Gravitational Waves," in *Les Théories Relativistes de la Gravitation,* pp. 147–153.

4. See especially Weber, *General Relativity and Gravitational Waves,* ch. 7, pp.

87–123; R. K. Sachs, in Dewitt and Dewitt, pp. 523–564; F. A. E. Pirani, "Gravitational Radiation," in Witten, *Gravitation: An Introduction to Current Research.* pp. 199–226.

5. Einstein, *The Meaning of Relativity,* p. 87.

6. Weber, *General Relativity and Gravitational Waves,* ch. 8, pp. 124–145; Weber, in Chiu and Hoffman, ch. 5, pp. 90–105; Wheeler, in *The Monist,* vol. 47 (Fall, 1962), pp. 49–55; Weber, in Møller, *Evidence for Gravitational Theories,* pp. 116–140.

7. Wheeler, *The Monist,* vol. 47 (Fall, 1962), pp. 49–55; Sidney Liebes, "Gravitational Lenses," *Phys. Rev.,* vol. 133 (1964), pp. B835–B844.

8. Wheeler, "Geons," *Phys. Rev.,* vol. 97 (1955), pp. 511–536; reprinted in *Geometrodynamics,* pp. 135–185; Wheeler, *Geometrodynamics,* pp. 25–31.

9. Wheeler, *Geometrodynamics,* pp. 27–31; Wheeler, *Rev. Mod. Phys.,* vol. 33 (1961), pp. 65, 71–76.

10. Wheeler, *Geometrodynamics,* pp. 135–140; also p. 28.

11. Frederick, J. Ernst, Jr., "Linear and Toroidal Geons," *Phys. Rev.,* vol. 105 (1957), pp. 1665–1670; *Rev. Mod. Phys.,* vol. 29 (1957), p. 496.

12. E. A. Power and John A. Wheeler, "Thermal Geons," *Rev. Mod. Phys.,* vol. 29 (1957), pp. 480–495; reprinted in *Geometrodynamics,* pp. 186–224.

13. James B. Hartle, *The Gravitational Geon,* A.B. senior thesis, Princeton University, 1960 (unpublished).

14. Wheeler, *The Monist,* vol. 47 (Fall, 1962), pp. 59–62; Wheeler, in Chiu and Hoffman, pp. 195–230; Kip S. Thorne, in Robinson et al., *Quasi-Stellar Sources and Gravitational Collapse,* pp. 83–92; University of Chicago Press, Chicago, 1965; Thorne, "Gravitational Collapse," *Scientific American,* vol. 217 (November, 1967), pp. 88–102; Misner and Sharp, *Phys. Rev.,* vol. 136 (1964), pp. B571–B576.

15. Descartes, *Discourse on Method,* Section 6.

16. Lindsay and Margenau, *Foundations of Physics,* pp. 48–55.

17. Wheeler, in Chiu and Hoffman, p. 66; *The Monist,* vol. 47 (Fall, 1962), p. 67.

18. Wheeler, *Geometrodynamics,* pp. 60f., 64; *The Monist,* vol. 47 (Fall, 1962), pp. 65–67.

19. Weber and Wheeler, *Rev. Mod. Phys.,* vol. 29 (1957), pp. 509–516.

20. Wheeler, *Geometrodynamics,* p. 75, footnote 53; *The Monist,* vol. 47 (Fall, 1962), pp. 62–72.

21. Hugh Everett III, "Relative State Formulation of Quantum Mechanics," *Rev. Mod. Phys.,* vol. 29 (1957), pp. 454–462; comments by Wheeler, pp. 463–465.

22. Sir Isaac Newton, Scholium to *Principia Mathematica,* Book I; reprinted in Danto and Morgenbesser, *Philosophy of Science,* pp. 322–329.

23. George Berkeley, *The Principles of Human Knowledge,* sections 109–116.

24. Ernst Mach, *The Science of Mechanics,* ch. 11, pp. 271–298, Open Court, Lasalle, Ill., 1960.

25. Bridgman, "The Mach Principle," in Mario Bunge, *The Critical Approach to Science and Philosophy* (Karl Popper Festschrift), pp. 224–233, Free Press, Glencoe, Ill., 1964.

26. Philip Morrison, "The Physics of the Large," in Colodny, *Beyond the Edge of Certainty,* pp. 130–133, Prentice-Hall, 1965.

27. Dennis W. Sciama, "On the Origin of Inertia," *Royal Astron. Soc. Monthly Notices,* vol. 113 (1953), pp. 34–42; Sciama, *The Unity of the Universe,* ch. 7–9, pp. 83–130, Anchor Books, Doubleday, Garden City, N.Y. 1961; cf. J. T. Fraser, *J. Franklin Inst.,* vol. 272 (1961), pp. 460–493.

28. See the various appendices in Dicke, "Experimental Relativity," in Dewitt and Dewitt, pp. 195–316; and elsewhere; also Brill, in Møller, *Evidence for Gravitational Theories,* pp. 51–69.

29. F. A. Kaempffer, "On Possible Realizations of Mach's Program," *Canadian J. Phys.,* vol. 36 (1958), pp. 151–159.

30. Parry Moon and Domina Eberle Spencer, "Mach's Principle," *Phil. of Sci.,* vol. 26 (1959), pp. 125–134.

31. V. W. Hughes, "Mach's Principle and Experiments on Mass Anisotropy," pp. 106–120 in Chiu and Hoffman; Dicke, *Phys. Rev. Letters,* vol. 7 (1961), pp. 359f.

32. Einstein, *The Meaning of Relativity,* p. 100.

33. Carl H. Brans, *Phys. Rev.,* vol. 125 (1962), pp. 388–396.

34. Törnebohm, *A Logical Analysis of the Theory of Relativity,* pp. 118–130.

35. Einstein, *The Meaning of Relativity,* pp. 100–108.

36. Joseph Callaway, "Mach's Principle and Unified Field Theory," *Phys. Rev.,* vol. 96 (1954), pp. 778–780.

37. W. Davidson, "General Relativity and Mach's Principle," *Royal Astron. Soc. Monthly Notices,* vol. 117 (1957), pp. 212–224.

38. G. J. Whitrow, in *The Monist,* vol. 47 (Fall, 1962), pp. 84–88.

39. Brans and Dicke, *Phys. Rev.,* vol. 124 (1961), pp. 925–935; Dicke, in Chiu and Hoffman, pp. 121–174; Dicke, in Møller, *Evidence for Gravitational Theories,* pp. 31–50.

40. Wheeler, "Mach's Principle as Boundary Condition for Einstein's Equations," in Chiu and Hoffman, pp. 303–349, esp. pp. 303–306; Wheeler, in Dewitt and Dewitt, pp. 364–366.

41. Wheeler, in Chiu and Hoffman, pp. 306–314; in Dewitt and Dewitt, pp. 366–369.

42. Einstein, *The Meaning of Relativity,* pp. 107f.

43. E. L. Hill, in Feigl and Maxwell, *Current Issues in the Philosophy of Science,* pp. 433f.

Chapter 18

1. Dicke, "Experimental Relativity," in Dewitt and Dewitt, pp. 165–316, and elsewhere in his writings; B. Bertotti, Dieter Brill and R. Krotkov, "Experiments in Gravitation," pp. 1–48, in Witten, *Gravitation: An Introduction to Current Research;* V. W. Hughes, "The Lyttleton-Bondi Universe and Charge Equality," pp. 259–278, in Chiu and Hoffman, *Gravitation and Relativity;* Dicke, "The Eötvös Experiment," *Scientific American,* December, 1961, pp. 84–94.

2. R. V. Pound and G. A. Rebka, Jr., *Phys. Rev. Letters,* vol. 4 (1960), p. 337; T. E. Cranshaw, in Møller, *Evidence for Gravitational Theories,* pp. 208–221; V. L. Ginzburg, "Artificial Satellites and Relativity," *Scientific American,* May 1959, pp. 149–160; Sergio deBenedetti, "The Mössbauer Effect," *Scientific American,* April, 1960.

3. Fred Hoyle, in Møller, *Evidence for Gravitational Theories,* pp. 141–173; Wheeler, in *The Monist* (Fall, 1962), pp. 43–48.

4. Chiu and Hoffman, introduction, pp. xxii, xxvii–xxxiii; Bunge, *Foundations of Physics,* pp. 232–235.

5. Wheeler, *Rev. Mod. Phys.,* vol. 33 (1961), p. 63.

6. J. L. Anderson, "Quantization of General Relativity," in Chiu and Hoffman, pp. 279–302; Bryce Dewitt, "The Quantization of Geometry," in Witten, pp. 266–381; Arnowitt, Deser, and Misner, "The Dynamics of General Relativity," in Witten, pp. 227–265; Misner, *Rev. Mod. Phys.,* vol. 29 (1957) pp. 497–509.

7. Wheeler, *Rev. Mod. Phys.,* vol. 34 (1962), p. 875.

8. Wheeler, *Geometrodynamics,* pp. 66–87; Wheeler, in Dewitt and Dewitt, pp. 501–520; Wheeler, "On the Nature of Quantum Geometrodynamics," *Ann. of Phys.,* vol. 2 (1957), pp. 604–614.

9. David Bohm, *Causality and Chance in Modern Physics,* ch. 4–5, Harper Torchbooks, Harper & Row, New York, 1961.

10. Wheeler, in Nagel, Suppes, and Tarski, p. 373.

11. Čapek, *Philosophical Impact,* pp. 224–241.

12. John G. Fletcher, "Geometrodynamics," in Witten, pp. 412–437.

13. Misner, "Mass as a Form of Vacuum," in McMullin, *The Concept of Matter,* pp. 609–612.

14. Grünbaum, "The Philosophical Retention of Absolute Space in Einstein's General Theory of Relativity," *Phil. Rev.,* vol. 66 (1957), pp. 525–534.

15. Synge, *Relativity: The General Theory,* pp. ix–xi.

The following bibliography is not intended to be a complete list of all items cited. In particular, it does not include individual articles of less than monograph length, whether they appear in journals or anthologies. These are individually cited in the references for the various chapters. However, it includes all books relevant for anyone wishing to make a serious study of the conceptual foundations of geometrodynamics, as well as the anthologies from which articles have been drawn. The books are listed in the alphabetical order of their authors. However, certain books, primarily the reports of summer conferences on general relativity, are not closely associated with any individual author or authors. These books are then listed by title at the end of the bibliography.

Abro, A. d', *The Evolution of Scientific Thought,* 2nd edition, Dover, New York, 1950.

Adler, Ronald; Bazin, Maurice; and Schiffer, Menahem, *Introduction to General Relativity,* McGraw-Hill, New York, 1965.

Baumrin, Bernard, ed., *Philosophy of Science—The Delaware Seminar,* Vol. 2, Interscience, New York, 1963.

Bergmann, Peter G., *Introduction to the Theory of Relativity,* Prentice-Hall, Englewood Cliffs, N.J., 1942.

Blake, Ralph M.; Ducasse, Curt J.; and Madden, Edward H., *Theories of Scientific Method,* University of Washington Press, Seattle, 1960.

Bohm, David, *Causality and Chance in Modern Physics,* Harper Torchbooks, Harper & Row, New York, 1961.

Bridgman, Percy W., *The Logic of Modern Physics,* Macmillan, New York, 1927.

——— *The Nature of Physical Theory,* Dover, New York, 1936.

——— *The Way Things Are,* Viking Press, New York, 1961.

Bunge, Mario, ed., *The Critical Approach to Science and Philosophy* (Essays in Honor of Karl Popper), Free Press, Glencoe, Ill., 1964.

——— *Foundations of Physics,* Springer-Verlag, New York, 1967.

——— *The Myth of Simplicity,* Prentice-Hall, New York, 1963.

——— *Metascientific Queries,* Charles C Thomas, Springfield, Illinois, 1959.

Burtt, Edwin A., *The Metaphysical Foundations of Modern Science,* Doubleday, New York, 1955.

Campbell, Norman R., *Foundations of Physics* (formerly entitled *Physics: The Elements*), Dover, New York, 1957.

Čapek, Milič, *The Philosophical Impact of Contemporary Physics,* Van Nostrand, Princeton, 1961.

Chiu, Hong-Yee, and Hoffman, William F., eds., *Gravitation and Relativity,* Benjamin, New York, 1964.

Clifford, William Kingdon, *The Common Sense of the Exact Sciences,* Knopf, New York, 1946.

Colodny, Robert G., ed., *Beyond the Edge of Certainty,* Prentice-Hall, Englewood Cliffs, N.J., 1965.

——— ed., *Frontiers of Science and Philosophy,* University of Pittsburgh Press, Pittsburgh, 1962.

Cornford, Francis M., *Plato's Cosmology* (includes a translation of Plato's *Timaeus*), Liberal Arts Press, New York, 1957.

Danto, Arthur, and Morgenbesser, Sidney, eds., *Philosophy of Science,* Meridian, New York, 1960.

Dewitt, Bryce, and Dewitt, C., eds., *Relativity, Groups, and Topology,* Gordon and Breach, New York, 1964.

Dijksterhuis, E. J., *The Mechanization of the World-Picture,* Clarendon Press, Oxford, 1961.

Duhem, Pierre, *The Aim and Structure of Physical Theory* (Philip Wiener, trans.), Princeton University Press, Princeton, 1954.

Dugas, René, *A History of Mechanics* (J. R. Maddox, trans.), Neuchâtel, Switzerland, 1955.

Eddington, Arthur S., *The Mathematical Theory of Relativity,* Cambridge University Press, London, 1960.

——— *The Nature of the Physical World,* Cambridge University Press, London, 1928.

Edelen, Dominic G. B., *The Structure of Field Space,* University of California Press, Berkeley, 1962.

Einstein, Albert, *The Meaning of Relativity,* 5th edition, Princeton University Press, Princeton, 1955.

——— *The Principle of Relativity,* Dover, New York, 1923.

Einstein, Albert, and Infeld, Leopold, *The Evolution of Physics,* Simon and Schuster, New York, 1938.

Feigl, Herbert, and Brodbeck, May, eds., *Readings in the Philosophy of Science,* Appleton-Century-Crofts, New York, 1953.

Feigl, Herbert, and Maxwell, Grover, eds., *Current Issues in the Philosophy of Science,* Holt, Rinehart, and Winston, New York, 1961.

Feigl, Herbert, and Sellars, Wilfred, eds., *Readings in Philosophical Analysis,* Appleton-Century-Crofts, New York, 1949.

Fock, V., *The Theory of Space, Time, and Gravitation,* Pergamon Press, New York, 1959.

Fraser, J. T., ed., *The Voices of Time,* Braziller, New York, 1966.

Gale, Richard M., ed., *The Philosophy of Time,* Anchor Books, Doubleday, Garden City, N.Y., 1967.

Gillispie, Charles C., *The Edge of Objectivity,* Princeton University Press, Princeton, 1960.

Gold, Thomas, ed., *The Nature of Time,* Cornell University Press, Ithaca, N.Y., 1967.

Goldstein, Herbert, *Classical Mechanics,* Addison-Wesley, Reading, Mass., 1950.

Good, I. J., ed., *The Scientist Speculates,* Capricorn Books, New York, 1965.

Graves, John C., *Singularities in Solutions of the Field Equations of General Relativity,* A.B. senior thesis, Princeton University, 1960 (unpublished).

Grünbaum, Adolf, *Philosophical Problems of Space and Time,* Knopf, New York, 1963.

Haldane, Elizabeth S., and Ross, G. R. T., *Philosophical Works of Descartes,* Vol. 1, Dover, New York, 1931 (used for all citations from Descartes' works).

Hanson, Norwood Russell, *Patterns of Discovery,* Cambridge University Press, London, 1958.

Harré, Romano, *An Introduction to the Logic of the Sciences,* Macmillan, London, 1960.

Hesse, Mary, *Forces and Fields,* Thomas Nelson & Sons, New York, 1961.

——— *Models and Analogies in Science,* Sheed & Ward, London, 1963.

Hlavaty, V., *Geometry of Einstein's Unified Field Theory,* Gröningen, Germany, 1957.

Infeld, Leopold, and Plebanski, Jerzy, *Motion and Relativity,* Pergamon Press, New York, 1960.

Jammer, Max, *Concepts of Force,* Harper Torchbooks, Harper & Row, New York, 1962.

——— *Concepts of Mass,* Harvard Universitv Press, Cambridge, Mass., 1961.

——— *Concepts of Space,* Harper & Row, New York, 1954.

Kemp Smith, Norman, *Studies in the Cartesian Philosophy,* Macmillan, London, 1902.

Koren, Henry J., ed., *Readings in the Philosophy of Nature,* Newman Press, Westminster, Maryland, 1961.

Körner, Stephen, ed., *Observation and Interpretation in the Philosophy of Physics* (Colston Papers No. 9), Dover, New York, 1962.

Koyré, Alexandre, *From the Closed World to the Infinite Universe,* Harper Torchbooks, Harper & Row, New York, 1958.

Kuhn, Thomas S., *The Structure of Scientific Revolutions,* University of Chicago Press, Chicago, 1962.

Landau, Lev, and Lifschitz, E., *The Classical Theory of Fields* (Morton Hammermesh, trans.), Addison-Wesley, Reading, Mass., 1951.

Lenzen, Victor F., *Causality in Natural Science,* Charles C Thomas, Springfield, Ill., 1954.

Levinson, Horace C., and Zeisler, Ernest B., *The Law of Gravitation in Relativity,* University of Chicago Press, Chicago, 1931.

Lichnerowicz, Andre, *Théories Relativistes de la Gravitation et de l'Électromagnétisme,* Masson et Cie., Paris, 1955.

Lindsay, Robert B., and Margenau, Henry, *Foundations of Physics,* Dover, New York, 1957.

Linsky, Leonard, ed., *Semantics and the Philosophy of Language,* University of Illinois Press, Urbana, 1952.

Mach, Ernst, *The Science of Mechanics,* Open Court, LaSalle, Ill., 1960.

McMullin, Ernan, ed., *The Concept of Matter,* University of Notre Dame Press, Notre Dame, Ind., 1963.

McVittie, G. C., *General Relativity and Cosmology,* University of Illinois Press, Urbana, 1962.

Møller, Christian, *The Theory of Relativity,* Clarendon Press, Oxford, 1952.

——— ed., *Evidence for Gravitational Theories,* Proceedings of the International School of Physics, Enrico Fermi School, course 20, Academic Press, New York, 1962.

Nagel, Ernest, *The Structure of Science,* Harcourt, Brace, & World, New York, 1961.

Nagel, Ernest; Suppes, Patrick; and Tarski, Alfred, eds., *Logic, Methodology, and Philosophy of Science,* Stanford University Press, Stanford, 1962.

Nash, Leonard K., *The Nature of the Natural Sciences,* Little, Brown, Boston, 1963.

Oliver, W. Donald, *Theory of Order,* Antioch Press, Yellow Springs, Ohio, 1951.

Palter, Robert M., *Whitehead's Philosophy of Science,* University of Chicago Press, Chicago, 1960.

Pap, Arthur, *An Introduction to the Philosophy of Science,* Free Press, Glencoe, Ill., 1962.

Pauli, Wolfgang, *Theory of Relativity* (G. Field, trans.), Pergamon Press, New York, 1958.

Popper, Karl R., *The Logic of Scientific Discovery,* Basic Books, New York, 1959.

——— *Conjectures and Refutations,* Basic Books, New York, 1962.

Quine, Willard Van Orman, *From a Logical Point of View,* Harvard University Press, Cambridge, Mass., 1953.

Rainich, G. Y., *Mathematics of Relativity,* John Wiley & Sons, New York, 1950.

Reichenbach, Hans, *The Philosophy of Space and Time* (Maria Reichenbach, trans.), Dover, New York, 1957.

——— *The Theory of Relativity and A Priori Knowledge,* University of California Press, Berkeley, 1965.

Schilpp, Paul A., ed., *Albert Einstein: Philosopher-Scientist* (Library of Living Philosophers series), Tudor Books, New York, 1949.

Schrödinger, Erwin, *Expanding Universes,* Cambridge University Press, London, 1956.

——— *Space-Time Structure,* Cambridge University Press, London, 1950.

Sciama, Dennis W., *The Unity of the Universe,* Anchor Books, Doubleday, Garden City, N.Y., 1961.

Scott, J. F., *The Scientific Work of Descartes,* Taylor and Francis, London, 1952.

Sellars, Wilfred, *Science, Perception, and Reality,* Humanities Press, New York, 1963.

Smart, J. J. C., *Philosophy and Scientific Realism,* Humanities Press, New York, 1963.

Synge, J. L., *Relativity: The General Theory,* North Holland, Amsterdam, 1960.

Tolman, Richard C., *Relativity, Thermodynamics, and Cosmology,* Clarendon Press, Oxford, 1934.

Tonnelat, M. A., *Les Principes de la Théorie Électromagnétique et de la Relativité,* Masson et Cie., Paris, 1959.

Törnebohm, Hakan, *A Logical Analysis of the Theory of Relativity,* Almqvist and Wiksell, Stockholm, 1952.

Toulmin, Stephen, *Foresight and Understanding,* Harper Torchbooks, Harper & Row, New York, 1963.

——— *Philosophy of Science,* Harper Torchbooks, Harper & Row, New York, 1960.

Veatch, Henry, *Intentional Logic,* Yale University Press, New Haven, Conn., 1952.

Wasserman, Robert, and Wells, Charles P., eds., *Relativistic Fluid Mechanics and Hydrodynamics,* Academic Press, New York, 1963.

Weber, Joseph J., *General Relativity and Gravitational Waves,* Interscience, New York, 1961.

Weyl, Hermann, *Philosophy of Mathematics and Natural Science,* Princeton University Press, Princeton, 1949.

——— *Space, Time, Matter,* Dover, New York, 1922.

Wheeler, John A., *Geometrodynamics,* Academic Press, New York, 1962.

Whitehead, Alfred North, *The Principle of Relativity,* Cambridge University Press, London, 1922.

Whittaker, Edmund, *From Euclid to Eddington,* Dover, New York, 1958.

——— *A History of the Theories of Aether and Electricity,* vol. 2, Harper Torchbooks, Harper & Row, New York, 1960.

Witten, Louis, ed., *Gravitation: An Introduction to Current Research,* John Wiley & Sons, New York, 1962.

Bern Jubilee of Relativity Theory, Helvetica Physica Acta, suppl. vol. 4, 1956.

Colloquie sur la Théorie de la Relativité, Centre Belge de Recherches Mathématiques, Louvain, 1960.

Minnesota Studies in the Philosophy of Science, vol. 1, University of Minnesota Press, Minneapolis, 1956.

Minnesota Studies in the Philosophy of Science, vol. 2, University of Minnesota Press, Minneapolis, 1959.

Minnesota Studies in the Philosophy of Science, vol. 3, University of Minnesota Press, Minneapolis, 1962.

Recent Developments in General Relativity, Macmillan, New York, 1962.

Scientific Papers Presented to Max Born, Hafner, New York, 1953.

Les Théories Relativistes de la Gravitation, Royaumont Conference, 1959, Centre National de la Recherche Scientifique, Paris, 1962.

Reviews of Modern Physics, vol. 29 (1957), is essentially a symposium on general relativity, with a great many articles cited in the references above.

The Monist, vol. 47 (Fall, 1962), is a valuable symposium on general relativity and cosmology.

Index